W9-CDM-656

Symbol	What the symbol represents	Page(s) where the symbol is defined
$S \cap T$	The intersection of S and T	99
$S + T$	The vector sum of S and T	99
$\mathrm{Sp}\{\alpha_1, \ldots, \alpha_k\}$	The linear span of $\alpha_1, \ldots, \alpha_k$	100
ϵ_k	The kth standard basis n-tuple	107
$\dim V$	The dimension of V	118
$[\gamma]_S$	The S-coordinates of the vector γ	126
$(\alpha \mid \beta)$	The inner product of α with β	159
$\|\alpha\|$	The length of α	162
$d(\alpha, \beta)$	The distance between α and β	162
α_S	The projection of α on S	176
$\alpha_{S\perp}$	The vector $\alpha - \alpha_S$	177
$S^\perp$	The orthogonal complement of S	181
A^*	The conjugate transpose of A	194 ; 327
$\mathfrak{N}_L$	The null space of the linear transformation L	203
$\mathfrak{R}_L$	The range space of the linear transformation L	203
$\nu(L)$	The nullity of L	204
$\rho(L)$	The rank of L	204
$\mu(L)$	A matrix representative of L	212
$\det A$	The determinant of the matrix A	257
A_{pq}	The submatrix formed by crossing out the pth row and q^{th} column of A	272
$\mathfrak{A}_{pq}$	The cofactor $(-1)^{p+q} \det A_{pq}$	272
$\mathrm{adj}\, A$	The adjugate of the matrix A	275
$p_A(x)$	$\det (xI_n - A)$, the characteristic polynomial of A	287
$\mathfrak{e}_\lambda$	The eigenspace associated with the eigenvalue λ of a given linear transformation	293
$\mathfrak{S}_\lambda$	The eigenspace associated with the eigenvalue λ of a given matrix	293
type $(n; r, s)$	A quadratic form whose symmetric representative has rank r and signature s	339

A Primer Of
Linear Algebra

A Primer Of Linear Algebra

Gerald L. Bradley

Department of Mathematics
Claremont Men's College

PRENTICE-HALL, INC.
Englewood Cliffs, New Jersey

Library of Congress Cataloging in Publication Data

BRADLEY, GERALD L
  A primer of linear algebra.

  Bibliography: p.
  1. Algebras, Linear. I. Title.
QA184.B7   512'.5   74-2025
ISBN 0-13-700328-5

© 1975 by Prentice-Hall, Inc., Englewood Cliffs, New Jersey

All rights reserved.
No part of this book may be reproduced
in any form or by any means without
permission in writing from the publisher.

10 9 8 7 6 5 4

Printed in the United States of America

PRENTICE-HALL INTERNATIONAL, INC., *London*
PRENTICE-HALL OF AUSTRALIA, PTY. LTD., *Sydney*
PRENTICE-HALL OF CANADA, LTD., *Toronto*
PRENTICE-HALL OF INDIA PRIVATE LIMITED, *New Delhi*
PRENTICE-HALL OF JAPAN, INC., *Tokyo*

To my parents,
who encouraged me to start this book,
and to my wife, Kay,
who nagged me into finishing it.

Contents

Chapter 2
Vector Spaces 74

Chapter 3
Real Inner Product Spaces 149

Chapter 4
Linear Transformations 195

Chapter 5
Determinants 255

Preface

Until recently, most linear algebra courses were populated almost exclusively by junior and senior mathematics majors. However, in the past few years, there has been a dramatic reevaluation of the place of linear algebra in the mathematics curriculum, and now, the subject is well on its way to assuming a role second in importance only to calculus. In fact, many schools now include a quarter or a semester of linear algebra as part of the fundamental freshman-sophomore calculus sequence. As a result, linear algebra is no longer the exclusive domain of mathematicians, and a typical class in the subject often includes majors from areas such as engineering, science, economics, and psychology as well as pure and applied mathematics. Many of these students have considerable aptitude for mathematics, but do not necessarily enjoy abstraction for its own sake. Such students are often confused by the traditional axiomatic approach to linear algebra and need to be guided carefully through the intricacies of formal arguments. Furthermore, these students legitimately demand to know what linear algebra is good for and are most attracted by those ideas that are well motivated and have interesting applications.

A Primer of Linear Algebra evolved out of notes written to meet the needs of this new breed of linear algebra student. For the basic development of ideas, I have assumed only that the reader has a sound background in high school algebra. However, there are a few examples and exercises based on ideas usually covered in an elementary calculus course. Whenever possible I have attempted to motivate key ideas in advance of the formal development by showing how these ideas are natural extensions of familiar properties. For instance, matrices are introduced as a convenient means for representing the gaussian elimination process, and concepts such as linear independence, basis, inner product, and projection are first discussed in the context of the plane and 3-space before being extended to general vector spaces. The text also contains many applications of linear algebra to other areas of interest, such as engineering, probability, numerical analysis, fourier analysis, linear economics, and differential equations. Finally, since students seem to have better intuition for the arithmetic and geometric features of linear algebra than for abstraction, I have used geometric arguments whenever possible and have developed computational methods to accompany key theoretical results. Formal proofs are also given but only when such arguments illustrate important ideas.

In my opinion, one of the most important objectives of any text is to relieve the instructor of the burden of lecturing on fundamentals so that he will have time for more important duties, such as discussing applications and advanced topics and conducting problem sessions. I have attempted to meet this goal by subdividing the subject matter of the text into approximately forty learning units, each of which is devoted to the development of a single idea and is intended to represent one day of class work. Ideally, the student should be able to read and understand the material in the text without a great deal of help from his instructor. Toward this end, explanations are given in detail, and many important ideas are examined from several different angles. Those proofs that are included are discussed carefully, and in order to help the reader identify the substance of a proof, the formal argument always begins with the word *Proof* and ends with the symbol ▮. Furthermore, examples are provided to dramatize each key point, and both theoretical and computational ideas are illustrated by solved problems. Each exercise set is closely integrated with the associated reading material, and the problems are graded from routine to difficult. Problems that are especially challenging are designated by an asterisk (*). Finally, each chapter ends with a supplementary exercise section which includes both review problems and exercises that outline the development of advanced ideas. For instance, the Gerschgorin Disk Theorem and several applications are developed in the supplementary exercises at the end of chapter 6.

Enough material is contained in this book for a one quarter or one semester introductory course. It also contains several optional sections and selected supplementary exercises that may be used as the basis for advanced work. All the usual topics are present, but as is the case with any text, there are certain characteristics that reflect the point of view or prejudices of the author. A few of these are listed below, each with a brief explanation:

1. The development of ideas begins with an analysis of systems of linear equations. In my opinion, this approach works better than plunging immediately into a study of abstract vector spaces. Moreover, by covering gaussian elimination first, the student is automatically supplied with a wealth of concrete examples that come in handy later in the course.

2. The first two sections of chapter 2 and the first section of chapter 3 illustrate how concepts such as linear independence, basis, and scalar product may be used to study the plane and 3-space. These sections are intended to help the student develop intuition for abstract ideas he will soon encounter, but they may be skipped if the instructor wants to save time or feels that such an introduction is not necessary for his class.

3. At the end of section 2.5, there is a computational supplement that shows how the reduction algorithm may be used to determine dependency relationships among n-tuples. In a sense, this material is off the main line of development, but I feel the digression is more than justified by the usefulness of the technique.

4. Many examples in chapters 2 and 4 deal with properties of the set of all polynomials of degree at most n. From the standpoint of later applications, it can be argued that this set is not as important as others. However, it has the advantage of being one of the few abstract vector spaces for which students have some natural intuition, mainly because of the obvious relationship between the polynomial $a_n t^n + \cdots + a_1 t + a_0$ and the n-tuple $(a_n, \ldots, a_1, a_0)$.

5. Perhaps the most controversial aspect of the order of topics is the early appearance of inner product spaces in chapter 3. Since many students in an introductory linear algebra course have already had some experience with the scalar product in the plane and 3-space, the extension of metric concepts to abstract vector spaces is a fairly natural process. Moreover, the geometric flavor of the theory of inner product spaces appeals to the student's intuition, and for this reason, chapter 3 provides a pause that refreshes the student who is weary of struggling with abstraction.

   If the instructor wishes to cover linear transformations immediately after chapter 2, there should be no real difficulty in skipping chapter 3. The discussion of inner product spaces can then be postponed until after chapter 4 or just before chapter 7 (see suggested schedule III.)

However, it should be noted that section 4.7 and certain exercises scattered throughout the last four chapters involve terminology developed in chapter 3.

Although I am the only one with writer's cramp as a result of this text, many others have contributed to it. First of all, I am indebted to all the teachers and others who have inspired me throughout my life, and I would especially like to thank Professors Olga Taussky, John Todd, and H. F. Bohnenblust for introducing me to the mysteries of linear algebra. I am also grateful to Professor Kenneth Hoffman for his many helpful comments and criticisms, and to Professors Robert Mosher, Courtney Coleman, Bob Borrelli, and Janet Myhre and to my student assistant, Richard Schwartz, all of whom read various versions of the manuscript and made important contributions to its final form. I am also indebted to the many students who, both wittingly and unwittingly, have helped me to clarify my thoughts and purify my exercise sets. Finally, I would like to thank Art Wester and Helena DeKeukelaere, the editors at Prentice-Hall whose efforts have caused my manuscript to become a text, and Judie Timko, my faithful secretary and typist, who gave up sleep and two weekends at the horse races to help me meet my deadline.

G. L. BRADLEY
Claremont Men's College

RELATIONS AMONG MAJOR TOPICS

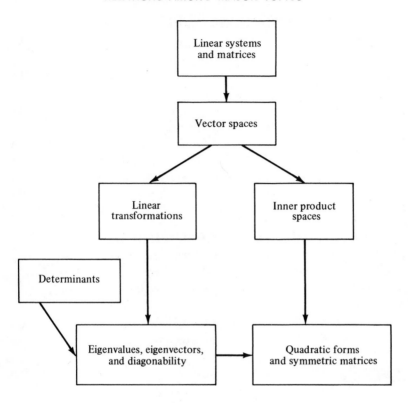

SUGGESTED SCHEDULES

I One Quarter (30 hours)

Required sections	Optional sections	Estimated time (hours)
1.1–1.6	1.7, 1.8	6
2.3–2.8	2.1, 2.2	8
4.1–4.6		7
5.2–5.3	5.1	2
6.1–6.3	6.4	3

Examinations and optional material ⟶
$$\begin{array}{r} 26 \\ +\ 4 \\ \hline 30 \end{array}$$

II One Semester (42 hours)

Required sections	Optional sections	Estimated time (hours)
1.1–1.7	1.8	7
2.3–2.8	2.1, 2.2	8
3.2–3.4	3.1, 3.5	4
4.1–4.6	4.7	7
5.1–5.3		3
6.1–6.3	6.4	4
7.1–7.3		4

Examinations and optional material ⟶
$$\begin{array}{r} 37 \\ +\ 5 \\ \hline 42 \end{array}$$

III One Semester (42 hours)

Required sections	Optional sections	Estimated time (hours)
1.1–1.7	1.8	7
2.3–2.8	2.1, 2.2	8
4.1–4.6		7
5.1–5.3		3
6.1–6.3	6.4	4
3.2–3.4	3.1, 3.5, 4.7	4
7.1–7.3		4

Examinations and optional material ⟶
$$\begin{array}{r} 37 \\ +\ 5 \\ \hline 42 \end{array}$$

A Primer Of
Linear Algebra

CHAPTER

1

Matrices and Systems of Linear Equations

1.1 MATRIX REPRESENTATION OF A LINEAR SYSTEM

Systems of linear equations play an important role not only in mathematics, but also in many other areas of scientific inquiry, such as physics, economics, biology, and engineering to name a few. Historically, the subject now known as linear algebra evolved out of the attempts of nineteenth century mathematicians to formalize the study of linear systems, and we shall find that such systems lie at the heart of even the most abstract topics in linear algebra. For this reason, it is important to have a general method for solving linear systems from the very beginning of our work, and the first three sections of this initial chapter are devoted to the development of such a method.

First, we must establish the basic vocabulary to be used in discussing systems of linear equations. Accordingly, suppose that a system of m linear equations in the n variables $x_1, x_2, \ldots, x_n$ is written in the form

$$
\begin{aligned}
a_{11}x_1 + a_{12}x_2 + \cdots + a_{1n}x_n &= b_1 \\
a_{21}x_1 + a_{22}x_2 + \cdots + a_{2n}x_n &= b_2 \\
&\vdots \\
a_{m1}x_1 + a_{m2}x_2 + \cdots + a_{mn}x_n &= b_m
\end{aligned}
\tag{1.1}
$$

where the mn coefficients a_{ij} on the left and the m constants b_k on the right are fixed scalars.* We shall refer to (1.1) as an $m \times n$ *linear system* and to any ordered n-tuple $(x_1, x_2, \ldots, x_n)$ whose components x_j satisfy (1.1) as a *solution* of that system. If the system has at least one solution, it is said to be *consistent*, while an *inconsistent* system is one which has no solution. The system is said to be *solved* if it is inconsistent or if it is consistent and all its solutions can be displayed or described in a meaningful fashion. In the special case where $b_1 = b_2 = \ldots = b_m = 0$, the resulting system is said to be *homogeneous*, and if at least one of the constants b_k is non-zero, the system is *non-homogeneous*. The zero n-tuple $(0, 0, \ldots, 0)$, which is a solution of any homogeneous $m \times n$ linear system, is often referred to as the *trivial solution* of such a system, while any other solution is said to be *non-trivial*.

EXAMPLE 1.   Consider the 3×2 linear systems

$$x_1 + 2x_2 = 0 \qquad\qquad x_1 + 2x_2 = 5$$

$$2x_1 + x_2 = 0 \quad \text{and} \quad 2x_1 + x_2 = 4$$

$$-x_1 + x_2 = 0 \qquad\qquad -x_1 + x_2 = 1$$

The system on the left is homogeneous, while that on the right is nonhomogeneous. Both systems are consistent, and it can be shown that the homogeneous system has only the trivial solution $(0, 0)$ and that the nonhomogeneous system has the unique solution $(1, 2)$. Sometimes a small change can convert a consistent linear system into one which has no solution. For instance, the system

$$x_1 + 2x_2 = 5$$

$$2x_1 + x_2 = 4$$

$$-x_1 + x_2 = 0$$

is inconsistent since the first two equations have the unique solution $x_1 = 1$, $x_2 = 2$ but the third equation is not satisfied because $-1 + 2 \neq 0$.

The reader may wonder why we chose to use the subscripted variables $x_1, x_2, \ldots, x_n$ to represent the unknowns in system (1.1) instead of representing each unknown by a different letter of the alphabet, as in the following

*Throughout this text, it may be assumed that any statement which involves the word "scalar" is valid for both real numbers and complex numbers. In more abstract treatments of linear algebra, it is shown that most of our results are valid if "scalar" is taken to mean an element of a mathematical system known as a field.

example:

$$2x + y + z = -3$$
$$-x + 2y - 3z = 5 \tag{1.2}$$

The latter method of labeling variables does have some advantages, but it clearly cannot be used to describe a system with a great many variables. Therefore, in order to insure uniformity throughout our work, we shall usually express even "small" systems such as (1.2) in the general form described in (1.1). Thus, system (1.2) becomes

$$2x_1 + x_2 + x_3 = -3$$
$$-x_1 + 2x_2 - 3x_3 = 5$$

Now that we have introduced enough terminology to be able to talk about linear systems, our next goal is to develop a method for solving them. Probably the most widely used systematic approach to solving a given linear system is the process known as either *elimination of variables* or *gaussian elimination* (after C. F. Gauss, a mathematical genius of the nineteenth century). This method is based on the observation that the solution set of a given linear system is not affected by any of the following procedures:

Type I.   Exchanging two equations.
Type II. Multiplying an equation by a non-zero scalar.
Type III. Adding a scalar multiple of one of the equations to   (1.3)
            another equation.

These manipulations are then used to "eliminate" variables and systematically reduce the given system to a "simple" form which is easier to solve.

EXAMPLE 2.   To solve the 2×3 linear system

$$x_1 + 2x_2 + x_3 = -8$$
$$3x_1 + 2x_2 - 5x_3 = -4 \tag{1.4}$$

by gaussian elimination, we first eliminate x_1 from the second equation by adding -3 times the first equation to the second. The resulting system looks like this:

$$x_1 + 2x_2 + x_3 = -8$$
$$- 4x_2 - 8x_3 = 20 \tag{1.5}$$

System (1.5) may be simplified even further if we multiply the second equation by $-\frac{1}{4}$ and then add -2 times the new second equation to the first equation:

$$x_1 \qquad - 3x_3 = 2$$
$$x_2 + 2x_3 = -5$$

(1.6)

Since we have used only the allowable manipulations listed in (1.3) in our reduction, it follows that the "simple" system displayed in (1.6) has the same solution set as (1.4). To solve both systems, we note that (1.6) can be rewritten as

$$x_1 = \quad 2 + 3x_3$$
$$x_2 = -5 - 2x_3$$

(1.7)

from which it follows that (x_1, x_2, x_3) is a solution if and only if there exists a scalar t such that

$$x_1 = 2 + 3t$$
$$x_2 = -5 - 2t$$
$$x_3 = t$$

(1.8)

Incidentally, an auxiliary variable such as t is called a *parameter*, and for this reason, the mode of representation used in (1.8) is known as a *parametric solution* of the given system.

If we look closely at example 2 or any other example of gaussian elimination, it becomes clear that the unknowns $x_1, x_2, \ldots, x_n$ serve only to mark a position in each equation and play no direct role in the actual reduction process. Therefore, a given $m \times n$ linear system can be solved with an economy of notation by simply removing the variables and performing appropriate manipulations on the rows of the array formed from the coefficients and right side constants of the system. For instance, the 3×3 system

$$x_1 + 5x_2 + 3x_3 = \quad 1$$
$$2x_1 - 2x_2 + \quad x_3 = \quad 4$$
$$-4x_1 + 0x_2 + 3x_3 = -1$$

(1.9)

can be represented by the 3×4 array

$$A = \begin{bmatrix} 1 & 5 & 3 & 1 \\ 2 & -2 & 1 & 4 \\ -4 & 0 & 3 & -1 \end{bmatrix}$$

The variable x_1 may be eliminated from the second equation of system (1.9) by adding -2 times the first equation to the second equation to form the system

$$x_1 + 5x_2 + 3x_3 = 1$$

$$0x_1 - 12x_2 - 5x_3 = 2$$

$$-4x_1 + 0x_2 + 3x_3 = -1$$

which is represented by the array

$$B = \begin{bmatrix} 1 & 5 & 3 & \vdots & 1 \\ 0 & -12 & -5 & \vdots & 2 \\ -4 & 0 & 3 & \vdots & -1 \end{bmatrix}$$

Thus, the reduction step in question may be represented by writing

$$\begin{bmatrix} 1 & 5 & 3 & \vdots & 1 \\ 2 & -2 & 1 & \vdots & 4 \\ -4 & 0 & 3 & \vdots & -1 \end{bmatrix} \xrightarrow{-2R_1 + R_2 \to R_2} \begin{bmatrix} 1 & 5 & 3 & \vdots & 1 \\ 0 & -12 & -5 & \vdots & 2 \\ -4 & 0 & 3 & \vdots & -1 \end{bmatrix}$$

where the notation $-2R_1 + R_2 \to R_2$ reminds us that the second row of B is obtained by adding -2 times the first row of A to its second row.

Rectangular arrays of numbers, which mathematicians call *matrices*, occur so often and in so many different contexts in linear algebra that a special body of notation and terminology is used when discussing them. For definiteness, consider the $m \times n$ matrix

$$A = \begin{bmatrix} a_{11} & a_{12} & \cdots & a_{1n} \\ a_{21} & a_{22} & \cdots & a_{2n} \\ \cdot & \cdot & & \cdot \\ \cdot & \cdot & & \cdot \\ \cdot & \cdot & & \cdot \\ a_{m1} & a_{m2} & \cdots & a_{mn} \end{bmatrix}$$

The integers m and n are called the *row* and *column dimensions* of A, respectively, and in the special case where these two numbers are the same, we shall say that A is a *square* matrix. The symbol a_{ij} is used to denote the entry located in the ith row and jth column of A. It is convenient to refer to a_{ij} as the (i, j) entry of A and to write $A = (a_{ij})$ to indicate the relationship between A and its general entry a_{ij}. Finally, any matrix formed from A by removing certain rows and columns is called a *submatrix* of A.

To illustrate these notational conventions, let

$$A = \begin{bmatrix} -1 & 2 & -5 & 8 \\ 3 & 7 & 0 & -4 \\ 5 & 6 & -2 & 1 \end{bmatrix} \quad \text{and} \quad B = \begin{bmatrix} 9 & -2 \\ 0 & 4 \end{bmatrix}$$

Then B is a square matrix, while A has row dimension 3 and column dimension 4. The $(1, 2)$ entry of B is -2, a_{32} is 6, and $b_{11} = 9$. The 3×3 array

$$\begin{bmatrix} -1 & -5 & 8 \\ 3 & 0 & -4 \\ 5 & -2 & 1 \end{bmatrix}$$

is the submatrix of A formed by removing the second column. Other submatrices of A include the following:

$$\begin{bmatrix} -1 & 2 & 8 \\ 5 & 6 & 1 \end{bmatrix}, \quad \begin{bmatrix} 3 & -4 \\ 5 & 1 \end{bmatrix}, \quad \text{and} \quad \begin{bmatrix} 2 & -5 \\ 7 & 0 \\ 6 & -2 \end{bmatrix}$$

An $m \times 1$ matrix C is also known as a *column vector*, and its jth component is denoted by c_j instead of c_{j1}. Similarly, a $1 \times n$ matrix R is called a *row vector*, and its entries are labeled r_i. For example,

$$C = \begin{bmatrix} -1 \\ 7 \\ 0 \end{bmatrix}$$

is a column vector, and $R = [-9 \ 3 \ 1 \ 5]$ is a row vector.

The special matrices and manipulations used in the matrix form of gaussian elimination also have names of their own. To be specific, the $m \times n$ matrix $A = (a_{ij})$ whose (i, j) entry is the coefficient of the jth variable in the ith equation of system (1.1) is called the *coefficient matrix* of that system. The $m \times (n + 1)$ matrix formed by joining the column vector of right side constants to the right of the coefficient matrix A is known as the *augmented matrix* of system (1.1). For instance, the coefficient matrix and augmented matrix of the 3×4 linear system

$$2x_1 - \ x_2 + 5x_3 - \ x_4 = 7$$

$$-x_1 + 4x_2 \qquad\quad + 2x_4 = 1$$

$$3x_2 - \ x_3 + 9x_4 = 0$$

are, respectively,

$$\begin{bmatrix} 2 & -1 & 5 & -1 \\ -1 & 4 & 0 & 2 \\ 0 & 3 & -1 & 9 \end{bmatrix} \quad \text{and} \quad \left[\begin{array}{cccc:c} 2 & -1 & 5 & -1 & 7 \\ -1 & 4 & 0 & 2 & 1 \\ 0 & 3 & -1 & 9 & 0 \end{array}\right]$$

Anything that is done to the equations of a given linear system is reflected in a change in the rows of the augmented matrix C of the system. In particular, the three basic manipulations described in (1.3) affect the rows of C in the following manner:

Type I.   Exchanges two rows of C.

Notation: "$R_1 \leftrightarrow R_3$" means "exchange rows 1 and 3."

Type II.   Multiplies a row of C by a non-zero scalar.

Notation: "$-5R_2 \rightarrow R_2$" means "multiply the second row by -5."

Type III.   Adds a scalar multiple of one row of C to another row.

Notation: "$-2R_3 + R_5 \rightarrow R_5$" means "add -2 times the third row to the fifth row."

Collectively, these row manipulations are called *elementary row operations*. If the $m \times n$ matrix B can be derived from the $m \times n$ matrix A by a finite sequence of elementary row operations, we say that B is *row equivalent* to A. It can be shown that B is row equivalent to A if and only if A is row equivalent to B (see exercise 11). Therefore, when discussing this relationship, we are justified in saying simply that A and B are row equivalent to each other.

We have already observed that the solution set of a given linear system is not affected by the manipulations listed in (1.3). The matrix form of this observation may be stated as follows:

> If the augmented matrices of two $m \times n$ linear systems are row equivalent to one another, then the systems either have the same solutions or they are both inconsistent.

In the following problem, we show how this principle may be used to solve a specific linear system.

PROBLEM.   Find all solutions of the 3×4 linear system

$$x_1 - 2x_2 + 2x_3 + 5x_4 = 0$$

$$3x_1 - 6x_2 + x_3 \qquad\qquad = 1 \qquad\qquad (1.10)$$

$$-x_1 + 2x_2 + x_3 + 4x_4 = 2$$

or prove that the system is inconsistent.

Solution: In order to reduce system (1.10) to a form in which its solutions will be apparent, we begin by using the first equation to eliminate the variable x_1 from the other two. In the matrix version of this reduction step, we perform two type 3 elementary row operations on the augmented matrix of the system:

$$
\left[\begin{array}{rrrr:r}
1 & -2 & 2 & 5 & 0 \\
3 & -6 & 1 & 0 & 1 \\
-1 & 2 & 1 & 4 & 2
\end{array}\right]
\xrightarrow[R_1+R_3 \to R_3]{-3R_1+R_2 \to R_2}
\left[\begin{array}{rrrr:r}
1 & -4 & 2 & 5 & 0 \\
0 & 0 & -5 & -15 & 1 \\
0 & 0 & 3 & 9 & 2
\end{array}\right]
$$

As a by-product of this reduction step, the variable x_2 has also been eliminated from the second and third equations. Next, we use the new second equation to eliminate the variable x_3 from the third equation. The matrix form of this step is as follows:

$$
\left[\begin{array}{rrrr:r}
1 & -4 & 2 & 5 & 0 \\
0 & 0 & -5 & -15 & 1 \\
0 & 0 & 3 & 9 & 2
\end{array}\right]
\xrightarrow[-3R_2+R_3 \to R_3]{-\frac{1}{5}R_2 \to R_2}
\left[\begin{array}{rrrr:r}
1 & -4 & 2 & 5 & 0 \\
0 & 0 & 1 & 3 & -\frac{1}{5} \\
0 & 0 & 0 & 0 & \frac{13}{5}
\end{array}\right]
$$

If the variables are restored to the matrix on the right of this reduction, we obtain the system

$$x_1 - 4x_2 + 2x_3 + 5x_4 = 0$$

$$0x_1 + 0x_2 + x_3 + 3x_4 = -\tfrac{1}{5}$$

$$0x_1 + 0x_2 + 0x_3 + 0x_4 = \tfrac{13}{5}$$

which is inconsistent, for the third equation clearly has no solution. Since the augmented matrix of this inconsistent system is row equivalent to that of system (1.10), we conclude that system (1.10) is also inconsistent.

Although we have talked a great deal about gaussian elimination in this section, we have yet to say what it really is. Therefore, in the next section, we provide a precise description of the logical "strategy" behind the elimination process, and in section 1.3, we translate our findings into matrix terms.

Exercise Set 1.1

1. In each of the following cases, either show that the given system is inconsistent or obtain a parametric solution (example 2 and the solved problem may help).

   (i) $x_1 + x_2 = 0$

       $2x_1 - x_2 = 0$

       $3x_1 + 2x_2 = 0$

   (ii) $x_1 - 2x_2 = 5$

       $2x_1 + x_2 = 3$

       $x_1 - x_2 = 1$

(iii) $x_1 - x_2 + x_3 = 5$ (iv) $x_1 \qquad + 5x_3 \qquad = 3$

$-x_1 + x_2 + 2x_3 = 7$ $x_2 - 7x_3 \qquad = 0$

$2x_1 + x_2 - x_3 = 4$ $x_4 = 8$

2. Answer the following questions about the 3×4 matrix

$$A = \begin{bmatrix} -2 & 4 & -1 & 2 \\ -3 & 0 & 3 & -6 \\ 7 & 8 & 5 & 10 \end{bmatrix}$$

(i) What is the $(2, 4)$ entry of A? What is a_{32}?

(ii) Write out the entries of the 2×3 submatrix formed by removing the first row and second column of A.

(iii) Which of the following arrays is *not* a submatrix of A?

$$B_1 = \begin{bmatrix} -2 \\ -3 \\ 7 \end{bmatrix}, \quad B_2 = \begin{bmatrix} 4 & -1 \\ 8 & 5 \end{bmatrix}, \quad B_3 = \begin{bmatrix} 4 & -1 \\ 3 & -6 \end{bmatrix},$$

$$B_4 = \begin{bmatrix} -3 & 3 & -6 \end{bmatrix}$$

3. In each of the following cases, fill in the entries of the matrix on the right by performing the indicated elementary row operations.

(i) $\begin{bmatrix} -3 & 7 & 2 & 8 \\ 5 & 1 & 7 & 2 \\ 1 & -1 & 0 & 1 \end{bmatrix} \xrightarrow[\substack{-5R_1 + R_2 \to R_2 \\ 3R_1 + R_3 \to R_3}]{R_1 \longleftrightarrow R_3} \begin{bmatrix} & & & \\ & & & \\ & & & \end{bmatrix}$

(ii) $\begin{bmatrix} 1 & -3 & 4 & 2 \\ -3 & 8 & 1 & 5 \\ -1 & 2 & 9 & 9 \end{bmatrix} \xrightarrow[\substack{-R_2 \to R_2 \\ R_2 + R_3 \to R_3}]{\substack{3R_1 + R_2 \to R_2 \\ R_1 + R_3 \to R_3}} \begin{bmatrix} & & & \\ & & & \\ & & & \end{bmatrix}$

4. (i) Write out the entries of the coefficient matrix and augmented matrix of the 3×4 non-homogeneous linear system

$$3x_1 + x_2 - 3x_3 + 4x_4 = 12$$

$$-x_1 + 2x_2 + 8x_3 + x_4 = -4$$

$$x_1 \qquad - 2x_3 + x_4 = 4$$

(ii) Show that the augmented matrix of the system in part (i) is row equivalent to the matrix

$$\begin{bmatrix} 1 & 0 & -2 & 1 & | & 4 \\ 0 & 1 & 3 & 1 & | & 0 \\ 0 & 0 & 0 & 0 & | & 0 \end{bmatrix}$$

(iii) Find a parametric solution for the system in part (i).

5. The augmented matrix of a certain linear system is known to be row equivalent to the matrix

$$\begin{bmatrix} 1 & -4 & 0 & 2 & \vdots & 0 \\ 0 & 0 & 1 & -3 & \vdots & -9 \\ 0 & 0 & 0 & 0 & \vdots & 4 \end{bmatrix}$$

Determine whether or not the system is consistent, and if it is, find all its solutions.

6. Show that a 4×3 linear system whose augmented matrix is row equivalent to the matrix

$$\begin{bmatrix} 1 & 0 & 0 & \vdots & -2 \\ 0 & 1 & 0 & \vdots & 5 \\ 0 & 0 & 1 & \vdots & 7 \\ 0 & 0 & 0 & \vdots & 0 \end{bmatrix}$$

has a unique solution. What is the solution?

7. A man has a pocket full of nickels, dimes, and pennies. If he has a total of 14 coins and 89 cents, how many coins of each type does he have?

*8. The graph of a certain quadratic polynomial $P(x) = ax^2 + bx + c$ passes through the point $(-1, 5)$ and has a horizontal tangent at $(1, 7)$. Find the coefficients a, b, and c.

*9. Show that the effect of each of the three elementary row operations can be reversed by an operation of the same type. For example, the effect of multiplying a row by 3 can be reversed by multiplying the same row by $\frac{1}{3}$.

*10. Show that the $m \times n$ matrix A is row equivalent to the matrix B if and only if B is row equivalent to A. *Hint:* Exercise 9 may help.

*11. Show that if A is row equivalent to B, and B is row equivalent to C, then A is also row equivalent to C.

1.2 THE GENERAL FORM OF GAUSSIAN ELIMINATION

The intuitive approach to gaussian elimination used in section 1.1 is fine for most purposes, but there are some important applications in which it is necessary to have a precise general description of the elimination process. In this section, we provide such a description, and in so doing, we develop terminology and record results which will be used throughout our work.

We begin by analyzing a kind of system that is so "simple" it can be solved immediately. Then after learning how to deal with such a system, we shall describe a step by step process for reducing a given linear system to a "simple" system with the same solution set. Unfortunately, in order to cover all possible cases which may arise, we must use rather formal language in our definition. For this reason, we advise the reader to use the example which follows definition 1.1 as a guide to understanding the details of the definition.

Definition 1.1.   An $m \times n$ *reduced linear system* is one with the following characteristics:

(i) The equations are arranged so that each of the first r $(0 \le r \le m)$ equations has at least one non-zero coefficient, but the last $m - r$ equations are trivial, that is, of the form $0x_1 + 0x_2 + \cdots + 0x_n = 0$.

(ii) For $i = 1, 2, \ldots, r$, the leading coefficient in the ith equation is a 1. The variable with which this coefficient is associated is called the ith *basic variable* of the system.

(iii) If $i < j$ and x_p and x_q are the ith and jth basic variables, respectively, then $p < q$.

(iv) The ith basic variable appears (with non-zero coefficient) in only the ith equation.

EXAMPLE 1.   The 4×5 linear system

$$x_1 - 3x_2 + 0x_3 - 2x_4 + 0x_5 = 7$$

$$0x_1 + 0x_2 + x_3 + 5x_4 + 0x_5 = 0$$

$$0x_1 + 0x_2 + 0x_3 + 0x_4 + x_5 = 9 \tag{1.11}$$

$$0x_1 + 0x_2 + 0x_3 + 0x_4 + 0x_5 = 0$$

is in reduced form. Note that $r = 3$, since the system has three non-trivial equations, and that the first, second, and third basic variables are x_1, x_3, and x_5, respectively.

Thanks to condition (iii) of the definition, the basic variables of any reduced system must lie on a jagged diagonal tending from the upper left corner to the lower right, and condition (iv) assures us that each basic variable appears in just one equation. These features become more apparent in system (1.11) when the trivial equation is left out and "gaps" are used in place of 0's:

$$x_1 - 3x_2 \quad - 2x_4 \quad = 7$$

$$x_3 + 5x_4 \quad = 0 \tag{1.12}$$

$$x_5 = 9$$

It is easy to obtain a parametric solution of a reduced system because each non-trivial equation in such a system can be rewritten so that its basic variable is expressed in terms of non-basic variables. For example, system

(1.12) can be rewritten as

$$x_1 = 7 + 3x_2 + 2x_4$$

$$x_3 = \qquad\qquad - 5x_4$$

$$x_5 = 9$$

and by identifying x_2 and x_4 with the parameters t_1 and t_2, respectively, we obtain the parametric solution

$$x_1 = 7 + 3t_1 + 2t_2$$

$$x_2 = \qquad t_1$$

$$x_3 = \qquad\qquad - 5t_2 \qquad\qquad (1.13)$$

$$x_4 = \qquad\qquad t_2$$

$$x_5 = 9$$

Since each reduced system has a parametric solution, we can make the following general statement.

Lemma 1.1.   An $m \times n$ reduced linear system is consistent.

From the parametric solution displayed in (1.13) we see that reduced system (1.11) has infinitely many solutions since each choice of t_1 and t_2 results in a different solution. For instance, with $t_1 = 1$ and $t_2 = 0$, we have the solution $(10, 1, 0, 0, 9)$, and with $t_1 = 0$, $t_2 = 1$ we have $(9, 0, 5, 1, 9)$. In general, an $m \times n$ reduced system has infinitely many solutions if at least one of its n variables is not basic, and it has a unique solution if all n variables are basic. These observations may be summarized as follows.

Lemma 1.2.   An $m \times n$ reduced linear system with r non-trivial equations has a unique solution if $r = n$ and infinitely many solutions if $r < n$.

We have already discussed a system with infinitely many solutions [system (1.12)], and the following is an example of a system with a unique solution.

EXAMPLE 2.   The 4×3 reduced linear system

$$\boxed{x_1} + 0x_2 + 0x_3 = -3$$
$$0x_1 + \boxed{x_2} + 0x_3 = 5$$
$$0x_1 + 0x_2 + \boxed{x_3} = 0$$
$$0x_1 + 0x_2 + 0x_3 = 0$$

has the unique solution $(-3, 5, 0)$.

In the special case where $m < n$, it is clearly impossible for an $m \times n$ reduced linear system to have n non-trivial equations, and we have the following result.

Lemma 1.3.   If $m < n$, an $m \times n$ reduced linear system must have infinitely many solutions.

Next, we shall describe a general procedure (known as the *reduction algorithm*) for using the manipulations listed in (1.3) to reduce a given linear system to a form which can be easily analyzed. To illustrate the main features of this algorithm, we shall couple a general description of its key steps with an analysis of the 3×4 system

$$3x_1 - 9x_2 + x_3 + 4x_4 = 6$$
$$-x_1 + 3x_2 + x_3 - 4x_4 = 2 \qquad (1.14)$$
$$-2x_1 + 6x_2 + 3x_3 - 10x_4 = 7$$

General Statement	*Example*
Step 1 Pick one of the left-most non-zero coefficients. This coefficient and the associated variable form the *pivot term*, and the equation in which this term appears is the *pivot equation*.	$3x_1 - 9x_2 + x_3 + 4x_4 = 6$ $-x_1 + 3x_2 + x_3 - 4x_4 = 2$ $-2x_1 + 6x_2 + 3x_3 - 10x_4 = 7$ pivot equation
Step 2 Exchange the pivot equation with the first equation, and then divide each coefficient in the pivot equation by the pivot coefficient.	$x_1 - 3x_2 - x_3 + 4x_4 = -2$ $3x_1 - 9x_2 + x_3 + 4x_4 = 6$ $-2x_1 + 6x_2 + 3x_3 - 10x_4 = 7$ $\dfrac{1}{-1}$ times the pivot equation

General Statement	*Example*

Step 3 Use the pivot term to eliminate the pivot variable from all equations "below" the pivot equation.

$$x_1 - 3x_2 - x_3 + 4x_4 = -2$$

$$4x_3 - 8x_4 = 12 \;\leftarrow$$

$$x_3 - 2x_4 = 3 \;\leftarrow$$

$2 \times$ pivot equation $+$ equation #3

$-3 \times$ pivot equation $+$ equation #2

At this point, the variable x_1 has been eliminated from all but the first equation, and we say that the first *stage* of the reduction process is complete. To enter the second stage, we simply ignore the first equation (the pivot equation of the first stage) and apply steps 1, 2, 3 to the $(m - 1) \times n$ subsystem formed from the other equations. In our example, we are led to consider the 2×4 system indicated next:

$$x_1 - 3x_2 - x_3 + 4x_4 = -2$$

$$0x_1 + 0x_2 + 4x_3 - 8x_4 = 12$$
$$0x_1 + 0x_2 + x_3 - 2x_4 = 3$$

subsystem to be reduced during the second stage

Since x_1 and x_2 have only 0 coefficients, the pivot variable of the second stage must be x_3. We choose $4x_3$ as the pivot term and carry out steps 2 and 3 to obtain the following system at the end of the second stage of reduction:

$$x_1 - 3x_2 - x_3 + 4x_4 = -2$$

$$0x_1 + 0x_2 + x_3 - 2x_4 = 3 \tag{1.15}$$

subsystem to be reduced during the third stage

$$0x_1 + 0x_2 + 0x_3 + 0x_4 = 0$$

In general, the process of reduction continues until a stage is reached (say, the kth stage) where it is impossible to find a non-zero coefficient in the $(m - k) \times n$ subsystem to be reduced during the $(k + 1)$st stage. At this point, we say that the *downward phase* of the reduction is complete. Note that there are only two ways for this terminal situation to occur:

1. At least one of the $m - k$ equations in the system to be reduced during the $(k + 1)$st stage has the form

$$0x_1 + 0x_2 + \cdots + 0x_n = d$$

where d is a non-zero scalar.

2. Either no equations or only trivial equations are left at the end of the kth stage.

If the first case occurs, the system must be inconsistent, for an equation of the form

$$0x_1 + 0x_2 + \cdots + 0x_n = d$$

clearly has no solution. Henceforth, we shall refer to such an equation as "bad." If none of the equations is "bad," the reduction process guarantees that the first variable in each equation has coefficient 1 and that this variable appears in no "lower" equation. We then complete the reduction by using the pivot term of each equation to eliminate the pivot variable from all "higher" equations. When this so-called *upward phase* of reduction is complete, the system which remains must be in reduced form and can thus be analyzed by the methods developed earlier in this section. The main features of the algorithm described above are displayed schematically in figure 1.1.

In our example, the downward phase of reduction is complete at the end of the second stage since the subsystem to be reduced during the third stage contains only the trivial equation

$$0x_1 + 0x_2 + 0x_3 + 0x_4 = 0$$

For the upward stage of reduction, we use the pivot term x_3 of the second equation in system (1.15) to eliminate x_3 from the first equation:

$$
\begin{array}{ccc}
x_1 - 3x_2 - x_3 + 4x_4 = -2 & & x_1 - 3x_2 \quad\quad + 2x_4 = 1 \\
& \text{becomes} & \\
x_3 - 2x_4 = 3 & & x_3 - 2x_4 = 3
\end{array}
$$

The parametric solution of the reduced system on the right is

$$
\begin{aligned}
x_1 &= 1 + 3t_1 - 2t_2 \\
x_2 &= \quad\ t_1 \\
x_3 &= 3 \quad\quad + 2t_2 \\
x_4 &= \quad\quad\quad\ t_2
\end{aligned}
\tag{1.16}
$$

and since our reduction process uses only the manipulations described in (1.3), we conclude that (1.16) also characterizes the solution set of system (1.14).

The following example shows what happens when the logical strategy described above is applied to an inconsistent system.

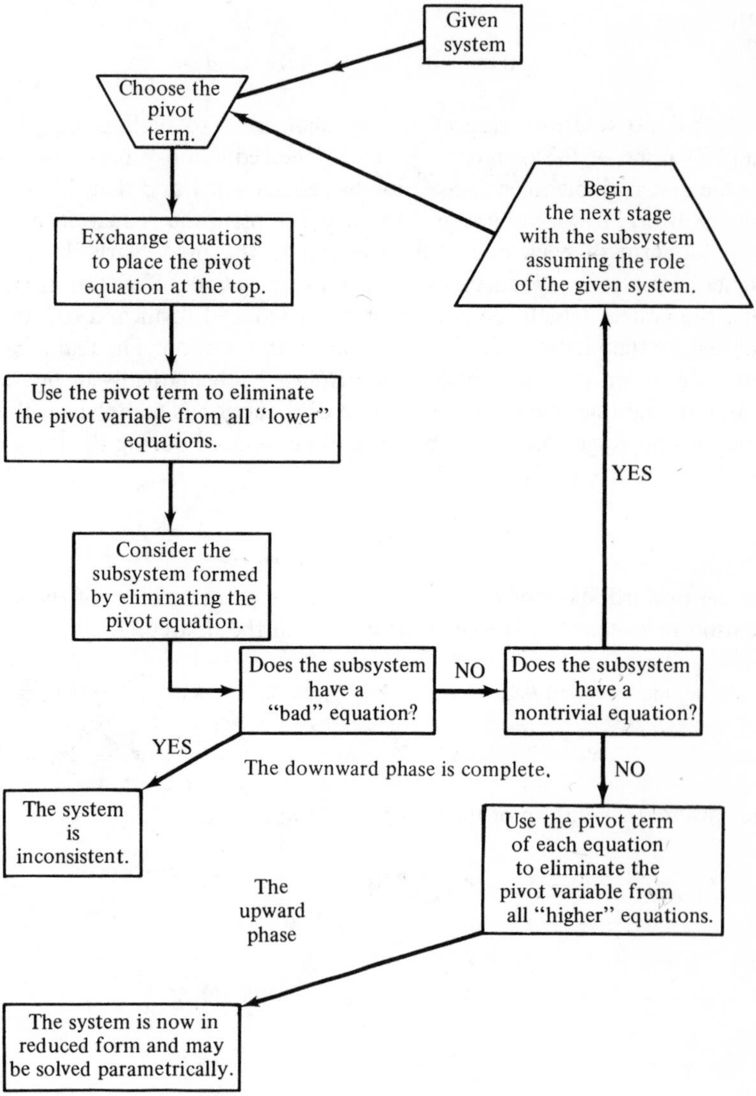

Given
system

Choose the
pivot
term.

Exchange equations
to place the pivot
equation at the top.

Begin
the next stage
with the subsystem
assuming the role
of the given system.

Use the pivot term to eliminate
the pivot variable from all "lower"
equations.

YES

Consider the
subsystem formed
by eliminating the
pivot equation.

| Does the subsystem have a "bad" equation? | NO | Does the subsystem have a nontrivial equation? |

The downward phase is complete.

YES

The system
is
inconsistent.

NO

The
upward
phase

Use the pivot term
of each equation
to eliminate the
pivot variable from
all "higher" equations.

The system is now in
reduced form and may
be solved parametrically.

The reduction algorithm

FIGURE 1.1

EXAMPLE 3.   Applying the reduction algorithm to the 4×4 linear system

$$2x_1 - 5x_2 + 4x_3 - 3x_4 = 7$$

$$-3x_1 + 7x_2 - x_3 + x_4 = -3$$

$$-x_1 + 2x_2 + 3x_3 - 2x_4 = 4 \qquad (1.17)$$

$$3x_1 - 8x_2 + 11x_3 - 8x_4 = 5$$

we obtain the following system at the end of the second stage:

$$x_1 - 2x_2 - 3x_3 + 2x_4 = -4$$

$$0x_1 + x_2 - 10x_3 + 7x_4 = -15$$

$$\boxed{0x_1 + 0x_2 + 0x_3 + 0x_4 = -13} \quad \text{``bad''} \atop \text{equation}$$

$$0x_1 + 0x_2 + 0x_3 + 0x_4 = 0$$

Since this system has a "bad" equation, it must be inconsistent, and so is system (1.17).

It is convenient to say that two $m \times n$ linear systems are *equivalent* to one another if they are both inconsistent or if they are both consistent and have the same set of solutions. Since the manipulations listed in (1.3) have no effect on the solution set of a given linear system, our algorithm shows that each such system must be equivalent either to a system with a "bad" equation or to one which is in reduced form. Using this observation in conjunction with lemmas 1.1, 1.2, and 1.3, we obtain the following important results.

Theorem 1.1.   When the reduction algorithm is applied to a consistent linear system, it results in the creation of an equivalent system in reduced form. However, if it is applied to an inconsistent system, the algorithm creates an equivalent system with a "bad" equation.

Theorem 1.2.   An $m \times n$ linear system which is equivalent to a reduced linear system with r non-trivial equations has a unique solution if $r = n$ and infinitely many solutions if $r < n$.

Theorem 1.3.   If $m < n$, a consistent $m \times n$ linear system must have infinitely many solutions.

These results are especially useful if we wish to make a general qualitative statement about the solution set of a system without actually finding the solutions. For instance, in chapter 2, our proof of a certain crucial result (lemma 2.1) requires us to show that a linear system with more variables than equations has more than one solution, and this is exactly what is stated in theorem 1.3.

In the next section, we conclude our introduction to linear systems by obtaining a matrix form of the reduction algorithm and translating the key results of this section into matrix terms.

Exercise Set 1.2

1. In each of the following cases, determine whether or not the given "simple" system is consistent, and if it is, identify its basic variables and provide a parametric solution.

(i) $x_1 - 2x_2 + 0x_3 + 0x_4 - 3x_5 = 1$ (ii) $x_1 + 3x_2 - 5x_3 = 9$

$\quad 0x_1 + 0x_2 + x_3 + 0x_4 + 4x_5 = 8$ $\quad 0x_1 + 0x_2 + 0x_3 = 0$

$\quad 0x_1 + 0x_2 + 0x_3 + x_4 - x_5 = 0$

(iii) $x_1 + 0x_2 + 2x_3 - x_4 = 5$ (iv) $x_1 + 0x_2 + 0x_3 = -1$

$\quad 0x_1 + x_2 - x_3 + 2x_4 = -1$ $\quad 0x_1 + x_2 + 2x_3 = 2$

$\quad 0x_1 + 0x_2 + 0x_3 + 0x_4 = 6$

(v) $0x_1 + x_2 - 3x_3 + 0x_4 = -2$ (vi) $x_1 + 0x_2 + 0x_3 = 5$

$\quad 0x_1 + 0x_2 + 0x_3 + x_4 = 7$ $\quad 0x_1 + x_2 + 0x_3 = 0$

$\quad 0x_1 + 0x_2 + 0x_3 + 0x_4 = 0$ $\quad 0x_1 + 0x_2 + x_3 = 2$

2. Each of the linear systems in the column on the left is equivalent to one of the reduced systems in the right column. Determine the correct pairings.

(1) $-4x_1 + x_2 + 2x_3 = 3$ (a) $x_1 - 2x_2 + 0x_3 = -11$

$\quad 3x_1 + x_2 + 7x_3 = -19$ $\quad 0x_1 + 0x_2 + x_3 = -3$

$\quad 0x_1 + 2x_2 + 5x_3 = -5$ $\quad 0x_1 + 0x_2 + 0x_3 = 0$

(2) $x_1 - x_2 - 4x_3 = 5$ (b) $x_1 + 5x_2 + 0x_3 = 1$

$\quad 3x_1 + 5x_2 - 4x_3 = 15$ $\quad 0x_1 + 0x_2 + 0x_3 = 0$

$\quad -x_1 - x_2 + 2x_3 = -5$ $\quad 0x_1 + 0x_2 + 0x_3 = 0$

(3) $4x_1 - 3x_2 - 15x_3 = 20$ (c) $x_1 + 0x_2 + 0x_3 = -1$

$\quad\quad x_1 + 6x_2 + 3x_3 = 5$ $\quad\quad 0x_1 + x_2 + 0x_3 = 5$

$\quad\quad 2x_1 - x_2 - 7x_3 = 10$ $\quad\quad 0x_1 + 0x_2 + x_3 = -3$

(4) $-2x_1 + 4x_2 + 9x_3 = -5$ (d) $x_1 + 0x_2 - 3x_3 = 5$

$\quad\quad x_1 - 2x_2 - 4x_3 = 1$ $\quad\quad 0x_1 + x_2 + x_3 = 0$

$\quad\quad -x_1 + 2x_2 + x_3 = 8$ $\quad\quad 0x_1 + 0x_2 + 0x_3 = 0$

3. In each of the following cases, either show that the given system is inconsistent, or find a parametric solution.

(i) $3x_1 + x_2 + x_3 + x_4 = 4$ (ii) $7x_1 - 5x_2 + 3x_3 - 3x_4 = 2$

$\quad -2x_1 - x_2 - x_3 + x_4 = 6$ $\quad -2x_1 + x_2 + x_3 + 8x_4 = -1$

$\quad\quad x_1 + x_2 + 0x_3 + x_4 = 1$ $\quad\quad 3x_1 - 3x_2 + 5x_3 + 13x_4 = 1$

(iii) $-3x_1 + 3x_2 + 4x_3 - 17x_4 = -8$

$\quad\quad 2x_1 - 2x_2 - x_3 + 8x_4 = 7$

$\quad\quad x_1 - x_2 + x_3 + x_4 = 5$

$\quad -x_1 + x_2 + 5x_3 - 13x_4 = 1$

(iv) $x_1 - 3x_2 + x_3 + x_4 = 4$

$\quad -7x_1 + x_2 + x_3 - 5x_4 = -6$

$\quad\quad x_1 + x_2 - x_3 + x_4 = 0$

$\quad\quad 4x_1 + 3x_2 - 2x_3 + x_4 = -11$

4. Show that the linear system

$$x_1 + x_2 - x_3 = a$$

$$3x_1 - x_2 + 2x_3 = b$$

$$x_1 - 3x_2 + 4x_3 = c$$

is consistent if and only if $2a - b + c = 0$.

5. For what real numbers t does the following system have a unique solution:

$$x_1 + 2x_2 + 2x_3 = 7$$

$$3x_1 + 5x_2 + x_3 = 6$$

$$-x_1 - 4x_2 + tx_3 = 20$$

6. The following is a list of statements which may or may not be true. In each case, either justify the statement by appealing to one of the results of this section or provide a specific example which shows that it is not always true.

(i) If an $m \times n$ reduced linear system has r non-trivial equations, then r cannot exceed n.

(ii) If $m > n$, a consistent $m \times n$ linear system has a unique solution.

(iii) In an $n \times n$ reduced linear system with a unique solution, the jth basic variable is x_j, for $j = 1, 2, \ldots, n$.

(iv) An $m \times n$ linear system must have a unique solution if it is equivalent to a reduced linear system with m non-trivial equations.

(v) The first basic variable of a reduced linear system must be x_1.

*7. The basic variables of a certain 3×5 reduced linear system are x_1, x_3, and x_5, and it is known that $(2, 0, -1, 0, 4)$, $(4, 1, 6, 1, 4)$, and $(7, 1, -1, 0, 4)$ are all solutions of the system.

(i) Obtain a parametric solution for the system.

(ii) Determine which (if any) of the following 5-tuples is a solution of the system:

$$\eta_1 = (-1, 0, 6, 1, 4), \quad \eta_3 = (-4, 3, 48, 7, 4)$$

$$\eta_2 = (1, 1, 0, 3, 4), \quad \eta_4 = (31, 5, 11, -2, 4)$$

1.3 THE MATRIX VERSION OF THE REDUCTION ALGORITHM

In section 1.1, we observed that the process of simplifying a given linear system can be carried out with an economy of notation by removing the variables and performing elementary row operations on the augmented matrix of the system. At that time, we could do little more than illustrate the matrix form of gaussian elimination with a few examples, for we had not yet analyzed the elimination process itself in any detail. We now have a general description of gaussian elimination, and the goal of this section is to translate the reduction algorithm and other important results of section 1.2 into matrix terms.

We begin by defining a "simple" type of matrix whose role in our description of the matrix version of the reduction algorithm is analogous to that played by reduced systems in section 1.2.

Definition 1.2. An $m \times n$ matrix E is said to be in *row echelon form* if it has the following characteristics.

(i) There is an index r $(0 \leq r \leq m)$ such that each of the first r rows of E contains at least one non-zero entry, but the last $m - r$ rows contain only zeroes.

(ii) The first non-zero entry in each of the first r rows is a 1 and is called the *leading entry* of its row. The column which contains the leading entry of the ith row is called the ith *basic column* of E.

(iii) If $i < j$, the ith basic column lies to the left of the jth basic column.

A row echelon matrix is said to be in *reduced row echelon form* if it satisfies the additional condition:

(iv) The leading entry of the ith row is the only non-zero entry in the ith basic column.

The number r of non-zero rows is called the *row rank* of E. Note that a reduced row echelon matrix must be in row echelon form, but not vice versa. For instance, each of the following matrices is in row echelon form, but only (i) and (iv) are in reduced row echelon form. To focus attention on the terminology introduced in definition 1.2, we have circled each leading entry and have placed an arrow above each basic column. Note also that (i) and (ii) have row rank 2, while the row rank of (iii) and (iv) is 3.

$$
\text{(i)} \begin{bmatrix} ① & 0 & -3 & 2 \\ 0 & ① & 7 & -9 \\ 0 & 0 & 0 & 0 \end{bmatrix}
\qquad
\text{(ii)} \begin{bmatrix} ① & -2 & 4 & 7 \\ 0 & 0 & ① & 3 \\ 0 & 0 & 0 & 0 \end{bmatrix}
$$

$$
\text{(iii)} \begin{bmatrix} 0 & ① & -2 & 5 \\ 0 & 0 & ① & 3 \\ 0 & 0 & 0 & ① \end{bmatrix}
\qquad
\text{(iv)} \begin{bmatrix} ① & 3 & 0 & -2 & 0 \\ 0 & 0 & ① & 0 & 0 \\ 0 & 0 & 0 & 0 & ① \end{bmatrix}
$$

Finally, to complete our list of new terms, if a row vector has only one non-zero entry and that entry appears in the right-most position, we shall say that both the row and its single non-zero entry are "bad." Observe that a "bad" equation is represented by a "bad" row vector; for example, the equation $0x_1 + 0x_2 + 0x_3 = 7$ is represented by $[0 \ \ 0 \ \ 0 \ \ 7]$.

Next, we shall describe a general method for using elementary row operations to place a given $m \times n$ matrix in reduced row echelon form. The logical strategy behind this reduction process is really the same as that used in the last section in connection with gaussian elimination. Therefore, instead of providing another general description of the steps and stages of the procedure, we shall content ourselves with working through the details of a typical reduction. The essential features of the general algorithm are given in schematic form in figure 1.2.

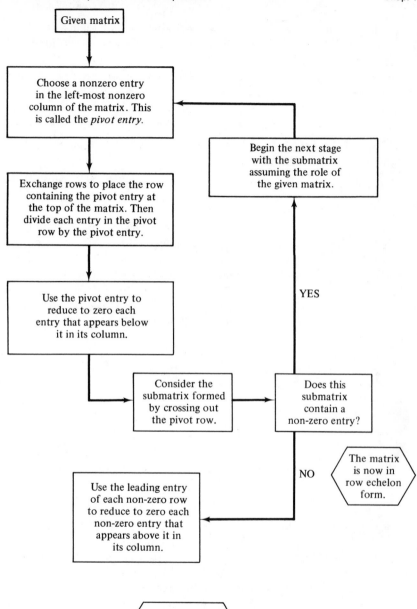

The algorithm for placing a matrix in echelon form

FIGURE 1.2

PROBLEM 1.   Use elementary row operations to place the following 3×5 matrix in reduced row echelon form:

$$A = \begin{bmatrix} -2 & -6 & 15 & -11 & 0 \\ -1 & -3 & 7 & -1 & 1 \\ 1 & 3 & -6 & -8 & 2 \end{bmatrix}$$

Solution: During the first stage of reduction, we use a type I elementary row operation to place a 1 in the upper left corner and then use two type III operations to reduce to 0 all entries of the first column which lie below the first row:

$$\begin{bmatrix} -2 & -6 & 15 & -11 & 0 \\ -1 & -3 & 7 & -1 & 1 \\ 1 & 3 & -6 & -8 & 2 \end{bmatrix} \xrightarrow[\substack{R_1+R_2 \to R_2 \\ 2R_1+R_3 \to R_3}]{R_1 \longleftrightarrow R_3} \begin{bmatrix} ① & 3 & -6 & -8 & 2 \\ 0 & 0 & 1 & -9 & 3 \\ 0 & 0 & 3 & -27 & 4 \end{bmatrix}$$

submatrix to be reduced
during the second stage

In the second stage, we use a type III elementary row operation to reduce to 0 all entries in the third column which lie below the second row:

$$\begin{bmatrix} ① & 3 & -6 & -8 & 2 \\ 0 & 0 & 1 & -9 & 3 \\ 0 & 0 & 3 & -27 & 4 \end{bmatrix} \xrightarrow{-3R_2+R_3 \to R_3} \begin{bmatrix} ① & 3 & -6 & -8 & 2 \\ 0 & 0 & ① & -9 & 3 \\ 0 & 0 & 0 & 0 & -5 \end{bmatrix}$$

submatrix to be reduced
during the third stage

The downward phase of reduction is then completed by using a type II elementary row operation to place a 1 in the lower right corner:

$$\begin{bmatrix} ① & 3 & -6 & -8 & 2 \\ 0 & 0 & ① & -9 & 3 \\ 0 & 0 & 0 & 0 & -5 \end{bmatrix} \xrightarrow{-\frac{1}{5}R_3 \to R_3} \begin{bmatrix} ① & 3 & -6 & -8 & 2 \\ 0 & 0 & ① & -9 & 3 \\ 0 & 0 & 0 & 0 & ① \end{bmatrix}$$

end of the downward phase

The matrix has now been placed in row echelon form, and to finish the reduction, we use three type III elementary row operations to reduce to 0 all entries in the third and fifth columns which lie above the leading entries of the seeond and third rows:

$$\begin{bmatrix} \textcircled{1} & 3 & -6 & -8 & 2 \\ 0 & 0 & \textcircled{1} & -9 & 3 \\ 0 & 0 & 0 & 0 & \textcircled{1} \end{bmatrix} \xrightarrow[\substack{-3R_3+R_2 \to R_2 \\ -2R_3+R_1 \to R_1 \\ 6R_2+R_1 \to R_1}]{} \begin{bmatrix} \textcircled{1} & 3 & 0 & -62 & 0 \\ 0 & 0 & \textcircled{1} & -9 & 0 \\ 0 & 0 & 0 & 0 & \textcircled{1} \end{bmatrix}$$

The matrix on the right is row equivalent to A and is in reduced row echelon form. To draw attention to its key features, we have circled the leading entries in each row and have indicated the basic columns with arrows. Note also that this matrix has row rank 3.

We shall refer to the procedure illustrated in problem 1 as *the matrix version of the reduction algorithm*. Using this procedure, we can always place a given matrix A in reduced row echelon form E. Morever, it is possible to show that A cannot be row equivalent to any reduced row echelon matrix other than E. These two statements may be combined as follows:

Theorem 1.4.   Each $m \times n$ matrix is row equivalent to one and only one $m \times n$ matrix in reduced row echelon form.

The proof of the uniqueness part of this theorem is rather complicated and will be omitted. However, a complete, formal proof may be found in almost any intermediate or advanced level text on linear algebra (e.g., see *Introduction to Matrices and Linear Transformations* by D. T. Finkbeiner, W. H. Freeman and Co., San Francisco, 2nd ed., 1966, pp. 126–127). Henceforth, we shall refer to the unique reduced row echelon matrix E to which A is row equivalent as the *reduced row echelon form of A*. This convention will help to simplify the language of some of our results.

Comparing definitions 1.1 and 1.2, we see that the augmented matrix of a reduced linear system is in reduced row echelon form. Thus, in order to solve a given $m \times n$ linear system, we carry out those elementary row operations which place the augmented matrix of the system in reduced row echelon form E. If E or any intermediate matrix in the reduction has a "bad" row, the system must be inconsistent. On the other hand, if the reduction produces no "bad" row, the system is consistent and may be solved by obtaining a parametric solution of the reduced linear system whose augmented matrix is E. This technique is illustrated in the following problems. Incidentally, in solving these problems, we write "$A \to B$" to symbolize the fact that A has been reduced to B by a finite sequence of elementary row operations. Only the end result of the reduction is displayed, and we invite the reader to test his understanding of the matrix version of the reduction algorithm by supplying the missing steps and stages.

PROBLEM 2.   Show that the system

$$-x_1 + x_2 - 5x_3 - 4x_4 = 2$$

$$3x_1 + 0x_2 + 9x_3 + 15x_4 = 1$$

$$x_1 + x_2 + x_3 + 6x_4 = 1$$

is inconsistent.

Solution: Applying the matrix version of the reduction algorithm to the augmented matrix of the given system, we obtain the following result at the end of the downward phase:

$$\begin{bmatrix} -1 & 1 & -5 & -4 & \vdots & 2 \\ 3 & 0 & 9 & 15 & \vdots & 1 \\ 1 & 1 & 1 & 6 & \vdots & 1 \end{bmatrix} \longrightarrow \begin{bmatrix} \textcircled{1} & 1 & 1 & 6 & \vdots & 1 \\ 0 & \textcircled{1} & -2 & 1 & \vdots & \frac{3}{2} \\ 0 & 0 & 0 & 0 & \vdots & 1 \end{bmatrix} \begin{matrix} \\ 2/_3 \\ 5/_3 \end{matrix}$$

"bad" row

Since the row echelon matrix on the right of this reduction has a "bad" row, the given system must be inconsistent.

PROBLEM 3.   Solve the 4 × 5 linear system

$$2x_1 - 4x_2 + 0x_3 + x_4 + 7x_5 = 11$$

$$x_1 - 2x_2 - x_3 + x_4 + 9x_5 = 12$$

$$-x_1 + 2x_2 + x_3 + 3x_4 - 5x_5 = 16$$

$$4x_1 - 8x_2 + x_3 - x_4 + 6x_5 = -2$$

Solution: The augmented matrix of the given system may be placed in reduced row echelon form as indicated below:

$$\begin{bmatrix} 2 & -4 & 0 & 1 & 7 & \vdots & 11 \\ 1 & -2 & -1 & 1 & 9 & \vdots & 12 \\ -1 & 2 & 1 & 3 & -5 & \vdots & 16 \\ 4 & -8 & 1 & -1 & 6 & \vdots & -2 \end{bmatrix} \longrightarrow \begin{bmatrix} \textcircled{1} & -2 & 0 & 0 & 3 & \vdots & 2 \\ 0 & 0 & \textcircled{1} & 0 & -5 & \vdots & -3 \\ 0 & 0 & 0 & \textcircled{1} & 1 & \vdots & 7 \\ 0 & 0 & 0 & 0 & 0 & \vdots & 0 \end{bmatrix}$$

Restoring the variables to the reduced row echelon form of the augmented matrix, we obtain a reduced linear system whose parametric solution is

$$x_1 = 2 + 2t_1 - 3t_2$$

$$x_2 = \qquad t_1$$

$$x_3 = -3 \qquad + 5t_2$$

$$x_4 = 7 \qquad - t_2$$

$$x_5 = \qquad t_2$$

and this also characterizes the solution set of the original system.

Later in our work, we shall encounter situations where it is necessary to solve several different systems which have the same coefficient matrix. The following solved problem illustrates how this can often be achieved by means of a single combined reduction.

PROBLEM 4. Solve the linear systems

$$x_1 - x_2 - 3x_3 + 8x_4 = 9 \qquad x_1 - x_2 - 3x_3 + 8x_4 = -2$$

$$3x_1 + 0x_2 - 3x_3 + 9x_4 = 6 \qquad 3x_1 + 0x_2 - 3x_3 + 9x_4 = -1$$

$$x_1 + x_2 + x_3 - 2x_4 = -5 \qquad x_1 + x_2 + x_3 - 2x_4 = 1$$

Solution: Since both systems have the same coefficient matrix, their augmented matrices differ only in the last column. Therefore, by performing the combined reduction

$$\begin{bmatrix} 1 & -1 & -3 & 8 & 9 & -2 \\ 3 & 0 & -3 & 9 & 6 & -1 \\ 1 & 1 & 1 & -2 & -5 & 1 \end{bmatrix} \longrightarrow \begin{bmatrix} ① & 0 & -1 & 3 & 2 & 0 \\ 0 & ① & 2 & -5 & -7 & 0 \\ 0 & 0 & 0 & 0 & 0 & ① \end{bmatrix}$$

"bad" entry

we find that the second system is inconsistent and that the first system has the parametric solution

$$x_1 = 2 + t_1 - 3t_2$$

$$x_2 = -7 - 2t_1 + 5t_2$$

$$x_3 = \qquad t_1$$

$$x_4 = \qquad t_2$$

Our final theorem is obtained by translating theorem 1.2 into matrix terms.

Theorem 1.5.   If the reduced row echelon form of the augmented matrix of a consistent $m \times n$ linear system has row rank r, then the system has a unique solution if $r = n$ and infinitely many solutions if $r < n$.

The system analyzed in problem 3 has infinitely many solutions since $r = 3$ and $n = 5$. The following problem illustrates how theorem 1.5 can be used to deduce that a system has a unique solution without actually finding it.

PROBLEM 5.   Show that the 4×3 linear system

$$-x_1 + x_2 + x_3 = 9$$

$$2x_1 + x_2 - x_3 = -10$$

$$3x_1 + 0x_2 - 2x_3 = -19$$

$$-x_1 + 2x_2 - 3x_3 = -10$$

has a unique solution.

Solution: At the end of the downward stage of reduction, the augmented matrix of the system is in row echelon form:

$$\begin{bmatrix} -1 & 1 & 1 & | & 9 \\ 2 & 1 & -1 & | & -10 \\ 3 & 0 & -2 & | & -19 \\ -1 & 2 & -3 & | & -10 \end{bmatrix} \longrightarrow \begin{bmatrix} ① & -1 & -1 & | & -9 \\ 0 & ① & -4 & | & -19 \\ 0 & 0 & ① & | & 5 \\ 0 & 0 & 0 & | & 0 \end{bmatrix}$$

Since the upward phase of reduction can neither create nor destroy a nonzero row (why not?), we see that the system is consistent and that $r = 3$. Thus, without even completing the reduction, we know that the system has a unique solution.

Note that an $n \times n$ linear system has a unique solution if and only if it is equivalent to a reduced system whose variables are all basic (see exercise 8). The coefficient matrix of such a reduced system must have a 1 in the (i, i) position for $i = 1, 2, \ldots, n$ and zeroes elsewhere. This special kind of matrix shall appear throughout our work and will be denoted by I_n or by just I in those cases where its dimension is not important. For example,

$$I_3 = \begin{bmatrix} 1 & 0 & 0 \\ 0 & 1 & 0 \\ 0 & 0 & 1 \end{bmatrix}$$

By combining these observations, we obtain the following result, which proves to be quite useful later in our work.

Corollary 1.1.   An $n \times n$ linear system has a unique solution if and only if its coefficient matrix is row equivalent to I_n.

It would not be honest to close our discussion of the reduction algorithm without conceding that there are other ways to solve linear systems. In fact, when solving a given system, it is usually possible to find a method of reduction which exploits special patterns within the system and thus requires fewer arithmetic steps than the algorithm. Nevertheless, the algorithmic approach still has two great advantages over any such free-lance method: it can be applied to any linear system, and its basic features are relatively easy to program for automatic computing. Most methods which appear to be shorter than the algorithm cannot be programmed as easily and may thus take much longer to execute.

This concludes our introduction to linear systems. Of all the new ideas introduced in these first three sections, perhaps the most important is the notion of matrix representation. In the sections which follow, we shall take a closer look at matrices, and in the process, shall introduce terminology and derive results to be used throughout our work.

Exercise Set 1.3

1. Determine which of the following matrices is in reduced row echelon form. For those that are in the desired form, find the row rank and identify the basic columns.

(i) $\begin{bmatrix} 1 & -3 & 0 & 4 \\ 0 & 0 & 1 & -2 \\ 0 & 0 & 0 & 0 \end{bmatrix}$
(ii) $\begin{bmatrix} 1 & 0 & 5 & 0 \\ 0 & 1 & 7 & 0 \\ 0 & 0 & 0 & -6 \end{bmatrix}$

(iii) $\begin{bmatrix} 0 & 1 & -2 & 0 & 5 & 0 & 0 \\ 0 & 0 & 0 & 1 & 9 & 0 & 0 \\ 0 & 0 & 0 & 0 & 0 & 1 & 0 \\ 0 & 0 & 0 & 0 & 0 & 0 & 0 \end{bmatrix}$
(iv) $\begin{bmatrix} 1 & 0 & 2 & 3 & 0 \\ 0 & 1 & 7 & 1 & 0 \\ 0 & 0 & 0 & 1 & 1 \\ 0 & 0 & 0 & 0 & 0 \end{bmatrix}$

2. In each of the following cases, find the reduced row echelon form of the given matrix.

(i) $\begin{bmatrix} 1 & 9 & 15 & -3 \\ 0 & 2 & 1 & 1 \\ 0 & 0 & 0 & 4 \\ 0 & 0 & 0 & 0 \end{bmatrix}$

(ii) $\begin{bmatrix} 1 & 0 & 0 & 9 \\ 0 & 1 & 5 & 0 \\ 0 & -2 & 1 & 0 \\ 3 & 0 & 0 & 7 \end{bmatrix}$

(iii) $\begin{bmatrix} 1 & -1 & -8 & 1 \\ 2 & 1 & -1 & 1 \\ -3 & 1 & 14 & 7 \end{bmatrix}$

(iv) $\begin{bmatrix} -3 & 1 & -18 & -5 & 4 & 4 \\ 1 & 1 & 2 & 3 & 1 & 1 \\ -1 & 1 & -8 & -1 & 0 & 0 \\ 1 & 2 & -1 & 4 & -5 & -5 \end{bmatrix}$

3. Use the matrix version of the reduction algorithm to solve each of the following linear systems.

(i) $-x_1 + 2x_2 + x_3 = 2$

   $x_1 - 2x_2 - 2x_3 = 7$

   $x_1 - 2x_2 + x_3 = 8$

(ii) $x_1 - x_2 + 10x_3 - 7x_4 = 0$

   $x_1 + 2x_2 + x_3 - x_4 = 0$

   $3x_1 + 5x_2 + 6x_3 - 5x_4 = 0$

(iii) $x_1 - x_2 + 3x_3 - 4x_4 + 2x_5 = 8$

   $3x_1 + x_2 + 5x_3 + 16x_4 + 0x_5 = 16$

   $-5x_1 + 2x_2 - 12x_3 - x_4 + 8x_5 = -7$

   $x_1 + x_2 + x_3 + 10x_4 + x_5 = 8$

(iv) $7x_1 + 14x_2 - 5x_3 = 13$

   $3x_1 + 6x_2 + 7x_3 = 33$

   $2x_1 + 4x_2 + 5x_3 = 23$

   $x_1 + 2x_2 + x_3 = 7$

4. In each of the following cases, determine whether or not the given linear system has a unique solution.

(i) $x_1 - x_2 + x_3 = 11$

   $2x_1 + 3x_2 - 2x_3 = 3$

   $x_1 + x_2 + x_3 = 5$

   $2x_1 + 5x_2 + 2x_3 = 1$

(ii) $13x_1 + 2x_2 - 5x_3 + x_4 = 17$

   $x_1 - x_2 + 14x_3 - 9x_4 = 2$

   $7x_1 + x_2 + 5x_3 - 19x_4 = 1$

5. (i) Show that two matrices are row equivalent to one another if and only if they both have the same reduced row echelon form.
   Hint: Use theorem 1.4.

   (ii) Show that two consistent $m \times n$ linear systems are equivalent if and only if their augmented matrices have the same reduced row echelon form.

6. Use the result of exercise 5 to determine which of the following matrices are row equivalent to one another.

$$A_1 = \begin{bmatrix} 5 & -2 & -25 & 12 \\ 1 & 1 & 2 & 1 \\ 3 & 7 & 26 & -1 \end{bmatrix}, \quad A_2 = \begin{bmatrix} 1 & -7 & 3 & 1 \\ -3 & 1 & 14 & -7 \\ 3 & 1 & -4 & 5 \end{bmatrix}$$

$$A_3 = \begin{bmatrix} 1 & 1 & 2 & 1 \\ 3 & 1 & -4 & 5 \\ -4 & 5 & 37 & -13 \end{bmatrix}, \quad A_4 = \begin{bmatrix} -3 & 1 & 14 & -7 \\ 2 & 1 & -1 & 3 \\ 4 & 3 & 3 & 5 \end{bmatrix}$$

7. The augmented matrix of a certain 3×4 linear system is known to be row equivalent to the matrix

$$\begin{bmatrix} -3 & 1 & -23 & 0 & \vdots & 11 \\ 2 & 1 & 12 & 15 & \vdots & 1 \\ 4 & -1 & 30 & 3 & \vdots & -13 \end{bmatrix}$$

Use this information to obtain a parametric solution for the system.

8. (i) Show that an $m \times n$ reduced linear system has a unique solution if and only if it has n basic variables.

   (ii) Show that a consistent $m \times n$ linear system has a unique solution if and only if the reduced row echelon form of its coefficient matrix has n basic columns.

   (iii) Describe the coefficient matrix of a 5×3 reduced linear system with a unique solution.

9. Show that an $m \times n$ linear system must be consistent if the reduced row echelon form of its coefficient matrix has row rank m.

10. Find a 4×5 reduced row echelon matrix E with all the following properties:
   1. The row rank is 3.
   2. Columns 1 and 2 are basic.
   3. The 4×4 non-homogeneous system whose augmented matrix is E is inconsistent.
   4. The 5-tuples $\eta_1 = (2, -5, 1, 0, 0)$ and $\eta_2 = (-1, -7, 0, 1, 0)$ are both solutions of the 4×5 homogeneous system whose coefficient matrix is E.

11. Show that I_n is the only $n \times n$ reduced row echelon matrix with row rank n.

12. For what values of λ does the linear system

$$x_1 + 0x_2 + 0x_3 = \lambda x_1$$

$$-3x_1 + 8x_2 + 4x_3 = \lambda x_2$$

$$7x_1 + 0x_2 + 9x_3 = \lambda x_3$$

have a unique solution?

1.4 THE THREE BASIC MATRIX OPERATIONS

We have already seen how matrices can be used to simplify the notation of the gaussian elimination process, and we shall encounter many more situations in which matrices play a useful role. The rest of this chapter is devoted to developing those properties of matrices which will be used most frequently throughout our work. In this section, we begin by defining and discussing three basic operations: matrix addition, multiplication of a matrix by a scalar, and multiplication of a matrix by a matrix.

Since a matrix is an array of numbers, it is natural to define the scalar multiple of a matrix and the sum of two matrices as entry by entry generalizations of ordinary multiplication and addition, respectively.

Definition 1.3. If $A = (a_{ij})$ and $B = (b_{ij})$ are $m \times n$ matrices, then the $m \times n$ matrix $C = (c_{ij})$ with $c_{ij} = a_{ij} + b_{ij}$ for all i and j is said to be the *sum* of A and B, and we write $C = A + B$.

Note that $A + B$ is defined only if A and B have the same dimensions.

EXAMPLE 1.   If

$$A = \begin{bmatrix} 1 & 3 & -5 \\ 0 & 2 & 6 \end{bmatrix} \quad \text{and} \quad B = \begin{bmatrix} -3 & 2 & 4 \\ 6 & -1 & 7 \end{bmatrix}$$

we have

$$A + B = \begin{bmatrix} (1) + (-3) & (3) + (2) & (-5) + (4) \\ (0) + (6) & (2) + (-1) & (6) + (7) \end{bmatrix} = \begin{bmatrix} -2 & 5 & -1 \\ 6 & 1 & 13 \end{bmatrix}$$

Definition 1.4. If $A = (a_{ij})$ is an $m \times n$ matrix and t is a scalar, then the $m \times n$ matrix $S = (s_{ij})$ with $s_{ij} = ta_{ij}$ for all i, j is called the *scalar multiple* of A by t, and we write $S = tA$.

EXAMPLE 2.   If

$$A = \begin{bmatrix} 5 & -1 \\ 3 & 0 \\ -2 & 4 \end{bmatrix}$$

we find that

$$3A = \begin{bmatrix} 3(5) & 3(-1) \\ 3(3) & 3(0) \\ 3(-2) & 3(4) \end{bmatrix} = \begin{bmatrix} 15 & -3 \\ 9 & 0 \\ -6 & 12 \end{bmatrix}$$

Matrices can also be multiplied together under certain circumstances, but what may seem to be the "natural" definition of a matrix product—namely, $AB = C$ if $c_{ij} = a_{ij}b_{ij}$ for all i and j—does not turn out to be as fruitful as the apparently unnatural definition given below.

Definition 1.5. If $A = (a_{ij})$ and $B = (b_{ij})$ are $m \times n$ and $n \times p$ matrices, respectively, then the $m \times p$ matrix $C = (c_{ij})$ with $c_{ij} = a_{i1}b_{1j} + a_{i2}b_{2j} + \ldots + a_{in}b_{nj}$ for all i and j is called the *product of A by B*, and we write $C = AB$.

Note that AB is defined only when the column dimension of A is the same as the row dimension of B. When this is the case, we say that A and B are *compatible for multiplication in the order AB*. The rule for finding the (i, j) entry of a product matrix is illustrated in two different ways in the following examples.

EXAMPLE 3.

The $(3, 2)$ entry of $C = AB$ is $(1)(4) + (5)(-5) = -21$

$$a_{31}b_{12} + a_{32}b_{22} = c_{32}$$

EXAMPLE 4. If

$$A = \begin{bmatrix} 3 & 2 \\ -1 & 4 \\ 0 & -2 \end{bmatrix} \text{ and } B = \begin{bmatrix} -1 & 0 \\ 3 & 4 \end{bmatrix}$$

then

$$AB = \begin{bmatrix} (3)(-1) + (2)(3) & (3)(0) + (2)(4) \\ (-1)(-1) + (4)(3) & (-1)(0) + (4)(4) \\ (0)(-1) + (-2)(3) & (0)(0) + (-2)(4) \end{bmatrix} = \begin{bmatrix} 3 & 8 \\ 13 & 16 \\ -6 & -8 \end{bmatrix}$$

The product BA is not even defined, since B has two columns and A has three rows.

Some of the reasons behind our definition of the matrix product cannot be properly discussed until later. Nevertheless, even at this stage of development, it is not difficult to find situations in which the product of two matrices is a natural consideration. Perhaps the most important immediate application of definition 1.5 comes from observing that a given $m \times n$ linear system can always be represented by the matrix equation $AX = \beta$, where A is the coefficient matrix of the system, X is a column vector of variables, and β is the column vector of right side constants. For instance, the 3×4 system

$$2x_1 - x_2 + 5x_3 + 3x_4 = 7$$

$$x_1 + 4x_2 - x_3 + 2x_4 = 1$$

$$9x_1 + 5x_2 + 0x_3 - x_4 = -3$$

can be represented by the equation

$$\begin{bmatrix} 2 & -1 & 5 & 3 \\ 1 & 4 & -1 & 2 \\ 9 & 5 & 0 & -1 \end{bmatrix} \begin{bmatrix} x_1 \\ x_2 \\ x_3 \\ x_4 \end{bmatrix} = \begin{bmatrix} 7 \\ 1 \\ -3 \end{bmatrix}$$

$$AX = \beta$$

By using the matrix equation $AX = \beta$ to represent a given linear system, we can employ algebraic methods to analyze the collection of all solution vectors X. This technique is used extensively in chapter 2, especially in section 2.8.

In addition to the applications we have already discussed, matrices are often used as a convenient means of organizing related bits of information, and in certain cases, interesting results can be obtained by forming an appropriate matrix product. Consider the following example.

EXAMPLE 5.   Four shoppers (S_1, S_2, S_3, S_4) live in a town where there are three supermarkets (M_1, M_2, M_3). In the table below, the shopping list of each person is displayed on the left, and the price per unit of each item at each supermarket is given in the array on the right:

| | Quantity desired by . . . | | | | $ Price per unit at . . . | | |
Item	S_1	S_2	S_3	S_4	M_1	M_2	M_3
potatoes	3	1	0	1	.08	.07	.08
tooth paste	1	0	1	0	.75	.72	.75
bubble gum	10	0	4	0	.05	.07	.05
minced yak	0	1	1	0	1.67	1.70	2.00
lettuce	2	3	0	2	.15	.13	.10

Thus, the first shopper (S_1) will spend

$$(3)(.08) + (1)(.75) + (10)(.05) + (0)(1.67) + (2)(.15) = \$1.79$$

if he goes to the first supermarket (M_1). Naturally, each shopper would like to go to the market where he will spend the least, but to check out all the possibilities, it is necessary to perform eleven more computations like the one already completed for S_1 at M_1. To help organize these computations, let Q denote the 4×5 matrix whose ith row is the shopping list of S_i and let C denote the 5×3 matrix whose jth column is the list of prices at M_j:

$$Q = \begin{bmatrix} 3 & 1 & 10 & 0 & 2 \\ 1 & 0 & 0 & 1 & 3 \\ 0 & 1 & 4 & 1 & 0 \\ 1 & 0 & 0 & 0 & 2 \end{bmatrix}, \qquad C = \begin{bmatrix} .08 & .07 & .08 \\ .75 & .72 & .75 \\ .05 & .07 & .05 \\ 1.67 & 1.70 & 2.00 \\ .15 & .13 & .10 \end{bmatrix}$$

Then, the total amount of money that shopper S_i would spend at super-market M_j appears as the (i, j) entry of the product matrix QC:

$$QC = \begin{array}{c} \\ \\ \\ \\ \end{array} \begin{matrix} M_1 & M_2 & M_3 \\ \begin{bmatrix} 1.79 & 1.89 & \boxed{1.69} \\ 2.20 & \boxed{2.16} & 2.38 \\ \boxed{2.62} & 2.70 & 2.95 \\ .38 & \boxed{.33} & .37 \end{bmatrix} & \begin{matrix} S_1 \\ S_2 \\ S_3 \\ S_4 \end{matrix} \end{matrix}$$

The smallest number in each row of QC is circled to indicate the least amount of money each shopper must spend to fill his needs. Thus, assuming that each market gives the same kind of trading stamps we can advise the first shopper to go to the third market and the third shopper to go to M_1, while S_2 and S_4 should share a ride to the second market.

When dealing with the matrix product, it is convenient to have a more compact form of the general term of a product matrix than the one given in definition 1.5. To describe such a form, we use a device known as the *summation* (or *sigma*) *notation*, in which the symbol

$$\sum_{k=p}^{n} a_k$$

is used to denote the sum $a_p + a_{p+1} + \cdots + a_n$. For example,

$$1^2 + 2^2 + 3^2 = \sum_{k=1}^{3} k^2$$

$$2x_1 + 3x_2 + 4x_3 + 5x_4 = \sum_{k=1}^{4} (k+1)x_k$$

The index k, which regulates the summation process, is called the *index of summation*, or more simply, the *running index*, and p is referred to as the *starting point* of the sum. When representing successive summations, we use the symbols

$$\sum_{i=1}^{n} \sum_{j=1}^{m} a_{ij} \quad \text{and} \quad \sum_{i=1}^{n} \left(\sum_{j=1}^{m} a_{ij} \right)$$

interchangeably. In both, it is understood that the inner summation (with running index j) is to be performed first. For instance

$$\sum_{i=1}^{3} \sum_{j=1}^{2} a_{ij} = \underbrace{(a_{11} + a_{12})}_{i=1} + \underbrace{(a_{21} + a_{22})}_{i=2} + \underbrace{(a_{31} + a_{32})}_{i=3}$$

It is important to remember that the running index is the only one which changes throughout the summation process, even if the general expression under the summation symbol involves other indices. An important application of this observation is the expansion

$$\sum_{k=1}^{n} a_{ik}b_{kj} = a_{i1}b_{1j} + a_{i2}b_{2j} + \cdots + a_{in}b_{nj}$$

which we recognize as the rule for finding the (i, j) entry of the product AB. Thus, definition 1.5 may be rewritten as follows.

Definition 1.5.   The *product* of the $m \times n$ matrix $A = (a_{ij})$ by the $n \times p$ matrix $B = (b_{ij})$ is the $m \times p$ matrix $C = (c_{ij})$, where $c_{ij} = \sum_{k=1}^{n} a_{ik}b_{kj}$ for all i and j.

This streamlined version of definition 1.5 makes it much easier to discuss the properties of matrix multiplication. This proves to be a great advantage in the next section, which is devoted to the development of the manipulative properties of all three matrix operations. Since the summation notation will play an important part in this development, it is appropriate to close this section by listing these properties of the notation which appear most frequently in our work.

1. $\displaystyle\sum_{i=1}^{n} a_i = \sum_{j=1}^{n} a_j$

2. $\sum_{i=1}^{n} (sa_i) = s \left(\sum_{i=1}^{n} a_i \right)$ for each scalar s

3. $\sum_{i=1}^{n} (a_i + b_i) = \sum_{i=1}^{n} a_i + \sum_{i=1}^{n} b_i$

4. $\sum_{i=1}^{n} \sum_{j=1}^{m} a_{ij} = \sum_{j=1}^{m} \sum_{i=1}^{n} a_{ij}$

5. $\sum_{k=1}^{n} a_{ik} \left(\sum_{l=1}^{p} b_{kl} c_{lj} \right) = \sum_{l=1}^{p} \left(\sum_{k=1}^{n} a_{ik} b_{kl} \right) c_{lj}$

for each pair of fixed indices i and j

Property 1 says that the summation process is independent of the symbol used for the running index, while properties 2, 3, 4, and 5 all express well-known features of scalar algebra. For example, we have

(3) $\sum_{i=1}^{3} (a_i + b_i) = (a_1 + b_1) + (a_2 + b_2) + (a_3 + b_3)$

$\qquad = (a_1 + a_2 + a_3) + (b_1 + b_2 + b_3) = \sum_{i=1}^{3} a_i + \sum_{i=1}^{3} b_i$

(4) $\sum_{i=1}^{3} \left(\sum_{j=1}^{2} a_{ij} \right) = (a_{11} + a_{12}) + (a_{21} + a_{22}) + (a_{31} + a_{32})$

$\qquad = (a_{11} + a_{21} + a_{31}) + (a_{12} + a_{22} + a_{32}) = \sum_{j=1}^{2} \left(\sum_{i=1}^{3} a_{ij} \right)$

Exercise Set 1.4

1. Consider the matrices

$$A = \begin{bmatrix} 1 & 3 & -1 \\ 2 & 0 & 4 \\ -7 & 1 & 5 \end{bmatrix}, \quad B = \begin{bmatrix} -4 & 2 \\ 0 & 1 \\ 1 & 3 \end{bmatrix}, \quad C = \begin{bmatrix} 1 & 3 & 8 \\ -1 & 5 & 3 \\ 0 & 0 & -7 \end{bmatrix}$$

In each of the following cases, either perform the indicated matrix operation or explain why the operation has no meaning:

(i) $A + C$ (ii) $A + B$ (iii) $5B$
(iv) $2A + 3C$ (v) AB (vi) BC
(vii) CA (viii) AC (ix) $C(AB)$
(x) $(CA)B$ (xi) $(A + C)B$ (xii) $AB + CB$

2. Let

$$A = \begin{bmatrix} 1 & y \\ -y & 2x \end{bmatrix} \quad \text{and} \quad B = \begin{bmatrix} x & -2x \\ y & 2 \end{bmatrix}$$

In each of the following cases, find all (real) values of x and y for which the given matrix equation is valid:

(i) $A + B = B + A$

(ii) $AB = BA$

(iii) $AB = 0$ where 0 is the 2×2 matrix whose entries are all 0

3. (i) Find a 3×3 matrix A and column vectors X and β so that the system

$$x_1 + x_2 - x_3 = 5$$

$$2x_1 + 0x_2 + 3x_3 = -7$$

$$4x_1 + 3x_2 - 2x_3 = 4$$

is represented by the matrix equation $AX = \beta$.

(ii) Find the matrix products AC and CA, where A is the matrix found in part (i) and

$$C = \begin{bmatrix} -9 & -1 & 3 \\ 16 & 2 & -5 \\ 6 & 1 & -2 \end{bmatrix}$$

(iii) Verify that $C(AX) = (CA)X = X$, and use this fact to show that $X = C\beta$ is the (unique) solution of the system displayed in (i).

4. Verify the following identities by expanding and rearranging the appropriate sums:

(i) $\displaystyle\sum_{k=1}^{4} a_{ik} \left(\sum_{l=1}^{3} b_{kl}c_{lj} \right) = \sum_{l=1}^{3} \left(\sum_{k=1}^{4} a_{ik}b_{kl} \right) c_{lj}$

(ii) $\displaystyle\sum_{k=1}^{3} a_{ik}(b_{kj} + c_{kj}) = \sum_{k=1}^{3} a_{ik}b_{kj} + \sum_{k=1}^{3} a_{ik}c_{kj}$

5. Define the *trace* of the $n \times n$ matrix $A = (a_{ij})$ to be the sum of the entries $a_{11}, a_{22}, \ldots, a_{nn}$. In summation notation form, we have

$$\operatorname{tr}(A) = \sum_{i=1}^{n} a_{ii}$$

Show that the trace function has the following properties:

(i) $\operatorname{tr}(sA) = s(\operatorname{tr} A)$ for each scalar s

(ii) $\operatorname{tr}(A + B) = \operatorname{tr}(A) + \operatorname{tr}(B)$

(iii) $\operatorname{tr}(AB) = \operatorname{tr}(BA)$

6. The *square* of an $n \times n$ matrix A is the product AA, and we denote this $n \times n$ matrix by A^2. Similarly, $A^3 = AA^2$, $A^4 = AA^3$, and so on, with the general rule being $A^k = AA^{k-1}$ for each positive integer k. Consider the matrices

$$C = \begin{bmatrix} -11 & -9 & 6 \\ 20 & 16 & -10 \\ 8 & 6 & -3 \end{bmatrix} \quad \text{and} \quad N = \begin{bmatrix} -22 & -3 & -25 \\ 40 & 6 & 44 \\ 16 & 3 & 16 \end{bmatrix}$$

(i) Verify that $C^2 = C$, and then show that $C^k = C$ for all k. A matrix with this property is said to be *idempotent*.

(ii) Verify that N^3 is the 3×3 matrix whose entries are all 0. Such a matrix is called *nilpotent*.

7. Three men (Arthur, Ben, and Charles) are dating four girls (Francine, Gertrude, Hattie, and Inger). The table on the left gives the number of times each man plans to take a girl on one of four different kinds of dates, while the table on the right indicates the number of units of pleasure (negative for displeasure) each girl shares with her companion during each type of date:

Type of Date	Number of dates planned by . . .			Units of pleasure shared by . . .			
	A	B	C	F	G	H	I
movie	3	0	2	-1	3	2	2
sports event	3	2	1	-1	2	2	-1
drive in the country	1	2	3	2	2	-2	3
dinner, dancing, etc.	2	5	4	5	1	4	3

(i) Form the 3×4 matrix T whose (i, j) entry is the total pleasure shared by the ith man and jth woman if they date each other exclusively; for example, the $(2, 3)$ entry of T represents the total pleasure shared by Ben and Hattie and has a numerical value of

$$(0)(2) + (2)(2) + (2)(-2) + (5)(4) = 20$$

(ii) Who shares the most pleasure with Ben? With Francine?

(iii) Assuming that everyone in the group wants to go steady, how should they pair up? Which girl is left without a mate? Explain your reasoning.

8. The *transpose* A^t of the $m \times n$ matrix $A = (a_{ij})$ is the $n \times m$ matrix whose (i, j) entry is a_{ji}. For example,

$$\begin{bmatrix} -3 & 7 & 1 \\ 5 & 0 & 1 \end{bmatrix}^t = \begin{bmatrix} -3 & 5 \\ 7 & 0 \\ 1 & 1 \end{bmatrix}$$

Show that the transpose has the following properties:

(i) $(A^t)^t = A$

(ii) $(sA)^t = sA^t$ for each scalar s

(iii) $(A + B)^t = A^t + B^t$

(iv) $(AB)^t = B^t A^t$

9. An $n \times n$ matrix A is said to be *symmetric* if $A = A^t$ and *skew-symmetric* if $A = -A^t$. For example, the matrix S displayed below is symmetric and K is skew-symmetric:

$$S = \begin{bmatrix} 3 & -7 & 5 \\ -7 & 0 & 2 \\ 5 & 2 & 1 \end{bmatrix}, \quad K = \begin{bmatrix} 0 & 9 & -5 \\ -9 & 0 & 3 \\ 5 & -3 & 0 \end{bmatrix}$$

(i) Show that each diagonal entry of a skew-symmetric matrix must be 0.

(ii) If S is symmetric and K is skew-symmetric, show that $S = K$ if and only if both are equal to O, the matrix of all zeroes.

(iii) If S_1 and S_2 are symmetric matrices, show that $S_1 + S_2$ is also symmetric. Show that $S_1 S_2$ is symmetric if and only if $S_1 S_2 = S_2 S_1$.

(iv) If K_1 and K_2 are skew-symmetric, what can be said about $K_1 + K_2$ and $K_1 K_2$?

10. If A is an $n \times n$ matrix, let $S = \frac{1}{2}(A + A^t)$ and $K = \frac{1}{2}(A - A^t)$.

(i) Show that S is symmetric and that K is skew-symmetric.

(ii) Verify that $A = S + K$, and then show that this is the *only* way A can be written as the sum of a symmetric and a skew-symmetric matrix.
   Hint: If $A = S_1 + K_1$, then $S - S_1 = K_1 - K$. Use exercise 9 to show that $S - S_1 = K_1 - K = O$.

11. An $n \times n$ *diagonal* matrix D is an array in which the (i, j) entry is 0 whenever $i \neq j$. Since any non-zero entries of D must appear on the main diagonal, it is convenient to write $D = \text{diag}(d_1, d_2, \ldots, d_n)$, where d_k is the (k, k) entry of D. For example,

$$\text{diag}(3, -7, 0) = \begin{bmatrix} 3 & 0 & 0 \\ 0 & -7 & 0 \\ 0 & 0 & 0 \end{bmatrix}$$

(i) If $D = \text{diag}(d_1, \ldots, d_n)$ and $F = \text{diag}(f_1, \ldots, f_n)$ are diagonal matrices, show that

(1) $sD = \text{diag}(sd_1, \ldots, sd_n)$ for each scalar s

(2) $D + F = \text{diag}(d_1 + f_1, \ldots, d_n + f_n)$

(3) $DF = FD = \text{diag}(d_1 f_1, \ldots, d_n f_n)$

(ii) If $A = (a_{ij})$ is an $n \times n$ matrix and $D = \text{diag}(d_1, \ldots, d_n)$, show that the (i, j) entry of AD is $d_j a_{ij}$. What is the (i, j) entry of DA?

12. Show that the jth column in the matrix product AB is A times the jth column of B. For example, the second column in the product

$$\begin{bmatrix} -3 & 5 & 2 \\ 8 & 7 & 1 \end{bmatrix} \begin{bmatrix} 2 & 1 \\ 1 & 5 \\ 0 & 2 \end{bmatrix}$$

is the vector

$$\begin{bmatrix} -3 & 5 & 2 \\ 8 & 7 & 1 \end{bmatrix} \begin{bmatrix} 1 \\ 5 \\ 2 \end{bmatrix} = \begin{bmatrix} 26 \\ 45 \end{bmatrix}$$

State and prove an analogous rule for finding the ith row of the product matrix AB.

1.5 THE ALGEBRA OF MATRICES

Since matrix addition and multiplication of a matrix by a scalar are entry by entry generalizations of ordinary addition and multiplication, it should come as no surprise to find that these matrix operations obey many of the familiar rules of scalar algebra. Some of these rules are contained in the following theorem.

Theorem 1.6. If A, B, and C are $m \times n$ matrices and r and s are scalars, then

(i) $A + B = B + A$ (commutativity of addition)

(ii) $A + (B + C) = (A + B) + C$ (associativity of addition)

(iii) $A + O = A = O + A$ where O is the $m \times n$ matrix containing only zeroes

(iv) $A + (-1)A = O = (-1)A + A$

(v) $r(A + B) = (rA) + (rB)$

(vi) $(r + s)A = (rA) + (sA)$

(vii) $r(sA) = (rs)A$

(viii) $(1)A = A$

Proof: In order to verify any one of these properties, it suffices to show that the matrix on one side of the proposed equation is identical to the matrix on the other side. For example, to prove (ii), we note that $A + (B + C)$ and $(A + B) + C$ are both $m \times n$ arrays and that

$$\underbrace{a_{ij} + (b_{ij} + c_{ij})}_{\substack{(i, j) \text{ entry of} \\ A + (B + C)}} = \underbrace{(a_{ij} + b_{ij}) + c_{ij}}_{\substack{(i, j) \text{ entry of} \\ (A + B) + C}}$$

by the associativity of scalar addition. The rest of the·proof follows similar lines and is left as an exercise. ∎

Henceforth, the $m \times n$ array which contains only 0's will be called the *zero matrix* and will always be denoted by O. In addition, we shall denote the scalar multiple $(-1)A$ by $-A$ and the matrix sum $B + (-A)$ by $B - A$. The reader may find it instructive to verify that this matrix "difference" operation has the following properties:

1. $A = -(-A)$
2. $A - B = -(B - A)$
3. $(A - B) - C = A - (B + C)$
4. $A - (B - C) = (A - B) + C$
5. $A + (B - C) = (A + B) - C$
6. $-(sA) = (-s)A$ for each scalar s

(1.18)

One of the most useful laws of scalar algebra states that if a and b are numbers and x is a scalar variable, then the equations $x + a = b$ and $a + x = b$ both have the unique solution $x = b - a$. The matrix analog of this result may be stated as follows.

Theorem 1.7. If A and B are $m \times n$ matrices, the equations $X + A = B$ and $A + X = B$ both have the unique solution $X = B - A$.

Proof: Since matrix addition is commutative, the two matrix equations in question are really the same, and we shall concentrate on showing that $B - A$ is the only solution of $X + A = B$. To this end, suppose C is a solution of this equation. Then, by adding $-A$ to each side of the equation $B = C + A$ and using properties (ii), (iii), and (iv) of theorem 1.6, we obtain

$$B - A = (C + A) - A = C + (A - A) = C + O = C$$

Thus, $B - A$ is the only solution of $X + A = B$, as claimed. ∎

The following problem shows how theorems 1.6 and 1.7 may be used to solve a matrix equation.

PROBLEM 1.   Solve the equation $5X + 3A = (2B - A) - 4X$, where A and B are constant $m \times n$ matrices.

Solution: Using properties (iv), (i), (vi), (vii), and (v) of theorem 1.6 successively, we can rewrite the given equation in the form

$$2B - A = (5X + 3A) + 4X = (3A + 5X) + 4X$$
$$= 3A + (5X + 4X) = 3A + 9X$$

According to theorem 1.7, this equation has the unique solution $9X = (2B - A) - 3A$, and by gathering like terms (justify this), we find that

$$9X = (2B - A) - 3A = 2B - (A + 3A) = 2B - 4A$$

Finally, applying properties (vii) and (v) of theorem 1.6, we obtain the unique solution

$$X = \tfrac{1}{9}(2B - 4A) = \tfrac{2}{9}B - \tfrac{4}{9}A$$

By and large, it is reasonably safe to deal with any algebraic problem involving matrix addition and multiplication of a matrix by a scalar as if it were the analogous problem of scalar algebra. However, matrix multiplication is a maverick operation which obeys some of the laws of scalar algebra but not others. To begin our discussion of the properties of this operation on a positive note, we prove that it satisfies the following familiar rules.

Theorem 1.8. If A, B, and C are matrices, then each of the following identities is valid whenever the appropriate products and sums are defined:
  (i) $A(sB) = s(AB) = (sA)B$ for each scalar s
  (ii) $A(B + C) = AB + AC$
  (iii) $(A + B)C = AC + BC$
  (iv) $A(BC) = (AB)C$

Proof: We shall prove (iv) in detail and leave the other parts to the reader as an exercise. Let $A = (a_{ij})$, $B = (b_{ij})$, and $C = (c_{ij})$ be $m \times n$, $n \times p$, and $p \times q$ matrices, respectively. Then $A(BC)$ and $(AB)C$ are both $m \times q$ matrices, and the proof will be complete as soon as we show that the (i, j) entry of $A(BC)$ equals that of $(AB)C$. To this end, let $D = BC$ and $F = AB$. Then we have

$$d_{kj} = \sum_{l=1}^{p} b_{kl}c_{lj} \quad \text{and} \quad f_{il} = \sum_{k=1}^{n} a_{ik}b_{kl}$$

and it follows that the (i, j) entries of $AD = A(BC)$ and $FC = (AB)C$ are, respectively,

$$\sum_{k=1}^{n} a_{ik}d_{kj} = \sum_{k=1}^{n} a_{ik}\left(\sum_{l=1}^{p} b_{kl}c_{lj}\right)$$

and

$$\sum_{l=1}^{p} f_{il}c_{lj} = \sum_{l=1}^{p}\left(\sum_{k=1}^{n} a_{ik}b_{kl}\right)c_{lj}$$

Applying property 5 of the summation notation, we find that

$$\underbrace{\sum_{k=1}^{n} a_{ik}\left(\sum_{l=1}^{p} b_{kl}c_{lj}\right)}_{(i,j)\ \text{entry of}\ A(BC)} = \underbrace{\sum_{l=1}^{p}\left(\sum_{k=1}^{n} a_{ik}b_{kl}\right)c_{lj}}_{(i,j)\ \text{entry of}\ (AB)C}$$

and our proof of identity (iv) is complete. ∎

The useful result which follows is an immediate consequence of theorem 1.8. We invite the reader to provide the details of the proof.

Corollary 1.2. If A, B, and C are matrices, then
  (i) $A(B - C) = AB - AC$
  (ii) $(A - B)C = AC - BC$
whenever these products and differences are defined.

The first three rules listed in theorem 1.8 are called *distributive laws*, while the fourth rule states that matrix multiplication is an *associative* operation. Note that the associative law makes it possible to remove the parentheses from a product involving three (or more) matrices; for instance, $A(BC)$ and $(AB)C$ can both be written as ABC. While these rules are certainly important, it is just as important to know what rules of scalar algebra are *not* satisfied by matrix multiplication. The three main differences between matrix multiplication and scalar multiplication are analyzed below:

1. *Matrix multiplication is not a commutative operation.* Specifically, AB is not necessarily equal to BA, even when both products are defined and have the same dimension.

EXAMPLE.

$$\begin{bmatrix} 1 & 5 \\ -7 & 2 \end{bmatrix}\begin{bmatrix} 3 & -6 \\ 4 & 9 \end{bmatrix} = \begin{bmatrix} 23 & 39 \\ -13 & 60 \end{bmatrix}$$

$$\begin{bmatrix} 3 & -6 \\ 4 & 9 \end{bmatrix}\begin{bmatrix} 1 & 5 \\ -7 & 2 \end{bmatrix} = \begin{bmatrix} 45 & 3 \\ -59 & 38 \end{bmatrix}$$

In scalar algebra, a product such as aba can also be written as a^2b or ba^2, since ordinary multiplication is commutative, and it is important to remember that such a rearrangement is not always valid in matrix algebra. Indeed, the matrix products ABA, A^2B, and BA^2 may all be different.

2. *The matrix equation $AX = B$ (or $XA = B$) may have no solution*, and even if it does, the solution may not be unique.

EXAMPLE.   Let

$$A = \begin{bmatrix} 0 & 0 \\ 1 & 1 \end{bmatrix}, \qquad B = \begin{bmatrix} 1 & 1 \\ 1 & 1 \end{bmatrix}, \qquad C = \begin{bmatrix} 0 & 0 \\ 2 & 3 \end{bmatrix}$$

There is no 2×2 matrix X such that $AX = B$, because no matter what X is chosen, the first row of AX must be [0 0] (verify this). On the other hand, there are many different ways to choose X so that $AX = C$, some of which are

$$\begin{bmatrix} 1 & 1 \\ 1 & 2 \end{bmatrix}, \qquad \begin{bmatrix} -1 & 5 \\ 3 & -2 \end{bmatrix}, \qquad \text{and} \qquad \begin{bmatrix} -2 & 3 \\ 4 & 0 \end{bmatrix}$$

This feature of matrix multiplication makes it impossible to define a general notion of matrix division.

3. *It is possible to have $AB = O$ even if neither A nor B is the zero matrix.*

EXAMPLE.

$$\begin{bmatrix} -4 & 2 \\ -2 & 1 \end{bmatrix}\begin{bmatrix} 1 & 3 \\ 2 & 6 \end{bmatrix} = \begin{bmatrix} 0 & 0 \\ 0 & 0 \end{bmatrix}$$

In scalar algebra, the fact that $ab = 0$ if and only if either $a = 0$ or $b = 0$ may be used for such purposes as showing that the scalar equation $x(x - a) = 0$ has exactly two solutions $x = 0$ and $x = a$. This technique is not generally valid in matrix algebra, for a matrix equation such as $X(X - A) = O$ may have solutions other than $X = O$ and $X = A$. For instance, the matrix

$$B = \begin{bmatrix} 4 & -6 \\ 2 & -3 \end{bmatrix}$$

is a solution of $X(X - I_2) = O$, since

$$B(B - I_2) = \begin{bmatrix} 4 & -6 \\ 2 & -3 \end{bmatrix}\begin{bmatrix} 3 & -6 \\ 2 & -4 \end{bmatrix} = \begin{bmatrix} 0 & 0 \\ 0 & 0 \end{bmatrix}$$

As a result of these features of matrix multiplication, certain matrix equations may be very hard to solve. In fact, it can happen that the only method of solution which offers any promise is the "direct" approach illustrated in the following problem.

PROBLEM 2. Show that the matrix equation $AX = I_2$ has no solution, where

$$A = \begin{bmatrix} -4 & 6 \\ 2 & -3 \end{bmatrix} \quad \text{and} \quad I_2 = \begin{bmatrix} 1 & 0 \\ 0 & 1 \end{bmatrix}$$

Solution: By computing the product matrix AX and equating its entries with the corresponding entries of I_2, we find that the equation $AX = I_2$ is equivalent to the non-homogeneous 4×4 linear system

$$\begin{aligned} -4x_{11} \qquad\quad + 6x_{21} \qquad\quad &= 1 \\ -4x_{12} \qquad\quad + 6x_{22} &= 0 \\ 2x_{11} \qquad\quad - 3x_{21} \qquad\quad &= 0 \\ 2x_{12} \qquad\quad - 3x_{22} &= 1 \end{aligned}$$

(1.19)

The reduction

$$\begin{bmatrix} -4 & 0 & 6 & 0 & \vdots & 1 \\ 0 & -4 & 0 & 6 & \vdots & 0 \\ 2 & 0 & -3 & 0 & \vdots & 0 \\ 0 & 2 & 0 & -3 & \vdots & 1 \end{bmatrix} \longrightarrow \begin{bmatrix} 1 & 0 & -\frac{3}{2} & 0 & \vdots & -\frac{1}{4} \\ 0 & 1 & 0 & -\frac{3}{2} & \vdots & 0 \\ 0 & 0 & 0 & 0 & \vdots & \text{①} \\ 0 & 0 & 0 & 0 & \vdots & 0 \end{bmatrix}$$

"bad" entry

tells us that system (1.19) is inconsistent and hence, that $AX = I_2$ has no solution.

It is important to note that the $n \times n$ matrix I_n, which has 1's on the main diagonal and 0's elsewhere, plays much the same role in matrix algebra as does the number 1 in scalar algebra. In particular, if A is an $m \times n$ matrix, it is easy to show that $I_m A = A = A I_n$ (see exercise 11), and for this reason, I_n is referred to as the nth order *identity* matrix. This formula is illustrated in the following example.

EXAMPLE.   If

$$A = \begin{bmatrix} -5 & 7 & 2 \\ 3 & 1 & -9 \end{bmatrix}$$

then

$$I_2 A = \begin{bmatrix} 1 & 0 \\ 0 & 1 \end{bmatrix} \begin{bmatrix} -5 & 7 & 2 \\ 3 & 1 & -9 \end{bmatrix} = \begin{bmatrix} -5 & 7 & 2 \\ 3 & 1 & -9 \end{bmatrix} = A$$

$$A I_3 = \begin{bmatrix} -5 & 7 & 2 \\ 3 & 1 & -9 \end{bmatrix} \begin{bmatrix} 1 & 0 & 0 \\ 0 & 1 & 0 \\ 0 & 0 & 1 \end{bmatrix} = \begin{bmatrix} -5 & 7 & 2 \\ 3 & 1 & -9 \end{bmatrix} = A$$

We know that each non-zero number c has a unique multiplicative inverse c^{-1} which satisfies $cc^{-1} = 1 = c^{-1}c$, but problem 2 shows that not every $n \times n$ matrix A has a multiplicative inverse. Specifically, it is not always possible to find a matrix A^{-1} such that $AA^{-1} = I_n = A^{-1}A$. However, some matrices do possess inverses, and in the next section, we examine the properties of such matrices.

Exercise Set 1.5

1. Let

$$A = \begin{bmatrix} -3 & 5 & 2 \\ 0 & 1 & 7 \\ 4 & -2 & 3 \end{bmatrix}, \quad B = \begin{bmatrix} 1 & 6 & 1 \\ 0 & -2 & 5 \\ -1 & 3 & 8 \end{bmatrix}, \quad C = \begin{bmatrix} 1 & 9 & -7 \\ 3 & 0 & 5 \\ -1 & 2 & 6 \end{bmatrix}$$

Verify the following equations:

(i) $(7A)C = 7(AC)$ (ii) $A(BC) = (AB)C$

(iii) $A(B + C) = AB + AC$ (iv) $A(B - C) = AB - AC$

(v) $(A + B)(A - B) = A^2 + BA - AB - B^2$

2. Use the results of this section to simplify each of the following matrix expressions as much as possible. You may assume that all sums, differences, and products are defined:

EXAMPLE: $(A - B)(A + B) - 2AB = A^2 - BA - AB - B^2$
  (i) $A(2B + C) - (A + B)C$
  (ii) $A[(3A - 7B) - (7A + 2B)] - (A - 3B)B$
  (iii) $(A + B)^3 - (A - B)^3$
  (iv) $[(2A - B)(A + B) - (3A^2 + 2B)] - (A^2 + I_n)B$
  (v) $[A(B + C)](B - C)$

3. Let

$$A = \begin{bmatrix} -1 & 2 & 1 \\ 0 & 5 & 8 \end{bmatrix}, \quad B = \begin{bmatrix} 0 & 6 & 3 \\ 1 & -4 & 7 \end{bmatrix}, \quad C = \begin{bmatrix} 2 & -5 \\ -6 & 15 \end{bmatrix}$$

In each of the following cases, either find a matrix X which solves the given equation or show that no such matrix exists.

(i) $2A - 3X = 7B$ (ii) $C(A + X) = B$

(iii) $CX = O$ but $X \neq O$

4. The following is a list of statements about $n \times n$ matrices. In each case, either show that the statement is generally true or find specific 2×2 matrices for which the statement is not true.
  (i) $A + B = 0$ if and only if $A = -B$.
  (ii) The matrix equation $X^2 = I_n$ is satisfied only when $X = I_n$ or $X = -I_n$.
  (iii) $(A - B)(A + B) = A^2 - B^2$ if and only if $AB = BA$.
  (iv) If $B = A^2 - 5A + 2I_n$, then $AB = BA$.
  (v) $AB = O$ if and only if $BA = O$.
  (vi) If $A^3 - 7A^2 + 5I_n = 0$, then $A^4 = 49A^2 - 5A - 35I_n$.
5. Verify the identities $A(B + C) = AB + AC$ and $A(B - C) = AB - AC$ in the special case where

$$A = \begin{bmatrix} a_{11} & a_{12} & a_{13} \\ a_{21} & a_{22} & a_{23} \end{bmatrix}, \quad B = \begin{bmatrix} b_{11} & b_{12} \\ b_{21} & b_{22} \\ b_{31} & b_{32} \end{bmatrix}, \quad C = \begin{bmatrix} c_{11} & c_{12} \\ c_{21} & c_{22} \\ c_{31} & c_{32} \end{bmatrix}$$

6. Let A, B, and C be $n \times n$ matrices which satisfy $AB = BA$ and $AC = CA$.
  (i) Show that $(AB)^k = A^k B^k$ for each positive integer k.
  (ii) Show that $A(B + C) = (B + C)A$.
*7. Let A be an $n \times n$ matrix, and let B and C be matrices such that $AB = I_n$ and $CA = I_n$. Use the associative property of matrix multiplication to show that $B = C$.
8. Let A be an $m \times n$ matrix and let $X = s_1 X_1 + s_2 X_2$, where s_1, s_2 are scalars and X_1, X_2 are column vectors which satisfy $AX_1 = O$ and $AX_2 = O$. Use the distributive laws of matrix multiplication (see theorem 1.8) to show that $AX = O$.

9. Verify in detail properties 3 and 4 of the difference operation [see (1.18)]:
   (3) $(A - B) - C = A - (B + C)$
   (4) $A - (B - C) = (A - B) + C$
10. Complete the proof of theorem 1.8 and its corollary.
    Hint: For instance, to prove $(A + B)C = AC + BC$, observe that

$$\underbrace{\sum_{k=1}^{n} (a_{ik} + b_{ik})c_{kj}}_{\substack{(i,j) \text{ entry of} \\ (A + B)C}} = \underbrace{\sum_{k=1}^{n} a_{ik}c_{kj}}_{\substack{(i,j) \text{ entry} \\ \text{of } AC}} + \underbrace{\sum_{k=1}^{n} b_{ik}c_{kj}}_{\substack{(i,j) \text{ entry} \\ \text{of } BC}}$$

and then explain why the sum of the expressions on the right is the (i,j) entry of $AC + BC$.

11. If A is an $m \times n$ matrix and I_n is the $n \times n$ identity matrix, verify that $AI_n = A$.
    Hint: What is the (i,j) entry of AI_n?

1.6 NON-SINGULAR MATRICES AND THE MATRIX INVERSION ALGORITHM

Matrices with inverses play a vital role in linear algebra, and in this section, we shall examine several important theoretical and computational questions associated with such matrices. We begin with the following definition.

Definition 1.6. An $n \times n$ matrix A is said to be *non-singular* (or *invertible*) if there exists a matrix A^{-1} such that $AA^{-1} = I_n = A^{-1}A$. If no such matrix exists, A is *singular*.

EXAMPLE 1.   The matrix

$$B = \begin{bmatrix} 5 & 2 \\ 3 & 1 \end{bmatrix}$$

is non-singular, for we have

$$\underset{B}{\begin{bmatrix} 5 & 2 \\ 3 & 1 \end{bmatrix}} \underset{B^{-1}}{\begin{bmatrix} -1 & 2 \\ 3 & -5 \end{bmatrix}} = \underset{I_2}{\begin{bmatrix} 1 & 0 \\ 0 & 1 \end{bmatrix}} = \underset{B^{-1}}{\begin{bmatrix} -1 & 2 \\ 3 & -5 \end{bmatrix}} \underset{B}{\begin{bmatrix} 5 & 2 \\ 3 & 1 \end{bmatrix}}$$

On the other hand, the matrix

$$A = \begin{bmatrix} -4 & 6 \\ 2 & -3 \end{bmatrix}$$

is singular, since the matrix equation $AX = I_2$ has no solution. (See problem 3 of section 1.5).

If A is an $n \times n$ non-singular matrix, it is easy to see that A^{-1} is the only matrix which satisfies the equation $AX = I_n = XA$, for if $AB = I_n = BA$, we find that

$$B = BI_n = B(AA^{-1}) = (BA)A^{-1} = I_nA^{-1} = A^{-1}$$

(see also exercise 7, section 1.5). Henceforth, A^{-1} will be referred to as *the inverse* of A. Our next theorem shows how A^{-1} is related to the inverses of several other non-singular matrices which are associated with A in one way or another.

Theorem 1.9.   If A and B are non-singular $n \times n$ matrices and s is a non-zero scalar, then A^{-1}, sA, A^t, and AB are all non-singular, and we have

(i) $(A^{-1})^{-1} = A$

(ii) $(sA)^{-1} = \dfrac{1}{s}A^{-1}$

(iii) $(A^t)^{-1} = (A^{-1})^t$

(iv) $(AB)^{-1} = B^{-1}A^{-1}$

Proof: In order to show that a given matrix C is non-singular, it is enough to find a solution of the equation $CX = I_n = XC$. For instance, AB is non-singular and has inverse $B^{-1}A^{-1}$ since

$$(AB)(B^{-1}A^{-1}) = ABB^{-1}A^{-1} = AI_nA^{-1} = AA^{-1} = I_n$$

and

$$(B^{-1}A^{-1})(AB) = B^{-1}A^{-1}AB = B^{-1}I_nB = B^{-1}B = I_n$$

The rest of the proof follows similar lines. Parts (i) and (ii) are easy to handle, and part (iii) is outlined in exercise 11. ∎

EXAMPLE 2.   Let

$$A = \begin{bmatrix} -1 & 5 \\ -2 & 9 \end{bmatrix} \quad \text{and} \quad B = \begin{bmatrix} 2 & 7 \\ 1 & 4 \end{bmatrix}$$

It can be shown that A and B are non-singular and that

$$A^{-1} = \begin{bmatrix} 9 & -5 \\ 2 & -1 \end{bmatrix} \quad \text{and} \quad B^{-1} = \begin{bmatrix} 4 & -7 \\ -1 & 2 \end{bmatrix}$$

Then, according to theorem 1.9, we have

(iii) $\begin{bmatrix} -1 & -2 \\ 5 & 9 \end{bmatrix}^{-1} = (A^t)^{-1} = (A^{-1})^t = \begin{bmatrix} 9 & 2 \\ -5 & -1 \end{bmatrix}$

(iv) $\begin{bmatrix} 3 & 13 \\ 5 & 22 \end{bmatrix}^{-1} = (AB)^{-1} = B^{-1}A^{-1} = \begin{bmatrix} 22 & -13 \\ -5 & 3 \end{bmatrix}$.

In section 1.5, we observed that the matrix equation $AX = B$ does not always have a solution and that it is possible to have $AC = O$ even when neither A nor C is the zero matrix. However, if A is non-singular, the equation $AX = B$ has the unique solution $X = A^{-1}B$, and in particular, $AC = O$ if and only if $C = A^{-1}O = O$. When this result is applied to the $n \times n$ linear system $AX = \beta$, we obtain the following theorem.

Theorem 1.10. The $n \times n$ linear system $AX = \beta$ has the unique solution $X = A^{-1}\beta$ if its coefficient matrix A is non-singular.

This theorem can be quite helpful if we need to solve several different systems with the same non-singular coefficient matrix. For example, it can be shown that

$$\begin{bmatrix} 5 & -1 & 3 \\ 2 & 0 & 1 \\ 3 & 2 & 1 \end{bmatrix}^{-1} = \begin{bmatrix} -2 & 7 & -1 \\ 1 & -4 & 1 \\ 4 & -13 & 2 \end{bmatrix}$$

Therefore, if a, b, and c are real numbers, a 3×3 linear system of the form

$$5x_1 - x_2 + 3x_3 = a$$

$$2x_1 + 0x_2 + x_3 = b \qquad (1.20)$$

$$3x_1 + 2x_2 + x_3 = c$$

has the unique solution

$$\begin{bmatrix} x_1 \\ x_2 \\ x_3 \end{bmatrix} = \begin{bmatrix} -2 & 7 & -1 \\ 1 & -4 & 1 \\ 4 & -13 & 2 \end{bmatrix} \begin{bmatrix} a \\ b \\ c \end{bmatrix} = \begin{bmatrix} -2a + 7b - c \\ a - 4b + c \\ 4a - 13b + 2c \end{bmatrix}$$

For instance, if $a = 3$, $b = 0$, $c = -2$ in (1.20), the corresponding unique solution is $x_1 = -4$, $x_2 = 1$, $x_3 = 8$, and in the case where $a = 9$, $b = -5$, $c = -4$, the solution is $x_1 = -49$, $x_2 = 25$, $x_3 = 93$.

Near the end of section 1.3, we showed that an $n \times n$ linear system has a unique solution if and only if its coefficient matrix is row equivalent to I_n. Combining this result with theorem 1.10, we find that a non-singular matrix must be row equivalent to I_n. The converse is also true, and we have the following useful criterion for non-singularity.

Theorem 1.11.   The $n \times n$ matrix A is non-singular if and only if it is row equivalent to I_n.

Proof: The remarks which precede the statement of the theorem constitute a proof of the "only if" part, but unfortunately, the "if" part is not easy to prove with the theoretical tools now at our disposal. Consequently, for the rest of this section, we shall simply assume that a matrix which is row equivalent to I_n must be non-singular, and then in section 1.7, we shall provide a formal proof of this fact. ∎

PROBLEM 1.   For what values of λ will the matrix

$$A_\lambda = \begin{bmatrix} 1 & 0 & 5 \\ 2 & 4 & -2 \\ -3 & 1 & \lambda \end{bmatrix}$$

be non-singular?

Solution: The reduction

$$\begin{bmatrix} 1 & 0 & 5 \\ 2 & 4 & -2 \\ -3 & 1 & \lambda \end{bmatrix} \longrightarrow \begin{bmatrix} 1 & 0 & 5 \\ 0 & 1 & -3 \\ 0 & 0 & \lambda + 18 \end{bmatrix}$$

enables us to deduce that A_λ will be row equivalent to I_3 unless $\lambda + 18 = 0$. Therefore, A_λ is non-singular when λ is any real number except -18.

In general, even if A and B are non-singular, the commutative law $AB = BA$ does not necessarily hold; for example, the matrices

$$\begin{bmatrix} 1 & -1 \\ 1 & 1 \end{bmatrix} \text{ and } \begin{bmatrix} 1 & 1 \\ 0 & 1 \end{bmatrix}$$

are non-singular, but

$$\begin{bmatrix} 1 & -1 \\ 1 & 1 \end{bmatrix}\begin{bmatrix} 1 & 1 \\ 0 & 1 \end{bmatrix} = \begin{bmatrix} 1 & 0 \\ 1 & 2 \end{bmatrix}$$

and

$$\begin{bmatrix} 1 & 1 \\ 0 & 1 \end{bmatrix}\begin{bmatrix} 1 & -1 \\ 1 & 1 \end{bmatrix} = \begin{bmatrix} 2 & 0 \\ 1 & 1 \end{bmatrix}$$

Nevertheless, our next result shows that there is at least one special case in which a product involving non-singular matrices is commutative.

Corollary 1.3. If A is non-singular and B satisfies $AB = I$, then B also satisfies $BA = I$. In other words, $B = A^{-1}$.

Proof: This follows immediately from the fact that a non-singular matrix has a unique inverse. ∎

Before non-singular matrices can be used effectively in our work, we must develop computational methods for determining whether or not a given matrix A is non-singular and for finding A^{-1} if it exists. Fortunately, both objectives can be accomplished within the framework of a single process, which derives its theoretical justification from theorem 1.11 and corollary 1.3. The general description of this process will have more meaning if we first demonstrate what we have in mind by analyzing a specific example. Accordingly, consider the matrix

$$A = \begin{bmatrix} 3 & 5 \\ 4 & 7 \end{bmatrix}$$

If A is non-singular, its inverse must satisfy the matrix equation $AX = I_2$. Comparing the corresponding entries in this equation, we obtain the 4×4 linear system

$$
\begin{array}{llll}
3x_{11} & + 5x_{21} & & = 1 \\
3x_{12} & & + 5x_{22} & = 0 \\
4x_{11} & + 7x_{21} & & = 0 \\
4x_{12} & & + 7x_{22} & = 1
\end{array}
\tag{1.21}
$$

which may be regarded as the following pair of 2×2 linear systems with coefficient matrix A:

$$
\begin{array}{ll}
3x_{11} + 5x_{21} = 1 & 3x_{12} + 5x_{22} = 0 \\
4x_{11} + 7x_{21} = 0 & 4x_{12} + 7x_{22} = 1
\end{array}
$$

first column of I_2 second column of I_2

These systems can then be solved simultaneously by performing the combined reduction

$$
\begin{bmatrix} 3 & 5 & | & 1 & 0 \\ 4 & 7 & | & 0 & 1 \end{bmatrix} \longrightarrow \begin{bmatrix} 1 & 0 & | & 7 & -5 \\ 0 & 1 & | & -4 & 3 \end{bmatrix}
$$

$$[A \; | \; I_2] \qquad\qquad [I_2 \; | \; B]$$

(see problem 4, section 1.3). The form of the reduced row echelon matrix $[I_2 \mid B]$ on the right of this reduction tells us that A is row equivalent to I_2 and hence that A is non-singular (theorem 1.11). Moreover, we also conclude that system (1.21) has the unique solution

$$x_{11} = 7, \quad x_{12} = -5$$
$$x_{21} = -4, \quad x_{22} = 3$$

and that

$$B = \begin{bmatrix} 7 & -5 \\ -4 & 3 \end{bmatrix}$$

is the unique solution of the equation $AX = I_2$. Since A is non-singular and $AB = I_2$, corollary 1.3 enables us to conclude that $B = A^{-1}$.

In general, if A is an $n \times n$ matrix, the equation $AX = I_n$ may be regarded as n separate $n \times n$ linear systems, each of which has A as its coefficient matrix. To solve these systems simultaneously, we perform the combined reduction

$$[A \mid I_n] \longrightarrow [E \mid B]$$

where E is the reduced row echelon form of A. We then have two cases to consider:

1. If E is any reduced row echelon matrix other than I_n, A must be singular (theorem 1.11).
2. If E is I_n, then A is non-singular (theorem 1.11). Moreover, the matrix B satisfies $AB = I_n$, and corollary 1.3 tells us that $B = A^{-1}$.

This process, which we shall call the *matrix inversion algorithm*, is illustrated in the following solved problems.

PROBLEM 2.   Determine whether or not the matrix

$$A = \begin{bmatrix} 3 & 8 & 3 \\ -1 & 1 & 10 \\ 1 & 2 & -1 \end{bmatrix}$$

is non-singular, and if it is, find its inverse.

Solution: The reduction

$$\left[\begin{array}{ccc|ccc} 3 & 8 & 3 & 1 & 0 & 0 \\ -1 & 1 & 10 & 0 & 1 & 0 \\ 1 & 2 & -1 & 0 & 0 & 1 \end{array}\right] \longrightarrow \left[\begin{array}{ccc|ccc} 1 & 0 & -7 & 0 & -\frac{2}{3} & \frac{1}{3} \\ 0 & 1 & 3 & 0 & \frac{1}{3} & \frac{1}{3} \\ 0 & 0 & 0 & 1 & -\frac{2}{3} & -\frac{11}{3} \end{array}\right]$$

$$[A \mid I_3] \qquad\qquad\qquad [E \mid B]$$

tells us that A is not row equivalent to I_3, and we conclude that A must be singular.

PROBLEM 3.   Show that the matrix

$$A = \begin{bmatrix} -1 & 3 & 0 & 5 \\ 2 & 1 & 4 & 0 \\ 0 & 2 & 1 & -1 \\ 1 & 0 & 1 & -19 \end{bmatrix}$$

is non-singular and find its inverse.

Solution: As the first step in the matrix inversion algorithm, we perform the reduction

$$\begin{bmatrix} -1 & 3 & 0 & 5 & | & 1 & 0 & 0 & 0 \\ 2 & 1 & 4 & 0 & | & 0 & 1 & 0 & 0 \\ 0 & 2 & 1 & -1 & | & 0 & 0 & 1 & 0 \\ 1 & 0 & 1 & -19 & | & 0 & 0 & 0 & 1 \end{bmatrix} \longrightarrow \begin{bmatrix} 1 & 0 & 0 & 0 & | & 134 & 44 & -223 & 47 \\ 0 & 1 & 0 & 0 & | & 40 & 13 & -66 & 14 \\ 0 & 0 & 1 & 0 & | & -77 & -25 & 128 & -27 \\ 0 & 0 & 0 & 1 & | & 3 & 1 & -5 & 1 \end{bmatrix}$$

$$[A \mid I_4] \qquad\qquad\qquad [I_4 \mid B]$$

From the form $[I_4 \mid B]$ of the reduced row echelon matrix on the right, we conclude that A is non-singular and that

$$A^{-1} = B = \begin{bmatrix} 134 & 44 & -223 & 47 \\ 40 & 13 & -66 & 14 \\ -77 & -25 & 128 & -27 \\ 3 & 1 & -5 & 1 \end{bmatrix}$$

This completes our survey of the fundamental properties of matrices. The material we have developed in these last three sections is of general interest and will be used throughout our work. However, there are other topics which are just as important as those we have discussed, but have a more narrow range of application, and we shall examine one such specialized topic in the section which follows.

Exercise Set 1.6

1. In each of the following cases, determine whether or not the given matrix is non-singular, and if it is, find its inverse.

(i) $\begin{bmatrix} 19 & 0 & 0 \\ 0 & -5 & 0 \\ 0 & 0 & 3 \end{bmatrix}$ (ii) $\begin{bmatrix} 5 & -3 & 1 \\ 0 & 1 & 4 \\ 0 & 0 & -6 \end{bmatrix}$

(iii) $\begin{bmatrix} -5 & -15 & 9 \\ 3 & 9 & 8 \\ 4 & 12 & -1 \end{bmatrix}$
(iv) $\begin{bmatrix} 5 & -2 & 3 \\ 0 & 1 & 7 \\ 2 & -1 & 0 \end{bmatrix}$

(v) $\begin{bmatrix} 5 & 7 & -1 & 9 \\ 1 & 1 & 1 & 1 \\ -1 & 2 & -10 & 3 \\ 7 & 9 & 1 & 11 \end{bmatrix}$
*(vi) $\begin{bmatrix} 1 & 0 & 1 & 0 \\ 1 & -1 & 1 & 3 \\ 5 & 2 & 1 & 1 \\ 2 & 0 & -3 & 9 \end{bmatrix}$

2. In each of the following cases, find all real numbers λ for which the given 3×3 matrix is non-singular.

(i) $\begin{bmatrix} -1 & 3 & \lambda \\ 1 & 1 & -5 \\ 5 & 2 & 2 \end{bmatrix}$
(ii) $\begin{bmatrix} \lambda & 1 & 1 \\ 1 & \lambda & 1 \\ 1 & 1 & \lambda \end{bmatrix}$

(iii) $\begin{bmatrix} 1 & 2 & \lambda \\ 3 & \lambda & 0 \\ \lambda & 0 & 1 \end{bmatrix}$
(iv) $\begin{bmatrix} -2 & \lambda & 3 \\ 1 & 2 & \lambda \\ 1 & 11 & 18 \end{bmatrix}$

3. A certain matrix A is known to be row equivalent to

$$B = \begin{bmatrix} -7 & 5 & 3 \\ 9 & 2 & -6 \\ 18 & 0 & -11 \end{bmatrix}$$

Is A singular or non-singular? Explain your reasoning.

4. If it is known that

$$A = \begin{bmatrix} a & 1 & 1 \\ -2 & 3 & b \\ -9 & c & 0 \end{bmatrix} \quad \text{and} \quad A^{-1} = \begin{bmatrix} 7 & d & 4 \\ e & 9 & -37 \\ 29 & -4 & f \end{bmatrix}$$

what are a, b, c, d, e, and f?

5. (i) Show that the matrix

$$A = \begin{bmatrix} 4 & -2 & 5 \\ 1 & 1 & -2 \\ 6 & 1 & -1 \end{bmatrix}$$

is non-singular, and find its inverse.

(ii) Use A^{-1} to find a general solution for each linear system of the form

$$4x_1 - 2x_2 + 5x_3 = a$$

$$x_1 + x_2 - 2x_3 = b$$

$$6x_1 + x_2 - x_3 = c \qquad \text{where } a, b, c \text{ are real numbers.}$$

(iii) Find the solutions which correspond to the following choices of a, b, c:

a	b	c	solution
0	0	0	
1	0	0	
2	−7	1	

6. Let A be an $n \times n$ matrix whose kth row contains only 0's. For example.

$$\begin{bmatrix} -3 & 2 & 5 \\ 0 & 0 & 0 \\ 7 & 1 & 9 \end{bmatrix}$$

is such a matrix.

(i) If B is an $n \times n$ matrix, show that the kth row of the product matrix AB also contains only 0's.

(ii) Explain why A must be singular.

(iii) Use a similar argument to show that an $n \times n$ matrix must be singular if one of its columns contains only 0's.

7. If A is an $n \times n$ non-singular matrix, show that A^k is also non-singular for each positive integer k and verify that $(A^k)^{-1} = (A^{-1})^k$.

8. The following is a list of statements regarding $n \times n$ nonsingular matrices. In each case, either prove that the statement is generally true or find 2×2 matrices for which it is false.

(i) If A and B are non-singular, then so is $A + B$.

(ii) If A and B are non-singular and $A^2B^2 = I_n$, then $(AB)^{-1} = BA$.

(iii) The only $n \times n$ non-singular reduced row echelon matrix is I_n.

(iv) If S is non-singular and symmetric (that is, $S = S^t$), then S^{-1} is also symmetric.

(v) If K is non-singular and skew-symmetric ($K = -K^t$), then K^{-1} is skew-symmetric.

(vi) I_n is the only $n \times n$ non-singular idempotent matrix (see exercise 6, section 1.4).

(vii) An $n \times n$ nilpotent matrix must be singular.

9. Let $D = \text{diag}(d_1, \ldots, d_n)$ be an $n \times n$ diagonal matrix (see exercise 11, section 1.4). Show that D is non-singular if and only if $d_j \neq 0$ for $j = 1, 2, \ldots, n$, and verify that $D^{-1} = \text{diag}(d_1^{-1}, \ldots, d_n^{-1})$ in the case where D is non-singular.

10. Find the inverse of the non-singular matrix

$$A = \begin{bmatrix} 3 & -5 & 2 \\ 1 & 4 & -1 \\ -10 & 1 & -2 \end{bmatrix}$$

and then solve the following matrix equations:

(i) $2AX = \begin{bmatrix} 1 & 1 \\ 0 & 1 \\ 2 & 0 \end{bmatrix}$ *(ii) $A^t X = A$

11. Complete the proof of part (iii) of theorem 1.9 by showing that

$$A^t(A^{-1})^t = I_n \quad \text{and} \quad (A^{-1})^t A^t = I_n$$

   Hint: Recall exercise 8, section 1.4.

*12. If $A_1, A_2, \ldots, A_m$ are all non-singular $n \times n$ matrices, show that the product $B = A_1 A_2 \ldots A_m$ is also non-singular and verify that $B^{-1} = A_m^{-1} A_{m-1}^{-1} \cdots A_2^{-1} A_1^{-1}$.

*13. Show that the $n \times n$ matrix A is non-singular if and only if the homogeneous linear system $AX = O$ has the unique solution $X = O$.
   Hint: For the "if" part, use theorem 1.11.

1.7   ELEMENTARY MATRICES

By definition, the matrix A is row equivalent to B if B can be derived from A by a finite sequence of elementary row operations. For certain theoretical purposes, it is useful to know that A is row equivalent to B if and only if there exists a non-singular matrix P such that $B = PA$. We shall prove this result and show how to actually construct P from the elementary row operations relating B to A.

We shall say that the $n \times n$ array P is an *elementary matrix* if it can be obtained by performing a single elementary row operation on I_n. There are three different kind of elementary matrices, one for each type of elementary row operation:

A *type I* elementary matrix (denoted by P_{kl}) is obtained by exchanging the kth and lth rows of I_n.

EXAMPLE.   Typical 4×4 elementary matrices of type 1 are

$$P_{13} = \begin{bmatrix} 0 & 0 & 1 & 0 \\ 0 & 1 & 0 & 0 \\ 1 & 0 & 0 & 0 \\ 0 & 0 & 0 & 1 \end{bmatrix} \quad \text{and} \quad P_{24} = \begin{bmatrix} 1 & 0 & 0 & 0 \\ 0 & 0 & 0 & 1 \\ 0 & 0 & 1 & 0 \\ 0 & 1 & 0 & 0 \end{bmatrix}$$

A *type II* elementary matrix is formed by multiplying the kth row of I_n by the non-zero scalar s. We denote such a matrix by $S_k(s)$.

EXAMPLE.   In the 3×3 case,

$$S_2(5) = \begin{bmatrix} 1 & 0 & 0 \\ 0 & 5 & 0 \\ 0 & 0 & 1 \end{bmatrix} \quad \text{and} \quad S_3(-\tfrac{1}{2}) = \begin{bmatrix} 1 & 0 & 0 \\ 0 & 1 & 0 \\ 0 & 0 & -\tfrac{1}{2} \end{bmatrix}$$

The *type III* elementary matrix $M_k(l; s)$ is obtained by adding s times the lth row of I_n to the kth row.

EXAMPLE.   For 3×3 matrices, we have

$$M_2(3; -\tfrac{1}{2}) = \begin{bmatrix} 1 & 0 & 0 \\ 0 & 1 & -\tfrac{1}{2} \\ 0 & 0 & 1 \end{bmatrix} \quad \text{and} \quad M_3(1; \sqrt{2}) = \begin{bmatrix} 1 & 0 & 0 \\ 0 & 1 & 0 \\ \sqrt{2} & 0 & 1 \end{bmatrix}$$

The effect of multiplying a given $m \times n$ matrix A on the left by an $m \times m$ elementary matrix E is the same as that achieved by performing the elementary row operation which corresponds to E. Thus, the reduction of A by a finite sequence of elementary row operations can be accomplished by multiplying A on the left by the corresponding sequence of elementary matrices. For example, the reduction step

$$\begin{bmatrix} -3 & 5 & 7 \\ 2 & -6 & 8 \\ 0 & 1 & 1 \end{bmatrix} \xrightarrow{R_1 \leftrightarrow R_2} \begin{bmatrix} 2 & -6 & 8 \\ -3 & 5 & 7 \\ 0 & 1 & 1 \end{bmatrix}$$

$$\qquad\quad A \qquad\qquad\qquad\qquad B_1$$

can be carried out by forming the product

$$\begin{bmatrix} 0 & 1 & 0 \\ 1 & 0 & 0 \\ 0 & 0 & 1 \end{bmatrix} \begin{bmatrix} -3 & 5 & 7 \\ 2 & -6 & 8 \\ 0 & 1 & 1 \end{bmatrix} = \begin{bmatrix} 2 & -6 & 8 \\ -3 & 5 & 7 \\ 0 & 1 & 1 \end{bmatrix}$$

$$\quad P_{12} \qquad\qquad A \qquad\qquad\qquad B_1$$

and the matrix B produced by the compound reduction

$$\begin{bmatrix} -3 & 5 & 7 \\ 2 & -6 & 8 \\ 0 & 1 & 1 \end{bmatrix} \xrightarrow[\substack{\frac{1}{2}R_1 \to R_1 \\ 3R_1 + R_2 \to R_2}]{R_1 \leftrightarrow R_2} \begin{bmatrix} 1 & -3 & 4 \\ 0 & -4 & 19 \\ 0 & 1 & 1 \end{bmatrix}$$

$$\qquad\quad A \qquad\qquad\qquad\qquad B$$

may be expressed as the product

$$\begin{bmatrix} 1 & 0 & 0 \\ 3 & 1 & 0 \\ 0 & 0 & 1 \end{bmatrix} \begin{bmatrix} \frac{1}{2} & 0 & 0 \\ 0 & 1 & 0 \\ 0 & 0 & 1 \end{bmatrix} \begin{bmatrix} 0 & 1 & 0 \\ 1 & 0 & 0 \\ 0 & 0 & 1 \end{bmatrix} \begin{bmatrix} -3 & 5 & 7 \\ 2 & -6 & 8 \\ 0 & 1 & 1 \end{bmatrix} = \begin{bmatrix} 1 & -3 & 4 \\ 0 & -4 & 19 \\ 0 & 1 & 1 \end{bmatrix}$$

$$\quad M_2(1;3) \qquad S_1(\tfrac{1}{2}) \qquad P_{12} \qquad\qquad A \qquad\qquad B$$

This product can be written more succinctly as $B = E_3 E_2 E_1 A$, where $E_1 = P_{12}$, $E_2 = S_1(\tfrac{1}{2})$, and $E_3 = M_2(1;3)$. Our example illustrates the general result contained in the following theorem.

Theorem 1.12. The $m \times n$ matrix A is row equivalent to B if and only if there exists a finite sequence of $m \times m$ elementary matrices $E_1, E_2, \ldots, E_k$ such that $B = E_k E_{k-1} \ldots E_1 A$.

Proof: This is just another way of saying that A is row equivalent to B if and only if B can be derived from A by a finite sequence of elementary row operations, the jth of which corresponds to E_j. ∎

In section 1.1, we observed that the effect of each elementary row operation can be reversed by an appropriate operation of the same type. The matrix version of this observation may be stated as follows.

Theorem 1.13. Elementary matrices are non-singular, and each type of elementary matrix has an inverse of the same type. To be specific,

$$P_{kl}^{-1} = P_{lk}$$

$$[S_k(s)]^{-1} = S_k(s^{-1})$$

$$[M_k(l;s)]^{-1} = M_k(l;-s)$$

Proof: Exercise.

EXAMPLE 1.   In the 3×3 case, we have

$$P_{13}^{-1} = P_{31} = P_{13} \qquad \begin{bmatrix} 0 & 0 & 1 \\ 0 & 1 & 0 \\ 1 & 0 & 0 \end{bmatrix}^{-1} = \begin{bmatrix} 0 & 0 & 1 \\ 0 & 1 & 0 \\ 1 & 0 & 0 \end{bmatrix}$$

$$[S_2(-3)]^{-1} = S_2(-\tfrac{1}{3}) \qquad \begin{bmatrix} 1 & 0 & 0 \\ 0 & -3 & 0 \\ 0 & 0 & 1 \end{bmatrix}^{-1} = \begin{bmatrix} 1 & 0 & 0 \\ 0 & -\tfrac{1}{3} & 0 \\ 0 & 0 & 1 \end{bmatrix}$$

$$[M_3(1;\sqrt{2})]^{-1} = M_3(1; -\sqrt{2}) \begin{bmatrix} 1 & 0 & 0 \\ 0 & 1 & 0 \\ \sqrt{2} & 0 & 1 \end{bmatrix}^{-1} = \begin{bmatrix} 1 & 0 & 0 \\ 0 & 1 & 0 \\ -\sqrt{2} & 0 & 1 \end{bmatrix}$$

If the matrix A is row equivalent to B, the "only if" part of theorem 1.12 tells us that there exist elementary matrices $E_1, E_2, \ldots, E_k$ such that $B = E_k \ldots E_1 A$. Since each E_j is nonsingular, we can write $A = E_1^{-1}E_2^{-1} \ldots E_k^{-1}B$ (verify this equation), and since each E_j^{-1} is an elementary matrix, the "if" part of theorem 1.12 enables us to conclude that B is row equivalent to A. The above argument can be reversed by simply exchanging the roles of A and B, and we have at last proved the following result, which was first stated without proof in section 1.1.

Theorem 1.14.   The matrix A is row equivalent to B if and only if B is row equivalent to A.

Theorem 1.14 can be used to show that matrices which are row equivalent to one another have a great deal in common. For instance, our next result shows that a non-singular matrix cannot be row equivalent to one which is singular.

Lemma 1.4.   Two $n \times n$ matrices which are row equivalent to one another must either both be singular or both be non-singular.

Proof: Let A be row equivalent to B, and suppose that $B = E_k E_{k-1} \ldots E_1 A$. If A is non-singular, then so is B since it is the product of a finite number of non-singular matrices (see exercise 12, section 1.6). For the same reason, if A is singular, the equation $A = E_1^{-1}E_2^{-1} \ldots E_k^{-1}B$ enables us to conclude that B is also singular (why?), and the proof is complete. ∎

Since each matrix A is row equivalent to a reduced row echelon matrix E, it follows from lemma 1.4 that A is non-singular if and only if E is non-singular. However, the only non-singular $n \times n$ reduced row echelon matrix is I_n, because all other $n \times n$ matrices in reduced row echelon form have at least one row of zeroes and must therefore be singular (see exercise 6, section 1.6). By combining these observations, we can finally complete the proof of theorem 1.11, which is so important that we state it again.

Theorem 1.11.   An $n \times n$ matrix is non-singular if and only if it is row equivalent to I_n.

Theorems 1.11, 1.12, and 1.14 can be used for many different theoretical purposes. For example, according to theorem 1.14, I_n is row equivalent to

A whenever A is row equivalent to I_n. Using this fact in conjunction with theorems 1.11 and 1.12, we see that A is non-singular if and only if there exists a finite sequence of elementary matrices $E_1, E_2, \ldots, E_k$ such that $A = E_k \ldots E_1 I_n$, and we have proved the following useful result.

Corollary 1.4. The $n \times n$ matrix A is non-singular if and only if it can be expressed as the product of a finite number of elementary matrices.

EXAMPLE 2.   If

$$A = \begin{bmatrix} 3 & 8 \\ 1 & 3 \end{bmatrix}$$

we find that the reduction

$$A = \begin{bmatrix} 3 & 8 \\ 1 & 3 \end{bmatrix} \xrightarrow[-3R_1 + R_2 \to R_2]{R_1 \leftrightarrow R_2} \begin{bmatrix} 1 & 3 \\ 0 & -1 \end{bmatrix} \xrightarrow[-3R_2 + R_1 \to R_1]{-R_2 \to R_2} \begin{bmatrix} 1 & 0 \\ 0 & 1 \end{bmatrix} = I_2$$

can also be written as the matrix equation

$$\underset{I_2}{\begin{bmatrix} 1 & 0 \\ 0 & 1 \end{bmatrix}} = \underset{E_4}{\begin{bmatrix} 1 & -3 \\ 0 & 1 \end{bmatrix}} \underset{E_3}{\begin{bmatrix} 1 & 0 \\ 0 & -1 \end{bmatrix}} \underset{E_2}{\begin{bmatrix} 1 & 0 \\ -3 & 1 \end{bmatrix}} \underset{E_1}{\begin{bmatrix} 0 & 1 \\ 1 & 0 \end{bmatrix}} \underset{A}{\begin{bmatrix} 3 & 8 \\ 1 & 3 \end{bmatrix}}$$

Thus, we have $I_2 = E_4 E_3 E_2 E_1 A$, and it follows that

$$A = E_1^{-1} E_2^{-1} E_3^{-1} E_4^{-1} I_2 = \begin{bmatrix} 0 & 1 \\ 1 & 0 \end{bmatrix} \begin{bmatrix} 1 & 0 \\ 3 & 1 \end{bmatrix} \begin{bmatrix} 1 & 0 \\ 0 & -1 \end{bmatrix} \begin{bmatrix} 1 & 3 \\ 0 & 1 \end{bmatrix}$$

Thanks to corollary 1.4, we can restate theorem 1.12 in the following form.

Corollary 1.5. The $m \times n$ matrix A is row equivalent to B if and only if there exists a non-singular matrix P such that $B = PA$.

EXAMPLE 3.   The matrix equation

$$\underset{B}{\begin{bmatrix} 1 & 0 & -3 \\ 0 & 1 & 1 \end{bmatrix}} = \underset{E_3}{\begin{bmatrix} 1 & -3 \\ 0 & 1 \end{bmatrix}} \underset{E_2}{\begin{bmatrix} 1 & 0 \\ -2 & 1 \end{bmatrix}} \underset{E_1}{\begin{bmatrix} 0 & 1 \\ 1 & 0 \end{bmatrix}} \underset{A}{\begin{bmatrix} 2 & 7 & 1 \\ 1 & 3 & 0 \end{bmatrix}}$$

can also be written as $B = PA$, where

$$P = E_3E_2E_1 = \begin{bmatrix} -3 & 7 \\ 1 & -2 \end{bmatrix}$$

As a final application of the results of this section, we shall now use corollary 1.5 to show that a square matrix with a "right" inverse is necessarily non-singular. Note that this is a stronger result than corollary 1.3 (see section 1.6), since we do not assume in advance that A is non-singular.

Corollary 1.6. If A is an $n \times n$ matrix and $AB = I_n$, then A is non-singular and $BA = I_n$. That is, $B = A^{-1}$.

Proof: Let E be the reduced row echelon form of A and let P be a non-singular matrix such that $A = PE$. Since the equation $AB = I_n$ can also be written as $AB = (PE)B = P(EB) = I_n$, it follows that $EB = P^{-1}$, and EB must be non-singular. Therefore, E must be I_n because any other reduced row echelon matrix would have at least one row of zeroes and so would EB (see exercise 6, section 1.6), but this could occur only if EB were singular (why?). Since the reduced row echelon form of A is I_n, A must be nonsingular, and corollary 1.3 enables us to conclude that $BA = I_n$ and $B = A^{-1}$. ∎

For the most part, the results obtained in this section will not be developed further. However, the non-singularity criterion established in theorem 1.11 will appear throughout our work, and theorem 1.12 will be used in chapter 5.

Exercise Set 1.7

1. Represent each of the following reduction steps as a matrix equation involving suitable elementary matrices.

(i) $\begin{bmatrix} 1 & 2 & -7 \\ 3 & 1 & 4 \end{bmatrix} \xrightarrow[\substack{-1/5R_2 \to R_2 \\ -2R_2 + R_1 \to R_1}]{-3R_1 + R_2 \to R_2} \begin{bmatrix} 1 & 0 & 3 \\ 0 & 1 & -5 \end{bmatrix}$

(ii) $\begin{bmatrix} 5 & -10 & 30 \\ 8 & 4 & -2 \\ -3 & 1 & 0 \end{bmatrix} \xrightarrow[\substack{-8R_1 + R_2 \to R_2 \\ 3R_1 + R_3 \to R_3}]{1/5R_1 \to R_1} \begin{bmatrix} 1 & -2 & 6 \\ 0 & 20 & -50 \\ 0 & -5 & 18 \end{bmatrix}$

(iii) $\begin{bmatrix} 3 & -5 & 7 \\ 4 & -3 & 1 \\ 1 & 2 & -6 \\ 5 & -1 & -5 \end{bmatrix} \xrightarrow[\substack{-3R_1 + R_3 \to R_3 \\ -5R_1 + R_4 \to R_4}]{\substack{R_1 \leftrightarrow R_3 \\ -4R_1 + R_2 \to R_2}} \begin{bmatrix} 1 & 2 & 6 \\ 0 & -11 & 25 \\ 0 & -11 & 25 \\ 0 & -11 & 25 \end{bmatrix}$

2. Show that the following matrices are both non-singular and represent each as the product of elementary matrices.

(i) $\begin{bmatrix} 7 & 5 \\ 4 & 3 \end{bmatrix}$

(ii) $\begin{bmatrix} 1 & -1 & 1 \\ 0 & 5 & 3 \\ -2 & 0 & -3 \end{bmatrix}$

3. Find the inverse of the 3×3 non-singular matrix A, where

$$A = \begin{bmatrix} -3 & 0 & 0 \\ 0 & 1 & 0 \\ 0 & 0 & 1 \end{bmatrix} \begin{bmatrix} 1 & 0 & 0 \\ 0 & 0 & 1 \\ 0 & 1 & 0 \end{bmatrix} \begin{bmatrix} 1 & 0 & -4 \\ 0 & 1 & 0 \\ 0 & 0 & 1 \end{bmatrix} \begin{bmatrix} 1 & 0 & 0 \\ 0 & 5 & 0 \\ 0 & 0 & 1 \end{bmatrix}$$

4. Find the 3×4 matrix B which satisfies the equation $A = E_3 E_2 E_1 B$, where

$$A = \begin{bmatrix} -7 & 2 & -5 & 8 \\ 9 & 1 & 0 & 1 \\ 3 & 1 & 2 & -1 \end{bmatrix}, \quad E_1 = P_{23}, \quad E_2 = S_3(\tfrac{1}{2}), \quad E_3 = M_2(1; -5)$$

5. Let

$$A = \begin{bmatrix} 2 & 1 & -4 & 11 \\ 1 & -2 & -7 & 3 \\ -3 & 1 & 11 & -14 \end{bmatrix} \quad \text{and} \quad B = \begin{bmatrix} 3 & -2 & -13 & 13 \\ 1 & 4 & 5 & 9 \\ -7 & 9 & 39 & -26 \end{bmatrix}$$

(i) Show that A and B have the same reduced row echelon form E.

(ii) Find non-singular matrices P_1 and P_2 such that $E = P_1 A$ and $E = P_2 B$.

(iii) Find a non-singular matrix P such that $A = PB$.

6. If A is row equivalent to B and B is row equivalent to C, show that A is row equivalent to C.

Hint: Let P_1, P_2 be non-singular matrices such that $B = P_1 A$ and $C = P_2 B$, and find a non-singular matrix P such that $C = PA$.

7. If A and B are $n \times n$ matrices which satisfy $AB = I_n$, show that they must both be non-singular.

8. Let A and B be $m \times n$ and $n \times m$ matrices, respectively.

(i) Let E be the reduced row echelon form of A. If $AB = I_m$, show that E must have row rank m.

Hint: Use the fact that $A = PE$ for some non-singular matrix P to show that EB is non-singular. Then observe that E cannot have a row of zeroes (why not?).

(ii) If $m > n$, show that AB cannot be I_m.

1.8 (OPTIONAL) SELECTED APPLICATIONS: KIRCHHOFF'S LAWS, THE LEONTIEF MODEL, AND MARKOV CHAINS.

In this section, we shall examine three of the many areas in which linear systems and matrices play a prominent role. Naturally, in the limited space available here, we can do little more than introduce the main ideas of each topic. However, the reader who wants to know more about these or other applications will find it interesting to check through the references listed in the applications section of the bibliography.

I. Engineering: Analysis of Circuits. A direct current electrical circuit is an arrangement of conductors, resistors, and batteries. In dealing with such a circuit, it is assumed that $E = IR$ (Ohm's law) is valid, where E is the voltage drop across a conductor, I is the current in the conductor, and R is the resistance. It is also assumed that the voltage drop caused by a current passing through several resistors connected in series is the sum of the voltage drops in each individual resistor. There are two important rules, known as *Kirchhoff's laws*, which apply to a circuit:

1. At any point in the circuit where two or more conductors are joined, the algebraic sum of the currents is zero. By convention, currents directed toward the junction are regarded as positive.
2. Around any closed path in the circuit, the algebraic sum of the voltage drops is zero.

If the value of each resistance and the voltage of each battery in the circuit are known, Kirchhoff's laws may be used to set up a system of linear equations in which the "unknowns" are the currents in the various sections of the circuit. We shall illustrate this process by analyzing the circuit diagrammed in figure 1.3. According to Kirchhoff's first law, the algebraic sum of the currents entering and leaving junction A (and junction B) must be zero, and it follows that

$$-I_1 + I_2 - I_3 = 0 \quad \text{(sum of currents at } A\text{)}$$

$$I_1 - I_2 + I_3 = 0 \quad \text{(sum of currents at } B\text{)}$$

In addition, the second law tells us that the sum of the voltage drops in the

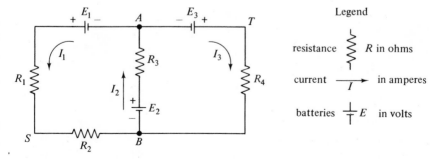

FIGURE 1.3

closed paths $ASBA$ and $ATBA$ must be zero. Thus, we have

$$E_1 + R_1I_1 + R_2I_1 + E_2 + R_3I_2 = 0$$

and

$$E_3 + R_4I_3 + E_2 + R_3I_2 = 0$$

where, for example, we have used Ohm's law to assert that the voltage drop associated with passing current I_1 through the first resistor is R_1I_1. Combining these four equations and rearranging terms, we obtain the linear system

$$-I_1 + I_2 - I_3 = 0$$

$$I_1 - I_2 + I_3 = 0$$

$$(R_1 + R_2)I_1 + R_3I_2 = -E_1 - E_2$$

$$R_3I_2 + R_4I_3 = -E_2 - E_3$$

For instance, if $R_1 = 3$, $R_2 = 1$, $R_3 = 6$, $R_4 = .6$, and $E_1 = 3$, $E_2 = 17$, $E_3 = 25$, the system is

$$I_1 - I_2 + I_3 = 0$$

$$-I_1 + I_2 - I_3 = 0$$

$$4I_1 + 6I_2 = -20$$

$$6I_2 + .6I_3 = -42$$

and the reduction

$$\begin{bmatrix} 1 & -1 & 1 & 0 \\ -1 & 1 & -1 & 0 \\ 4 & 6 & 0 & -20 \\ 0 & 6 & .6 & -42 \end{bmatrix} \longrightarrow \begin{bmatrix} 1 & 0 & 0 & 4 \\ 0 & 1 & 0 & -6 \\ 0 & 0 & 1 & -10 \\ 0 & 0 & 0 & 0 \end{bmatrix}$$

enables us to conclude that the currents in the circuit are $I_1 = 4$, $I_2 = -6$, $I_3 = -10$, where the negative signs show that I_2 and I_3 actually flow in the direction opposite to that indicated in the diagram.

The circuit we have just discussed is quite simple, but even the most complex circuits can be analyzed by similar methods.

II. Economics: The Leontief Model of an Interacting Economy. In recent years, techniques based on linear algebra, such as linear programming and game theory, have begun to play a prominent role in the social sciences.

The example we shall consider here is a mathematical model developed in the 1950's by Wassily Leontief, whose work in this area has since earned him a Nobel Prize. In its simplest form, the Leontief model deals with an economy involving n industries $I_1, \ldots, I_n$ that interact in the sense that part of the output of each industry is purchased by each of the others. For instance, the steel industry interacts in this fashion with the machine tool and automobile industries, among others.

To analyze this kind of situation, we need the following notation:

x_i is the value of the output of industry I_i.

z_{ij} is the value of the portion of the output of industry I_i, which is purchased by industry I_j.

y_i is the consumer demand for the output of I_i.

It is reasonable to expect z_{ij} to be related in some way to x_j, and we shall assume that this relationship is linear—namely, $z_{ij} = p_{ij}x_j$, where p_{ij} is a suitable constant. After all the other industries have made their purchases, the value of the output of industry I_i which remains is

$$x_i - \sum_{j=1}^{n} z_{ij} = x_i - \sum_{j=1}^{n} p_{ij}x_j$$

Under ideal market conditions, this value should equal the consumer demand y_i, and when this occurs for each index i, we have the matrix equation

$$(I_n - P)X = Y$$

where

$$P = \begin{bmatrix} p_{11} \cdots p_{1n} \\ \vdots \qquad \vdots \\ p_{n1} \cdots p_{nn} \end{bmatrix}, \qquad X = \begin{bmatrix} x_1 \\ \vdots \\ x_n \end{bmatrix}, \qquad \text{and} \qquad Y = \begin{bmatrix} y_1 \\ \vdots \\ y_n \end{bmatrix}$$

Note that the entries of P, X, and Y must be non-negative for the model to be realistic.

When the output distribution matrix P and the consumer demand vector Y are known, then a vector X is said to be an *equilibrium solution* of the model if it satisfies $(I_n - P)X = Y$ and its entries are all non-negative numbers. In using a model such as this, a researcher is likely to make certain assumptions (based on experience or other research) about P and Y in order to gain information about the corresponding equilibrium solutions. For instance, he might be interested in knowing what conditions to impose on P in order to guarantee the existence of an equilibrium solution no matter how the vector Y is chosen. This certainly occurs when the matrix $I_n - P$ is non-singular and $(I_n - P)^{-1}$ contains only non-negative entries, for in this case, if Y is any vector with non-negative entries, then $X = (I_n - P)^{-1} Y$ is an equilibrium solution. Unfortunately, in general, even when $I_n - P$ is non-singular, the matrix $(I_n - P)^{-1}$ will have negative as well as non-negative entries, and the analysis becomes more difficult. Additional questions which arise naturally in connection with the Leontief model and its generalizations are analyzed in many texts on linear economics. For instance, the reader will find an especially thorough treatment in *Linear Economic Theory*, by Daniel Vandermeulen, Prentice-Hall Inc., Englewood Cliffs, New Jersey, 1971.

III. Probability Theory: Markov Chains.

Definition 1.7. Consider a physical or mathematical system which is observed at regular time intervals. Such a system is said to satisfy the *Markov condition* if its state at the time of the kth observation depends only on its state during the $(k - 1)$st observation. The associated sequence of observations is called a *Markov chain*.

Many situations which arise in the physical and social sciences can be analyzed by referring to a probabilistic model which satisfies the Markov condition. One example is the so-called Ehrenfest diffusion model, which is used to analyze the tendency of gas molecules to diffuse across a permeable membrane, and another is the Hardy model of genotype distributions in a mixed population. These and other important models involving Markov chains are discussed in detail in several texts listed in the bibliography. For example, see *Elements of the Theory of Markov Processes and Their Applications*, by A. T. Bharucha-Reid, McGraw-Hill, New York, 1960.

A rigorous and complete treatment of the theory of Markov chains lies beyond the scope of this text. Instead, we shall focus attention on certain features of Markov chains that involve the tools of matrix analysis. These features are illustrated in the solution of the following problem.

PROBLEM. In Styxburg, there are two political parties, Democratic (D) and Republican (R), and it has been observed that at each election, the following voting pattern occurs.

1. Of those who voted D in the last election, 75% will vote D again and 25% will vote R.
2. Of those who voted R in the last election, 80% will do so again, while 20% will vote D.

In 1972, 60% of the electorate voted Democratic and 40% voted Republican. If this pattern continues, who will win the next election (in 1974)? What percentage of the total vote will each party receive in the 1976 election?

Solution: In the situation described above, the states are S_1, the Democratic vote, and S_2, the Republican vote. Let p_{ij} denote the likelihood (probability) that someone who voted for the S_i party in a given election will vote for the S_j party in the next election. For instance, p_{12} is the proportion of those who voted Democratic last time and will vote Republican this time. The values of the p_{ij} are displayed in the array below:

$$\begin{array}{c} \\ S_1 \\ S_2 \end{array} \begin{array}{cc} S_1 & S_2 \end{array} \\ \begin{bmatrix} \frac{3}{4} & \frac{1}{4} \\ \frac{1}{5} & \frac{4}{5} \end{bmatrix}$$

To answer the first question, note that since the Democrats received $\frac{3}{5}$ of the vote in 1972 and the Republicans received $\frac{2}{5}$, then in 1974, the Democrats will receive $(\frac{3}{5})(\frac{3}{4}) = \frac{9}{20}$ of the vote of former Democrats and $(\frac{2}{5})(\frac{1}{5}) = \frac{2}{25}$ from former Republicans. Therefore, the total Democratic proportion will be

$$(\tfrac{3}{5})(\tfrac{3}{4}) + (\tfrac{2}{5})(\tfrac{1}{5}) = \tfrac{53}{100}$$

We find that the Republicans will receive

$$(\tfrac{3}{5})(\tfrac{1}{4}) + (\tfrac{2}{5})(\tfrac{4}{5}) = \tfrac{47}{100}$$

of the total vote, and it follows that the Democrats will win the 1974 election. Note that if we set

$$P_1 = \begin{bmatrix} \frac{3}{4} & \frac{1}{4} \\ \frac{1}{5} & \frac{4}{5} \end{bmatrix} \quad \text{and} \quad X_0 = [\tfrac{3}{5}, \tfrac{2}{5}]$$

then the desired proportions appear as the components of the product matrix

$$X_1 = X_0 P_1 = [\tfrac{3}{5}, \tfrac{2}{5}] \begin{bmatrix} \frac{3}{4} & \frac{1}{4} \\ \frac{1}{5} & \frac{4}{5} \end{bmatrix} = [\tfrac{53}{100}, \tfrac{47}{100}]$$

Before we can predict the outcome of the 1976 election, we need to determine the probability that a person who voted for the S_i party two elections ago will vote for the S_j party this time. Denote this probability by $p_{ij}^{(2)}$. To find $p_{11}^{(2)}$, we note that there are exactly two ways for someone who voted Democratic two elections ago to vote Democratic again in the present election: namely, he could vote in either the pattern DDD or the pattern DRD. This and the various other voting patterns are depicted in the transition diagram labeled figure 1.4. As indicated in this diagram, the proportions of those who vote in the patterns DDD and DRD are $(\frac{3}{4})(\frac{3}{4}) = \frac{9}{16}$ and $(\frac{1}{4})(\frac{1}{5}) = \frac{1}{20}$, respectively, and we find that

$$p_{11}^{(2)} = (\tfrac{3}{4})(\tfrac{3}{4}) + (\tfrac{1}{4})(\tfrac{1}{5}) = \tfrac{49}{80}$$

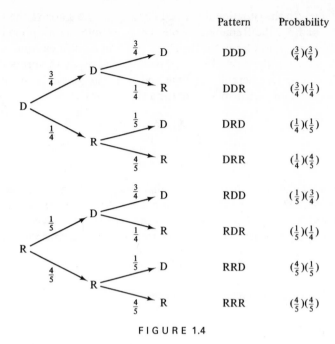

FIGURE 1.4

Similarly, the other three desired proportionalities may be found as follows:

$$p_{12}^{(2)} = (\tfrac{3}{4})(\tfrac{1}{4}) + (\tfrac{1}{4})(\tfrac{4}{5}) = \tfrac{31}{80} \quad (DDR \text{ and } DRR)$$

$$p_{21}^{(2)} = (\tfrac{1}{5})(\tfrac{3}{4}) + (\tfrac{4}{5})(\tfrac{1}{5}) = \tfrac{31}{100} \quad (RDD \text{ and } RRD)$$

$$p_{22}^{(2)} = (\tfrac{1}{5})(\tfrac{1}{4}) + (\tfrac{4}{5})(\tfrac{4}{5}) = \tfrac{69}{100} \quad (RDR \text{ and } RRR)$$

Finally, we conclude that in 1976, the Democrats will receive

$$(\tfrac{3}{5})p_{11}^{(2)} + (\tfrac{2}{5})p_{21}^{(2)} = (\tfrac{3}{5})(\tfrac{49}{80}) + (\tfrac{2}{5})(\tfrac{31}{100}) = \tfrac{983}{2000}$$

of the vote, and the Republicans will receive

$$(\tfrac{3}{5})p_{21}^{(2)} + (\tfrac{2}{5})p_{22}^{(2)} = (\tfrac{3}{5})(\tfrac{31}{80}) + (\tfrac{2}{5})(\tfrac{69}{100}) = \tfrac{1017}{2000}$$

Thus, the Republicans will win the 1976 election with approximately 51% of the total vote.

Note that the matrix

$$P_2 = (p_{ij}^{(2)}) = \begin{bmatrix} \tfrac{49}{80} & \tfrac{31}{80} \\ \tfrac{31}{100} & \tfrac{69}{100} \end{bmatrix}$$

which represents the voting proportions after two elections, can be expressed as $P_2 = P_1^2$, and the result of the 1976 election may be found by forming the product vector

$$X_2 = X_0 P_2 = [\tfrac{3}{5}, \tfrac{2}{5}] \begin{bmatrix} \frac{49}{80} & \frac{31}{80} \\ \frac{31}{100} & \frac{69}{100} \end{bmatrix} = [\tfrac{983}{2000}, \tfrac{1017}{2000}]$$

Returning to the general case, suppose we have a system with n states $S_1, S_2, \ldots, S_n$, which satisfies the Markov condition. Then we make the following definition.

Definition 1.8. The (one-step) *transition probability matrix* associated with a given Markov chain is the $n \times n$ array $P_1 = (p_{ij})$, where p_{ij} is the probability of going from the state S_i to the state S_j in a single interval of time.

Thanks to the probabilistic origins of the p_{ij}, the matrix $P_1 = (p_{ij})$ must have the following properties:

1. $0 \le p_{ij} \le 1$ for all i, j

   $\qquad\qquad\qquad\qquad\qquad\qquad\qquad\qquad\qquad$ (1.22)

2. $\sum\limits_{j=1}^{n} p_{ij} = 1$ for $i = 1, 2, \ldots, n$

Any array whose entries satisfy the two properties listed in (1.22) is called a *Markov matrix*. The terms *probability matrix* and *stochastic matrix* are also applied to such an array. Similarly, the n-tuple $X = (x_j)$ is called a *probability vector* if its components satisfy $0 \le x_j \le 1$ and $\sum_{j=1}^{n} x_j = 1$.

Let $p_{ij}^{(k)}$ denote the probability of going from the state S_i to the state S_j in exactly k intervals of time. Then it can be shown that P_k, the $n \times n$ matrix whose (i, j) entry is $p_{ij}^{(k)}$, is a Markov matrix and that $P_k = P_1^k$. Finally, if $X_0 = (x_j)$ is the probability vector whose components x_j represent the initial condition of the system, then the vector $X_k = X_0 P^k$ represents the condition of the system at the end of k intervals of time.

In analyzing a system with these characteristics, one is usually concerned with answering the following questions, among others:

1. Does the Markov chain terminate after a finite number of steps? In other words, is there a k such that $X_n = X_k$ for all $n \ge k$?
2. Is there a "steady state" behavior for the Markov chain? That is, do the entries of X_n approach a limiting value as n tends to infinity?
3. Do the answers to questions 1 and 2 depend in any way on the choice of the initial condition vector X_0?

The answers to these questions depend a great deal on the nature of the system in question. However, in the case where P_1 or some power P_1^k has only positive entries, it can be shown that a probability vector $Y = (y_j)$ exists for which

1. $\lim\limits_{n \to \infty} X_n = Y$
2. Y is independent of the choice of X_0.
3. $YP_1 = Y$

The reader who is interested in knowing more about Markov chains will find an excellent survey by Janet Myhre in Chapter 2 of *Markov Chains and Monte Carlo Calculus in Polymer Science*, Dekker Publication Corp., New York.

Exercise Set 1.8

1. Consider the electrical circuit diagrammed below:

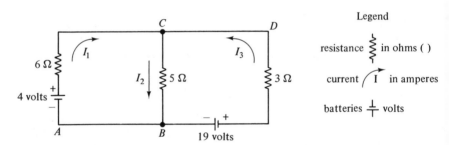

(i) Use Kirchhoff's laws to show that

$$I_1 - I_2 + I_3 = 0$$

$$6I_1 + 5I_2 \qquad = -4$$

$$5I_2 + 3I_3 = -19$$

and then find I_1, I_2, and I_3.

(ii) If we change the voltage supplied by the two batteries but leave the resistors alone, is it possible to create a circuit in which the currents are $I_1 = .5$, $I_2 = 1$, $I_3 = .5$?

2. Guy Goodfellow is dating three girls, Annabelle (*A*), Bertha (*B*), and Christine (*C*). He has a date with one of them each night, and after each date, the likelihood of his choosing a particular girl for his next date is given in the table below:

probability of next
date with . . .

		A	B	C
if last date was with . . .	A	$\frac{1}{3}$	0	$\frac{2}{3}$
	B	0	$\frac{1}{2}$	$\frac{1}{2}$
	C	$\frac{3}{4}$	$\frac{1}{4}$	0

For example, if his last date was with Christine, he will not ask her for the next date, but 75% of the time he will ask Annabelle and 25% he will ask Bertha.

(i) Suppose he begins dating on Sunday, and on that particular day, it is 20% likely that he will ask Annabelle for a date, 40% that he will ask Bertha and 40% that he will ask Christine. Who is most likely to be his date on Monday? On Tuesday?

(ii) What is the transition matrix P_1 for Guy's dating chain? Find a probability vector $Y = (y_j)$ which satisfies $YP_1 = Y$.

Roughly speaking, each component of Y represents the "ultimate" likelihood that Guy will date the corresponding girl. With whom is he most likely to "go steady" in the long run? Explain your reasoning.

3. Suppose that in analyzing a certain segment of the economy that involves three interacting industries, an economist assumes the existence of the following output distribution matrix P and consumer demand vector Y:

$$P = \begin{bmatrix} .2 & .2 & .3 \\ .1 & .5 & .7 \\ .4 & .3 & .1 \end{bmatrix} \quad \text{and} \quad Y = \begin{bmatrix} 1.0 \\ 2.1 \\ 1.3 \end{bmatrix}$$

(i) Compute the matrix $(I_3 - P)^{-1}$.

(ii) Show that an equilibrium solution exists for this particular model and find it.

SUPPLEMENTARY EXERCISES

1. An $n \times n$ *magic square* matrix is an array of non-negative integers arranged so that each row and column adds up to the same number S. For example, the following are magic square matrices:

$$\begin{bmatrix} 1 & 0 & 5 \\ 5 & 1 & 0 \\ 0 & 5 & 1 \end{bmatrix} \quad \text{and} \quad \begin{bmatrix} 16 & 2 & 3 & 13 \\ 5 & 11 & 10 & 8 \\ 9 & 7 & 6 & 12 \\ 4 & 14 & 15 & 1 \end{bmatrix}$$

$$S = 6 \qquad\qquad\qquad S = 34$$

If A and B are $n \times n$ magic squares, with sums S_1 and S_2, respectively, show that
  (i) $A + B$ is a magic square matrix with sum $S_1 + S_2$.
  (ii) AB is a magic square matrix with sum $S_1 S_2$.

2. A *rational lattice point* in the plane is a point (x, y) whose coordinates x and y are both rational numbers; for example, $(\frac{1}{2}, -\frac{1}{3})$ and $(-5, 0)$.
  (i) If the graph of the quadratic polynomial $p(x) = ax^2 + bx + c$ passes through the rational lattice points $(\frac{1}{2}, 1)$, $(3, 0)$, and $(-5, -32)$, what are a, b, and c?
  (ii) In general, show that the quadratic polynomial $q(x) = a_2 x^2 + a_1 x + a_0$ has rational coefficients if and only if its graph passes through at least three rational lattice points.

3. A man, driving down a straight, level road at the constant speed v_0 is forced to apply his brakes to avoid hitting a cow, and the car comes to a stop 3 sec later and s_1 ft from the point where the brakes were applied. Continuing on his way, he increases his speed by 20 ft/sec to make up time, but is again forced to hit the brakes, and this time it takes him 5 sec and s_2 ft to come to a full stop.

   Assuming that his brakes supplied a constant deceleration d each time they were used, find d and determine v_0, s_1, and s_2.

4. Let A be an $n \times n$ matrix, and let λ be a fixed scalar. Also, let X be a non-zero vector which satisfies $AX = \lambda X$.
  (i) Show that $A^k X = \lambda^k X$ for each positive integer k.
     Hint: First show that $A^2 X = \lambda^2 X$, and then use induction.
  (ii) If A is non-singular, show that $\lambda \neq 0$ and that $A^{-1} X = \lambda^{-1} X$.
     Hint: First, note that $X = \lambda(A^{-1} X)$. Can either λ be 0 or $A^{-1} X$ be the zero vector?

5. Let P be a non-singular $n \times n$ matrix, and let A and B be $n \times n$ matrices which are related by the equation $B = P^{-1}AP$ (such matrices are said to be *similar*; see section 4.6).
  (i) Show that $B^2 = P^{-1}A^2 P$. Find an analogous relationship involving B^k and A^k.
  (ii) If A is non-singular, show that B is also non-singular and that $B^{-1} = P^{-1}A^{-1}P$.

6. Let A be an $n \times n$ matrix which satisfies the equation $A^3 - 7A^2 + A + 2I_n = 0$.
  (i) Show that $I_n = -\frac{1}{2}(A^2 - 7A + I_n)A$.
  (ii) Use part (i) to conclude that A is non-singular, and find scalars a, b, c such that $A^{-1} = aA^2 + bA + cI_n$.

7. Let $X = (x_{ij})$ be an $m \times n$ matrix whose entries x_{ij} are all differentiable functions of the real variable t. Then we shall write dX/dt to denote the $m \times n$ matrix (x'_{ij}) whose (i, j) entry is the derivative of x_{ij}. For example,

$$\text{if} \quad X = \begin{bmatrix} e^t & \sin t & t^2 \\ 3t & e^{-t} & 4 \end{bmatrix} \quad \text{then} \quad \frac{dX}{dt} = \begin{bmatrix} e^t & \cos t & 2t \\ 3 & -e^{-t} & 0 \end{bmatrix}$$

Prove each of the following statements.
  (i) If A is a constant matrix, then $dA/dt = 0$.
  (ii) If s is a scalar, then $d(sX)/dt = s(dX/dt)$.

(iii) If f is a differentiable function of t and A is a constant matrix, then $d[f(t)A]/dt = f'(t)A$.

(iv) If X and Y are both $m \times n$ matrices, then, $d(X + Y)/dt = dX/dt + dY/dt$.

(v) If A is an $m \times n$ constant matrix and X is an $n \times p$ matrix, then $d(AX)/dt = A(dX/dt)$.

(vi) If X and Y are $m \times n$ and $n \times p$ matrices, respectively, then $d(XY)/dt = X(dY/dt) + (dX/dt)Y$.

8. Let

$$X = \begin{bmatrix} t & e^t \\ 1 & t \end{bmatrix}$$

(i) Find X^2 and X^{-1}.

(ii) Show that $d(X^2)/dt$ is not the same as $2X(dX/dt)$.

(iii) Is $d(X^{-1})/dt$ the same as $-X^{-2}(dX/dt)$?

9. Note that the system of linear differential equations

$$2\frac{dx_1}{dt} + 5\frac{dx_2}{dt} = t$$

$$\frac{dx_1}{dt} + 3\frac{dx_2}{dt} = 7\cos t$$

can be expressed as $A(dX/dt) = \beta$, where

$$A = \begin{bmatrix} 2 & 5 \\ 1 & 3 \end{bmatrix}, \qquad \frac{dX}{dt} = \begin{bmatrix} \dfrac{dx_1}{dt} \\ \dfrac{dx_2}{dt} \end{bmatrix}, \qquad \beta = \begin{bmatrix} t \\ 7\cos t \end{bmatrix}$$

(i) Find A^{-1} and write out the system symbolized by the matrix equation $dX/dt = A^{-1}\beta$.

(ii) Solve the given system subject to the initial conditions $x_1(0) = 0 = x_2(0)$.

10. Let A, B be $n \times n$ matrices, and let $C = AB$. If C is non-singular, show that both A and B must also be non-singular.

2

Vector Spaces

2.1 GEOMETRIC VECTORS

A physical or mathematical quantity such as work, area, or energy which can be adequately described in terms of magnitude alone is called a *scalar*. A quantity which requires both a magnitude and a direction for its description is said to be a *vector;* for example, velocity and force. In this section, we shall discuss the algebra of vector quantities and in so doing, shall provide a preview of important ideas which will occupy our attention later in this chapter.

If O and P are points in space, we shall write $\overrightarrow{OP}$ to indicate the directed line segment whose initial point is O and whose terminal point is P. On the printed page, the segment will appear as an "arrow" from O to P and its length will be denoted by $|\overrightarrow{OP}|$. When the end points are not an important consideration, we shall indicate the segment by a single letter of the Greek alphabet—say, α—and shall denote its length by $|\alpha|$. These notational conventions are illustrated in figure 2.1. Such a directed line segment is called a *geometric vector*, and it is customary to represent many different kinds of vector quantities in this form. For instance, a force F of 3 lbs which acts in

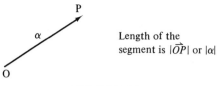

FIGURE 2.1

a direction inclined 45° to the horizontal may be represented by a geometric vector 3 units long which points in the appropriate direction:

In dealing with geometric vectors, we shall not distinguish between two vectors which point in the same direction and have the same magnitude. Accordingly, we begin our development of the algebra of geometric vectors with the following definition.

Definition 2.1.   Two geometric vectors are said to be *equivalent* if they point in the same direction and have the same magnitude. If α is equivalent to β, we shall write $\alpha = \beta$.

For instance, in figure 2.2, we have $\alpha = \beta$, but neither γ nor ρ is equivalent to α, for the reasons indicated.

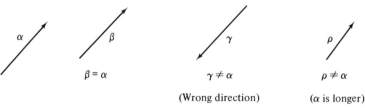

FIGURE 2.2

In the work which follows, we shall deal with equivalent vectors as if they were identically equal. Specifically, when examining relationships within a collection of vectors, we shall assume that it is always possible to move a given vector to a new position which is parallel to its initial position. This policy involves no real loss of generality and allows us to quickly focus attention on the essential features of vector algebra.

Next, we shall establish rules for adding geometric vectors and multiplying them by scalars. Experimental evidence indicates that the effect of applying two forces F_1 and F_2 at a point O is the same as that achieved by applying a single "resultant" force (denoted by $F_1 + F_2$) whose magnitude and direction may be determined by referring to the diagram below:

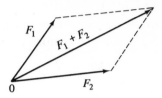

We shall use the following form of this so-called "parallelogram law" as the basis for our definition of vector addition.

Definition 2.2.   To add the geometric vector β to the vector α, place the end of β on the tip of α. Then the vector sum $\alpha + \beta$ is the geometric vector extending from the end of α to the tip of β.

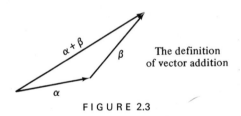

The definition of vector addition

FIGURE 2.3

In physics, it is common practice to denote by $2F$ the force which points in the same direction as a given force F but has twice the magnitude. Similarly, the velocity whose magnitude is only $\frac{1}{3}$ that of a given velocity V and which points in the opposite direction is denoted by $-\frac{1}{3}V$. The general rule for multiplying a given geometric vector by a scalar is provided by the following definition.

Definition 2.3.   Let α be a geometric vector and s be a real number. Then the *scalar multiple* of α by s is a vector (denoted by $s\alpha$) whose magnitude is $|s|$ times that of α and which points in the direction of α if $s > 0$, and in the direction opposite to that of α if $s < 0$.

This rule is illustrated in figure 2.4.

It is convenient to have names for certain special vectors and combinations of vectors which appear frequently in vector algebra. The vector which

FIGURE 2.4

has magnitude zero will be called the *zero vector* and will always be denoted by θ. If α is a non-zero vector, we shall denote by $-\alpha$ the vector whose magnitude is the same as that of α but which points in the opposite direction. Note that $-\alpha = (-1)\alpha$. Finally, we shall denote the vector sum $\alpha + (-\beta)$ by $\alpha - \beta$. Note that $\alpha - \beta$ may be obtained by the geometric construction indicated below:

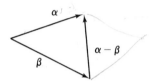

The following theorem provides a list of the most important properties of vector algebra.

Theorem 2.1.   If α, β, γ are vectors and s, t are scalars, then

  (i) $\alpha + \beta = \beta + \alpha$ (commutativity of addition)
 (ii) $\alpha + (\beta + \gamma) = (\alpha + \beta) + \gamma$ (associativity of addition)
(iii) $\alpha + \theta = \alpha$
 (iv) $\alpha - \alpha = \theta$
  (v) $(s + t)\alpha = s\alpha + t\alpha$
 (vi) $s(\alpha + \beta) = s\alpha + s\beta$
(vii) $(st)\alpha = s(t\alpha)$

Proof: Each of these properties may be derived by constructing a suitable geometric argument. For instance, the proof for the associativity of addition is suggested by the following diagram:

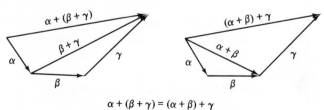

$$\alpha + (\beta + \gamma) = (\alpha + \beta) + \gamma$$

Associativity of vector addition

In the following solved problem, we use vector methods to prove a familiar theorem of plane geometry. To test his understanding of vector algebra, the reader should justify each step in the solution of this problem by referring to the appropriate part of theorem 2.1.

PROBLEM 1.   Show that the line segment joining the midpoints of two sides of a triangle is parallel to the third side and has half its length.

Solution: In the figure below, P and Q are the midpoints of sides AB and BC, respectively, of triangle ABC:

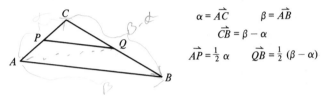

$$\alpha = \vec{AC} \qquad \beta = \vec{AB}$$
$$\vec{CB} = \beta - \alpha$$
$$\vec{AP} = \tfrac{1}{2}\,\alpha \qquad \vec{QB} = \tfrac{1}{2}\,(\beta - \alpha)$$

If vectors are assigned as indicated in the figure, we have

$$\beta = \vec{AB} = \vec{AP} + \vec{PQ} + \vec{QB} = \tfrac{1}{2}\alpha + \vec{PQ} + \tfrac{1}{2}(\beta - \alpha) = \vec{PQ} + \tfrac{1}{2}\beta$$

Solving, we find that $\vec{PQ} = \tfrac{1}{2}\,\beta$, and the segment PQ has the desired properties. (Why?)

According to definition 2.3, two vectors are parallel if and only if one is a scalar multiple of the other. This rather simple observation can be converted into the following important principle.

Theorem 2.2. (The principle of planar linear independence.)   If α and β are non-zero vectors which are not parallel, the vector equation $x\alpha + y\beta = \theta$ is satisfied only when x and y are both 0.

Proof: If $x = y = 0$, then $x\alpha + y\beta$ is obviously the zero vector. Conversely, suppose that $x\alpha + y\beta = \theta$. If x is not 0, then we must have $\alpha = -(y/x)\beta$, which means that α is a scalar multiple of β. However, this would contradict the assumption that α and β are not parallel, and we conclude that x must be 0. Finally, if $x = 0$, then $y = 0$ also (why?), and the proof is complete. ∎

The principle of planar linear independence is a useful tool for determining geometric proportionalities. Consider the following illustration.

PROBLEM 2.   Use vector methods to show that the medians of a triangle meet at a single point located two-thirds the distance from each vertex to the midpoint of the opposite side.

Solution: Let A, B, C be the vertices of an arbitrary triangle, and let Q and R be the midpoints of sides AB and BC, respectively:

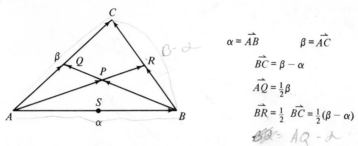

$$\alpha = \overrightarrow{AB} \qquad \beta = \overrightarrow{AC}$$
$$\overrightarrow{BC} = \beta - \alpha$$
$$\overrightarrow{AQ} = \tfrac{1}{2}\beta$$
$$\overrightarrow{BR} = \tfrac{1}{2}\ \overrightarrow{BC} = \tfrac{1}{2}(\beta - \alpha)$$

The medians AR and BQ intersect at P, and since this point lies on both medians, there exist scalars k and l such that $\overrightarrow{BP} = k\overrightarrow{BQ}$ and $\overrightarrow{AP} = l\overrightarrow{AR}$ (why?). Expressing $\overrightarrow{BP}$ and $\overrightarrow{AP}$ in terms of α and β, we find that

$$\overrightarrow{BP} = k\overrightarrow{BQ} = k(\tfrac{1}{2}\beta - \alpha)$$

and

$$\overrightarrow{AP} = l\overrightarrow{AR} = l[\alpha + \tfrac{1}{2}(\beta - \alpha)] = l(\tfrac{1}{2}\alpha + \tfrac{1}{2}\beta)$$

Substituting in the equation $\overrightarrow{AB} + \overrightarrow{BP} = \overrightarrow{AP}$, we have

$$\alpha + k(\tfrac{1}{2}\beta - \alpha) = l(\tfrac{1}{2}\alpha + \tfrac{1}{2}\beta)$$

and it follows that

$$(1 - k - \tfrac{1}{2}l)\alpha + (\tfrac{1}{2}k - \tfrac{1}{2}l)\beta = \theta$$

Finally, since α and β are not parallel, theorem 2.2 enables us to conclude that k and l must satisfy the equations

$$1 - k - \tfrac{1}{2}l = 0$$
$$\tfrac{1}{2}k - \tfrac{1}{2}l = 0$$

and we find that $k = l = \tfrac{2}{3}$.

Since this argument may be repeated for any other pair of medians (say, with AR and CS, where S is the midpoint of AB), it follows that each pair of medians intersect at P (why?), and the solution is complete.

The principle of planar linear independence is the first of many important new concepts we shall encounter in this chapter. In the next section, we shall continue our study of the properties of vectors by showing how vector methods may be used in analytic geometry.

Exercise Set 2.1

1. In the figure below, points P, Q, R, S are the midpoints of the sides of the quadrilateral $ABCD$.

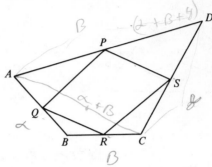

(i) Use vector methods to show that $\overrightarrow{QR} = \frac{1}{2}\overrightarrow{AC} = \overrightarrow{PS}$.

(ii) Show that $PQRS$ is a parallelogram.

2. In the figure below, P is the point where the diagonals of parallelogram $ABCD$ intersect one another:

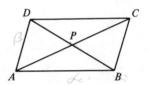

Let $\alpha = \overrightarrow{AB}$ and $\beta = \overrightarrow{AD}$, and let s and t be scalars such that $\overrightarrow{AP} = s\overrightarrow{AC}$ and $\overrightarrow{BP} = t\overrightarrow{BD}$.

(i) Use vector algebra to show that

$$s(\alpha + \beta) = \overrightarrow{AP} = \alpha + t(\beta - \alpha)$$

(ii) Use theorem 2.2 to show that $s = t = \frac{1}{2}$, and give a geometric interpretation of your result.

3. In the figure below, E is a point on side AD of parallelogram $ABCD$, and P is the point where EC intersects diagonal DB.

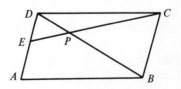

(i) If E is the midpoint of AD, show that $\overrightarrow{BP} = \frac{2}{3}\overrightarrow{BD}$.

(ii) More generally, if $\overrightarrow{AE} = r\overrightarrow{AD}$, where $0 \le r \le 1$, show that $\overrightarrow{BD} = (2 - r)\overrightarrow{BP}$.

4. Complete the proof of theorem 2.1.

5. If α and β are vectors, show that the equation $\alpha + x = \beta$ has the unique solution $x = \beta - \alpha$.

6. Let α_1, α_2, α_3 be non-zero vectors which do not lie in the same plane. If x, y, z are scalars such that $x\alpha_1 + y\alpha_2 + z\alpha_3 = \theta$, show that $x = y = z = 0$.

7. Show that any two vectors α and β which satisfy the equation

$$(\alpha + 3\beta) + 3(2\alpha - \beta) = 5\alpha + 6\beta$$

must be parallel. Which vector is "longer"?

8. Let β_1, β_2, and β_3 be non-zero vectors which satisfy $\beta_1 + \beta_2 + \beta_3 = \theta$ and suppose that $\alpha_1 = 3\beta_1 - 2\beta_2$ and $\alpha_2 = 5\beta_1 - \beta_3$.

   (i) Express $\alpha_3 = 2\alpha_1 + 3\alpha_2$ in terms of β_2 and β_3.

   (ii) If $\alpha_3 = \theta$, show that β_1, β_2, and β_3 must be parallel to one another.

2.2   VECTOR METHODS IN COORDINATE GEOMETRY

In plane analytic geometry results are obtained by first imposing a coordinate system on the plane and then using algebraic methods to analyze geometric relationships. In the classical approach to this subject, the plane is coordinatized by associating each point P with an ordered pair (a, b), where a and b are real numbers which fix the position of P with respect to two perpendicular reference lines, called *axes*. For example, the point associated with $(2, -3)$ is located 2 units to the right of the vertical axis and 3 units below the horizontal axis. Coordinates can also be introduced by vector methods, and the object of this section is to explore this approach. We begin by calling attention to an important property of planar vectors.

Theorem 2.3.   If α_1 and α_2 are non-zero vectors which are not parallel, then each vector α which lies in the same plane can be expressed in the form $\alpha = a_1\alpha_1 + a_2\alpha_2$, where a_1 and a_2 are uniquely determined real numbers.

Proof: It is easy to see that such numbers exist (see figure 2.5), and to show that they are uniquely determined, suppose that $\alpha = b_1\alpha_1 + b_2\alpha_2$. Then the equation $b_1\alpha_1 + b_2\alpha_2 = a_1\alpha_1 + a_2\alpha_2$ can be rewritten as $(b_1 - a_1)\alpha_1 + (b_2 - a_2)\alpha_2$, and the principle of planar linear independence

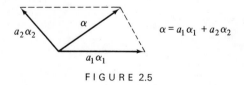

FIGURE 2.5

enables us to conclude that $b_1 - a_1 = 0$ and $b_2 - a_2 = 0$. In other words, $b_1 = a_1$ and $b_2 = a_2$, as claimed. ∎

Henceforth, we shall say that the ordered set $S = \{\alpha_1, \alpha_2\}$ is a *basis* for the plane if α_1 and α_2 are non-zero vectors which are not parallel. If $\alpha = a_1\alpha_1 + a_2\alpha_2$, we shall refer to the ordered pair (a_1, a_2) as the *S-coordinates* of α. Note that if (a_1, a_2) and (b_1, b_2) are the S-coordinates of vectors α and β, respectively, then

1. $\alpha = \beta$ if and only if $a_1 = b_1$ and $a_2 = b_2$.
2. If t is a real number, the vector $t\alpha$ has S-coordinates (ta_1, ta_2).   (2.1)
3. The vector sum $\alpha + \beta$ has S-coordinates $(a_1 + b_1, a_2 + b_2)$.
4. The vector $\alpha - \beta$ has S-coordinates $(a_1 - b_1, a_2 - b_2)$.

Part (1) of this list of observations follows immediately from the uniqueness statement of theorem 2.3, while the other three parts may be derived by using the first part in conjunction with the equations

$$t\alpha = t(a_1\alpha_1 + a_2\alpha_2) = (ta_1)\alpha_1 + (ta_2)\alpha_2$$

$$\alpha + \beta = (a_1\alpha_1 + a_2\alpha_2) + (b_1\alpha_1 + b_2\alpha_2) = (a_1 + b_1)\alpha_1 + (a_2 + b_2)\alpha_2$$

$$\alpha - \beta = \alpha + (-\beta) = (a_1 - b_1)\alpha_1 + (a_2 - b_2)\alpha_2$$

So far, we have shown that each vector in the plane has unique coordinates with respect to a fixed basis, and our next goal is to use this mode of representation for planar vectors to obtain a coordinatization of individual points in the plane. To this end, let O be a fixed reference point and let $S = \{\alpha_1, \alpha_2\}$ be a basis for the plane. Then we shall say that the point P has coordinates (a, b) if the vector $\overrightarrow{OP}$ has S-coordinates (a, b)—that is, if $\overrightarrow{OP} = a\alpha_1 + b\alpha_2$. The point O is called the *origin* or *center of coordinates*, while α_1 and α_2 are said to be *reference vectors*, and the lines through O whose directions are determined by α_1 and α_2 are called the *first* and *second coordinate axes*, respectively (see figure 2.6).

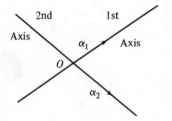   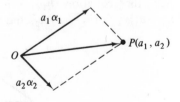

FIGURE 2.6

In the special case where α_1 and α_2 both have length 1 and α_2 is 90° in advance of α_1 (see figure 2.7), the resulting coordinatization of the plane is said to be *Cartesian* (after René Descartes, the man who first developed analytic geometry). Thus, whenever we speak of the *Cartesian coordinates* of a point, it should be understood that we are tacitly assuming the plane has been coordinatized in this special way.

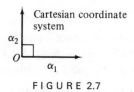

FIGURE 2.7

As our first application of vector methods in coordinate geometry, we shall characterize the line through two given points P_1 and P_2 in the plane. Accordingly, suppose that the plane has been coordinatized with center of coordinates O and reference vectors α_1 and α_2, and let P_1 and P_2 have coordinates (a_1, a_2) and (b_1, b_2), respectively. Then if $P(x, y)$ is an arbitrary point on the line (see figure 2.8), there exists a real number t such that $\overrightarrow{P_1P} = t\overrightarrow{P_1P_2}$ (why?), and we have

$$\overrightarrow{OP} = \overrightarrow{OP_1} + \overrightarrow{P_1P} = \overrightarrow{OP_1} + t\overrightarrow{P_1P_2}$$
$$= \overrightarrow{OP_1} + t(\overrightarrow{OP_2} - \overrightarrow{OP_1}) = t\overrightarrow{OP_2} + (1 - t)\overrightarrow{OP_1}$$

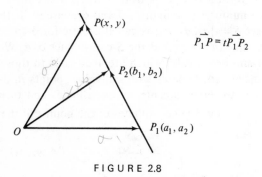

FIGURE 2.8

Expressing each of these vectors in terms of α_1 and α_2, we obtain

$$x\alpha_1 + y\alpha_2 = t(b_1\alpha_1 + b_2\alpha_2) + (1 - t)(a_1\alpha_1 + a_2\alpha_2)$$
$$= [tb_1 + (1 - t)a_1]\alpha_1 + [tb_2 + (1 - t)a_2]\alpha_2$$

and since $\{\alpha_1, \alpha_2\}$ is a basis, the uniqueness criterion established in theorem 2.3 enables us to conclude that

$$x = tb_1 + (1 - t)a_1$$
$$y = tb_2 + (1 - t)a_2$$
(2.2)

As t ranges over all real numbers, the corresponding points $P(x, y)$ range along the line in question. In this sense, (2.2) provides a complete characterization of the line.

PROBLEM 1.   With respect to a certain coordinatization of the plane, the points P_1 and P_2 have coordinates $(-5, 7)$ and $(1, 3)$. Show that the point $P(x, y)$ lies on the line through P_1 and P_2 if and only if $2x + 3y = 11$.

Solution: Substituting in (2.2), we obtain the parametric equations

$$x = t + (1 - t)(-5) = 6t - 5$$
$$y = 3t + (1 - t)(7) = -4t + 7$$

Thus, we have

$$\frac{x + 5}{6} = t = \frac{y - 7}{-4}$$

and it follows that $-4x - 20 = 6y - 42$ or $2x + 3y = 11$, as desired.

With only minor modifications, the approach we have used in coordinatizing the plane can also be used to coordinatize 3-space. Briefly, to obtain this coordinatization, we must first show that if $\alpha_1, \alpha_2, \alpha_3$ are non-zero vectors which do not lie in the same plane, then each vector α in 3-space can be expressed in the form $\alpha = a_1\alpha_1 + a_2\alpha_2 + a_3\alpha_3$, where the a_j's are uniquely determined real numbers (see figure 2.9 and exercise 9). If the α's have this property, the set $S = \{\alpha_1, \alpha_2, \alpha_3\}$ is called a basis for 3-space, and we refer to the ordered triple (a_1, a_2, a_3) as the S-coordinates of α. We leave it to the reader to state and prove rules which are analogous to those listed in (2.1). Finally, to associate coordinates with individual points in 3-space, we first choose a point O to serve as center of coordinates and then designate the ordered triple (a, b, c) as the coordinates of the point P if the vector $\overrightarrow{OP}$ has

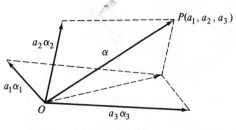

FIGURE 2.9

S-coordinates (a, b, c). This method of labeling points is illustrated in figure 2.9.

Next, suppose that 3-space has been coordinatized as indicated in the above discussion, with center of coordinates O and reference vectors α_1, α_2, and α_3. To obtain a parametric characterization of the line through the points $P_1(a_1, a_2, a_3)$ and $P_2(b_1, b_2, b_3)$, we proceed exactly as in the analogous planar problem. We leave it to the reader to show that $P(x, y, z)$ lies on the line in

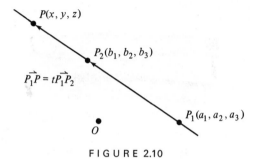

FIGURE 2.10

question if and only if there exists a real number t such that

$$x = tb_1 + (1 - t)a_1$$
$$y = tb_2 + (1 - t)a_2 \qquad (2.3)$$
$$z = tb_3 + (1 - t)a_3$$

Vector methods can also be used to characterize a plane in 3-space. To be specific, assume that we have the same coordinate system as before (origin O and reference vectors α_1, α_2, α_3), and let $P(x, y, z)$ be a general point on the plane determined by the three non-collinear points, $P_1(a_1, a_2, a_3)$, $P_2(b_1, b_2, b_3)$, and $P_3(c_1, c_2, c_3)$:

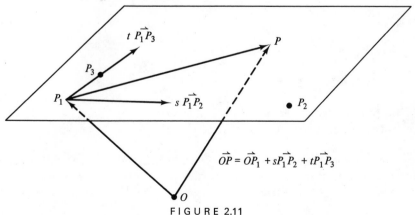

FIGURE 2.11

Since P_1, P_2, P_3, and P all lie in the same plane, there exist unique scalars s and t such that $\overrightarrow{P_1P} = s\overrightarrow{P_1P_2} + t\overrightarrow{P_1P_3}$, and we have

$$\overrightarrow{OP} = \overrightarrow{OP_1} + \overrightarrow{P_1P} = \overrightarrow{OP_1} + (s\overrightarrow{P_1P_2} + t\overrightarrow{P_1P_3})$$

$$= \overrightarrow{OP_1} + s(\overrightarrow{OP_2} - \overrightarrow{OP_1}) + t(\overrightarrow{OP_3} - \overrightarrow{OP_1}) \qquad (2.4)$$

$$= (1 - s - t)\overrightarrow{OP_1} + s\overrightarrow{OP_2} + t\overrightarrow{OP_3}$$

Expressing these vectors in terms of the reference vectors α_1, α_2, and α_3 and comparing coordinates, we find that

$$x = (1 - s - t)a_1 + sb_1 + tc_1$$
$$y = (1 - s - t)a_2 + sb_2 + tc_2 \qquad (2.5)$$
$$z = (1 - s - t)a_3 + sb_3 + tc_3$$

As s and t both range over all real numbers, the points $P(x, y, z)$ whose coordinates satisfy equations (2.5) range over the plane in question. Therefore, the plane is completely determined by these equations.

PROBLEM 2.   Find the equation of the plane in 3-space which contains the points $P_1(-2, 1, 1)$, $P_2(3, 0, -4)$, and $P_3(1, 1, 0)$.

Solution: The point $P(x, y, z)$ lies on the plane in question if there exist real numbers s and t such that

$$x = (1 - s - t)(-2) + s(3) + t(1) = 5s + 3t - 2$$
$$y = (1 - s - t)(1) + s(0) + t(1) = -s + 0t + 1 \qquad (2.6)$$
$$z = (1 - s - t)(1) + s(-4) + t(0) = -5s - t + 1$$

Applying the reduction algorithm to this system of three equations in the unknowns s and t, we obtain

$$\begin{bmatrix} 5 & 3 & | & x + 2 \\ -1 & 0 & | & y - 1 \\ -5 & -1 & | & z - 1 \end{bmatrix} \longrightarrow \begin{bmatrix} 1 & 0 & | & 1 - y \\ 0 & 1 & | & z - 5y + 4 \\ 0 & 0 & | & x - 10y + 3z + 9 \end{bmatrix}$$

and it follows that system (2.6) is solvable if and only if $x - 10y + 3z + 9 = 0$. Therefore, the equation of the plane is $x - 10y + 3z = -9$.

The approach we have used in coordinatizing the plane and 3-space depends more on the algebraic properties of vectors than on any intrinsic geometric properties of these spaces. Indeed, we shall soon see that the results of these first two sections can be extended to situations which appear to have little if anything in common with the plane and 3-space.

Exercise Set 2.2

1. In this exercise, assume the existence of a fixed coordinatization of the plane in which O is the origin, and answer the following questions:
   (i) What are the coordinates of the point P_3 located $\frac{2}{3}$ the distance from $P_1(-1, 7)$ to $P_2(8, -4)$?
   (ii) Show that the point $P(x, y)$ lies on the line through P_1 and P_2 if and only if

$$x = -1 + 9t$$
$$y = 7 - 11t$$

   for a real number t.
   (iii) Show that $P(x, y)$ lies on the line segment between P_1 and P_2 if and only if x and y satisfy the parametric equations of part (ii) for t such that $0 \leq t \leq 1$.
   (iv) Does the point $P_4(2, -1)$ lie on the boundary of triangle OP_1P_2? Explain.
   (v) Is the point $P_5(3, 2)$ inside or outside triangle OP_1P_2? What about $P_6(1, 5)$? Explain.

2. The *center* C of triangle $P_1P_2P_3$ is the point of intersection of its three medians (see problem 2, section 2.1). Let $\alpha_1 = \overrightarrow{P_1P_2}$ and $\alpha_2 = \overrightarrow{P_1P_3}$. Show that $S = \{\alpha_1, \alpha_2\}$ is a basis for the plane and that C has S-coordinates $(\frac{1}{3}, \frac{1}{3})$.

3. Let P_1, P_2, P_3 be points which lie in the same plane, and let $\alpha = P_1P_2$ and $\beta = P_1P_3$.
   (i) Show that the point P lies in the same plane as P_1, P_2, and P_3 if and only if there exist scalars s and t such that $P_1P = s\alpha + t\beta$.
   (ii) Show that P lies in the triangle whose vertices are P_1, P_2, and P_3 whenever $P_1P = s\alpha + t\beta$, where s and t are scalars such that $s \geq 0$, $t \geq 0$, and $s + t \leq 1$.

In exercises 4–6, assume the existence of a fixed coordinatization of 3-space in which O is the origin.

4. (i) Show that the equation of the line through the points $P_1(-1, 1, 5)$ and $P_2(6, -3, 0)$ is

$$\frac{x + 1}{7} = \frac{y - 1}{-4} = \frac{z - 5}{-5}$$

   (ii) Find the coordinates of the point where the line in part (i) intersects the plane $3x + 3y + 2z = 5$.
   Hint: Parameterize the equation of the line.

5. Find a parameterization of the type displayed in (2.5) for the plane which contains the point $P_1(-1, 3, 7)$ and the line

$$\frac{x - 1}{2} = \frac{y + 1}{2} = \frac{z}{3}$$

What is the equation of this plane?
Hint: Find two points in addition to P_1 which lie in the plane.

6. Find the equation of the line which passes through the origin O and is parallel to the line through the points $P_1(-1, 1, 0)$ and $P_2(1, 0, -7)$.

7. It is known that the point Q_1 is the midpoint of one side of a certain triangle $P_1P_2P_3$ and that C is the point where the medians of the triangle meet. If the plane is coordinatized in such a way that P_1 is at the origin and Q_1 and C have coordinates $(3, -5)$ and $(1, 1)$, respectively, what are the coordinates of P_2 and P_3?

8. Let $S = \{\alpha_1, \alpha_2\}$ and $T = \{\beta_1, \beta_2\}$ be two bases for the plane, and suppose that $\beta_1 = 5\alpha_1 - 3\alpha_2$ and $\beta_2 = -3\alpha_1 + 2\alpha_2$.

  (i) Express α_1 in the form $\alpha_1 = b_1\beta_1 + b_2\beta_2$, and find a similar expression for α_2.

  (ii) If the S-coordinates of the point P are $(-9, 3)$, what are its T-coordinates? Assume that the origin is the same for both coordinatizations.

*9. Let $\alpha_1, \alpha_2, \alpha_3$ be non-zero vectors in 3-space which do not all lie in the same plane, and suppose that

$$a_1\alpha_1 + a_2\alpha_2 + a_3\alpha_3 = b_1\alpha_1 + b_2\alpha_2 + b_3\alpha_3$$

where the a's and b's are real numbers. Show that $a_j = b_j$ for $j = 1, 2, 3$. *Hint:* Modify the proof of theorem 2.3.

2.3   VECTOR SPACES

At first glance, matrices and geometric vectors seem to have very little in common, but if we compare theorems 1.6 and 2.1, it becomes apparent that these two mathematical entities obey essentially the same manipulative rules. Indeed, there are numerous other mathematical and physical systems in which it is natural to add quantities and multiply them by scalars and which have essentially the same mathematical behavior as matrices and geometric vectors. It is the primary objective of this chapter to investigate such systems, but since they are so similar in nature, it would be a waste of time and effort to study them individually. Instead, we shall expedite matters by defining and analyzing an abstract model based on the fundamental features these systems have in common. Then, whenever a system under consideration is found to be a member of this class, its properties can be determined quite painlessly by simply referring to the list of properties of the model.

We begin by giving a formal definition of the abstract model we plan to investigate.

Definition 2.4. A *real vector space* $\{V; \oplus; \odot\}$ is a collection V of elements together with two operations, $\oplus$ and $\odot$, which obey the following rules (known as vector space axioms):

VS 1. If α and β are elements of V, then $\alpha \oplus \beta$ is a uniquely determined element of V.

VS 2. $\alpha \oplus \beta = \beta \oplus \alpha$ for all α, β in V.

VS 3. $\alpha \oplus (\beta \oplus \gamma) = (\alpha \oplus \beta) \oplus \gamma$ for all α, β, γ in V.

VS 4. There exists a unique element θ_v in V such that $\alpha \oplus \theta_v = \alpha = \theta_v \oplus \alpha$ for each α in V.

VS 5. For each α in V, there exists a unique element β of V such that $\alpha \oplus \beta = \theta_v = \beta \oplus \alpha$.

VS 6. If α is an element of V and t is a real number, then $t \odot \alpha$ is a uniquely determined element of V.

VS 7. $t \odot (\alpha \oplus \beta) = (t \odot \alpha) \oplus (t \odot \beta)$ for each real number t and elements α, β of V.

VS 8. $(s + t) \odot \alpha = (s \odot \alpha) \oplus (t \odot \alpha)$ for each α in V and all real numbers s and t.

VS 9. $(st) \odot \alpha = s \odot (t \odot \alpha)$ for each α in V and real numbers s and t.

VS 10. $1 \odot \alpha = \alpha$ for each α in V.

The elements of the set V are called *vectors*, and the operations $\oplus$ and $\odot$ are known as *vector addition* and *multiplication by a scalar*, respectively. VS 1 and VS 6 are often called "closure axioms" and tell us that the sum of two vectors and a scalar multiple of a given vector are always vectors. Axioms VS 2 and VS 3 say that vector addition is an operation which is both commutative and associative, while VS 7, VS 8, and VS 9 are various kinds of distributivity laws. The element θ_v whose existence is guaranteed by VS 4 is called the *zero vector*, and the vector β of VS 5 is known as the *additive inverse* of α.

If the word "real" is replaced by "complex" in axioms VS 6, VS 7, VS 8, and VS 9, we obtain a model system known as a *complex vector space*, and a similar modification may be used to define a vector space over any other scalar field. Most of the results of this chapter apply to any linear vector space, but for simplicity, we choose to concentrate on examples and applications in which the scalar field is the set of real numbers. Finally, to eliminate unnecessary verbiage from our discussion of definition 2.4, we shall henceforth refer to any system whose elements and operations satisfy definition 2.4 as simply a *vector space*, leaving out the modifier "real."

The attentive reader has probably already observed that the properties found in definition 2.4 are the same as those obtained for matrices in theorem 1.6 and for geometric vectors in theorem 2.1. Therefore, it should come as no surprise to find that the system of all geometric vectors and the system of all $m \times n$ real matrices are both vector spaces. For future reference, we shall now list these vector spaces along with several others which recur frequently throughout our work. We ask the reader to take special note of the terminology and notation developed in these examples.

Examples of Vector Spaces

EXAMPLE 1.   The system of all geometric vectors together with the operations described in definitions 2.2 and 2.3.

EXAMPLE 2.   The system of all forces which act at a point P in space together with the operations of force addition (resolution) and multiplication of a force by a scalar.

EXAMPLE 3.   The collection of all $m \times n$ real matrices, with the operations of matrix addition and multiplication of a matrix by a scalar.

We shall denote this real vector space by $_mR_n$, and the analogous complex vector space comprised of all $m \times n$ complex matrices by $_mC_n$.

EXAMPLE 4.   The collection of all ordered n-tuples of real numbers, together with the operations of componentwise addition and multiplication by a scalar:

$$(a_1, a_2, \ldots, a_n) + (b_1, b_2, \ldots, b_n) = (a_1 + b_1, a_2 + b_2, \ldots, a_n + b_n)$$

$$t(a_1, a_2, \ldots, a_n) = (ta_1, ta_2, \ldots, ta_n)$$

We shall denote this system by R_n and the corresponding system of all ordered n-tuples of complex numbers by C_n. Note that R_n may be regarded as a special case of the system described in example 3. Specifically, it is either $_1R_n$ or $_nR_1$, depending on whether we represent n-tuples in row or column form. The difference between column n-tuples and row n-tuples is largely superficial, and in this text, we shall distinguish between them only if it is necessary to avoid confusion.

EXAMPLE 5.   The system $F[a, b]$ comprised of all real-valued functions defined on the interval $[a, b]$ together with the familiar operations of functional addition and multiplication of a function by a scalar:

$$(f + g)(x) = f(x) + g(x)$$

$$(rf)(x) = rf(x)$$

To show that this system meets the conditions of definition 2.4, it is necessary to know only the rudiments of function theory, and we shall leave the verification to the reader. Note that the zero vector of $F[a, b]$ is the function f_0 such that $f_0(x) = 0$ for $a \leq x \leq b$, and the additive inverse of a given function f is the function g, where $g(x) = -f(x)$ for $a \leq x \leq b$.

EXAMPLE 6.   The system $C[a, b]$ of all functions continuous on $[a, b]$, with the operations defined in example 5 (see exercise 5).

Since each function of $C[a, b]$ is also contained in $F[a, b]$ and since both vector spaces have the same operations, it is natural to refer to $C[a, b]$ as a *subspace* of $F[a, b]$. This relationship will be explored in depth in section 2.4.

EXAMPLE 7.   The system P_n of all real polynomials of degree at most n, together with the operations defined in example 5.

To be specific, $p(t)$ belongs to P_n if and only if there exist real numbers $a_0, a_1, \ldots, a_n$ (some or all of which may be zero) such that $p(t) = a_n t^n + a_{n-1} t^{n-1} + \ldots + a_1 t + a_0$. Note that if $q(t) = b_n t^n + \ldots + b_1 t + b_0$ is another such polynomial and r is a scalar, the operations defined in example 5 assume the following convenient forms:

$$(p + q)(t) = (a_n + b_n)t^n + (a_{n-1} + b_{n-1})t^{n-1} + \ldots + (a_1 + b_1)t + (a_0 + b_0)$$

$$(rp)(t) = (ra_n)t^n + (ra_{n-1})t^{n-1} + \ldots + (ra_1)t + (ra_0)$$

The zero vector in this system is the zero polynomial, $p_0(t) = 0t^n + 0t^{n-1} + \ldots + 0t + 0$, and the additive inverse of $p(t)$ is the polynomial $-a_n t^n - a_{n-1} t^{n-1} - \ldots - a_1 t - a_0$.

EXAMPLE 8.   The system P_∞ of all real polynomials of all degrees, together with the operations defined in example 5.

At times we shall find it convenient to refer to the vector space V as a *function space* if its elements are general functions (e.g., examples 5 and 6) or as a *polynomial space* if its elements are polynomials, as in examples 7 and 8.

EXAMPLE 9.   The system $\{R^+; \oplus, \odot\}$, where R^+ is the set of all positive real numbers and the operations $\oplus$ and $\odot$ are defined as follows:

$$x \oplus y = xy \quad \text{for all } x, y \text{ in } R^+$$

$$r \odot x = x^r \quad \text{for each real number } r \text{ and all } x \text{ in } R^+$$

Note that axioms VS 1 and VS 6 of definition 2.4 are immediate consequences of the definitions of $\oplus$ and $\odot$, respectively. To verify axioms VS 3 and VS 9, for example, we observe that

VS 3. $(x \oplus y) \oplus z = (xy)z = x(yz) = x \oplus (y \oplus z)$ since multiplication of real numbers is associative.

VS 9. $(rs) \odot x = x^{rs} = (x^r)^s = s \odot (r \odot x)$ by the rules governing exponentiation within the real number system.

The verification of the remaining six properties is left as an exercise.

In order to prove that a given system is a vector space, it is necessary to show that each of the ten axioms of definition 2.4 is satisfied within the system. Thus, to show that a system is *not* a vector space, it suffices to prove that at least one of these properties is violated. Usually, if a system is not a vector space, one of the closure axioms (VS 1 and VS 6) will not be satisfied, but there are many systems which pass this test and are still not vector spaces. These statements are illustrated in the following two examples.

EXAMPLE 10.   Neither of the following collections of ordered 2-tuples is a vector space with respect to the operations defined in example 4:

(i) The set of all (x, y) such that $y = x^2$.
(ii) The set of all (x, y) such that $2x + y \geq 0$.

The system described above in (i) violates both closure axioms. For example, the set contains both $(2, 4)$ and $(3, 9)$ but does not contain their sum $(5, 13)$ or the scalar multiple $3(2, 4) = (6, 12)$. The second system is closed under addition (verify this fact), but not under multiplication by a scalar. For instance, $(1, 1)$ satisfies the given conditions, but $-1(1, 1) = (-1, -1)$ does not. The geometric configurations that correspond to these two systems are depicted in figure 2.12.

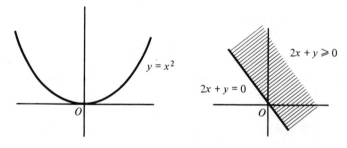

FIGURE 2.12

Incidentally, it can be shown (see exercise 11) that a collection of ordered 2-tuples (x, y) forms a vector space with respect to the operations of R_2 only if it falls in one of these three categories:

1. It is all of R_2.
2. It contains only the zero vector $(0, 0)$.
3. The coordinates of each of its points satisfy an equation of the form $Ax + By = 0$.

Geometrically, the points in the third case may be represented by a line through the origin, as illustrated in figure 2.13.

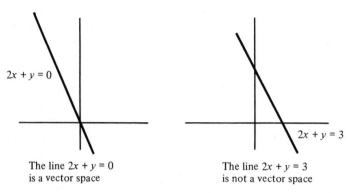

The line $2x + y = 0$
is a vector space

The line $2x + y = 3$
is not a vector space

FIGURE 2.13

EXAMPLE 11.   Let $\hat{R}$ denote the set of all non-negative real numbers and let $\oplus$ and $\odot$ be defined as follows:

$$x \oplus y = x \quad \text{for all } x, y \text{ in } \hat{R}$$
$$r \odot x = 0 \quad \text{for each real number } r \text{ and each } x \text{ in } \hat{R}$$

The system $\{\hat{R}; \oplus, \odot\}$ is not a vector space, since property 2 is not satisfied. For example, $1 \oplus 2 = 1$ and $2 \oplus 1 = 2$ and thus, $1 \oplus 2 \neq 2 \oplus 1$.

Note that axioms VS 1 and VS 6 are satisfied within this system, since $x \oplus y$ and $r \odot x$ are always non-negative real numbers whenever x and y are non-negative.

Our examples show that many different systems satisfy the conditions of definition 2.4 and can thus be analyzed by referring to the properties of our abstract model. Indeed, the main reason we used such formal language and abstract notation in stating definition 2.4 was to draw attention to the general nature of our model, but now that we have made this point, it is time to introduce more flexible terminology and notation. Accordingly, we shall henceforth denote the vector space $\{V; \oplus, \odot\}$ by writing V and shall use $+$ instead of $\oplus$ and juxtaposition of symbols rather than $\odot$. We shall denote the zero vector by θ instead of θ_v and the additive inverse of a given vector α by $-\alpha$. Finally, we shall write $\alpha - \beta$ in place of the cumbersome expression $\alpha + (-\beta)$. Using these notational conventions, we shall now show that vectors obey many of the same algebraic laws as the real numbers.

Theorem 2.4.   All vector spaces have the following properties:

(i) Given vectors α and β, the vector γ satisfies $\alpha + \gamma = \beta$ if and only if $\gamma = \beta - \alpha$.

(ii) $\alpha + \alpha = \alpha$ if and only if $\alpha = \theta$.

(iii) $0\alpha = \theta$ for each vector α.

(iv) $s\theta = \theta$ for each scalar s.

(v) $(-s)\alpha = -(s\alpha) = s(-\alpha)$ for each vector α and scalar s.

(vi) $-(\alpha + \beta) = (-\alpha) + (-\beta)$ for all vectors α and β.

(vii) If $s\alpha = \theta$, then either $s = 0$ or $\alpha = \theta$.

(viii) If $s\alpha = s\beta$ and $s \neq 0$, then $\alpha = \beta$.

(ix) If $s\alpha = r\alpha$ and $\alpha \neq \theta$, then $s = r$.

Proof: We shall prove parts ii and v in detail and then leave the rest of the proof as an exercise. Incidentally, the list of properties is arranged so that each part can be proved by using only the axioms of a vector space and certain parts of the theorem which appear earlier in the list.

(ii)   Certainly we must have $\theta + \theta = \theta$ (VS 4). Conversely, if $\alpha + \alpha = \alpha$, the existence of an additive inverse (VS 5) guarantees that

$$(\alpha + \alpha) + (-\alpha) = \alpha + (-\alpha) = \theta.$$

Moreover, axioms VS 3, VS 5, and VS 4 tell us that

$$(\alpha + \alpha) + (-\alpha) = \alpha + (\alpha + (-\alpha)) = \alpha + \theta = \alpha$$

and it follows that $\alpha = \theta$.

(v)   Let s be a scalar and let α be a vector in V. Combining axiom VS 8 with the result of part (iii), we find that

$$(-s)\alpha + s\alpha = (-s + s)\alpha = 0\alpha = \theta$$

and since the additive inverse of the vector $s\alpha$ is unique (VS 5), it follows that $(-s)\alpha = -(s\alpha)$. Similarly, using VS 5 and VS 4 in conjunction with part (iv), we obtain the equation

$$s(-\alpha) + s\alpha = s(-\alpha + \alpha) = s\theta = \theta$$

which enables us to deduce that $s(-\alpha) = -s(\alpha)$. ∎

Thanks to theorem 2.4, we now have a fairly complete list of properties to use in manipulating vector expressions and solving vector equations. In the next section, we shall continue our study of vector spaces by investigating the conditions under which a subset of a vector space is a vector space in its own right.

Exercise Set 2.3

1. Show that the collection of all ordered 3-tuples (x_1, x_2, x_3) whose components satisfy $3x_1 - x_2 + 5x_3 = 0$ forms a vector space with respect to the usual operations of R_3 (see example 4). Note that axioms VS 2, VS 3, VS 7, VS 8, VS 9, and VS 10 are all inherited from R_3 by the system in question.

2. Show that the collection of all functions f such that $f(1) = 0$ forms a vector space with respect to the operations defined in example 5.
3. Show that the collection of all 2×3 real matrices A whose (1, 1) entry is 0 forms a vector space with respect to the usual operations in ${}_2R_3$ (see example 3).
4. None of the systems described below is a vector space, but each satisfies some of the axioms of definition 2.4:
   (i) S_1 is the system of all 3-tuples (x_1, x_2, x_3) such that $2x_1 - x_2 + 5x_3 = 1$, together with the operations of R_3.
   (ii) S_2 is the system of all functions f such that $f(x) = f^2(x)$ for all real numbers x, together with the operations defined in example 5.
   (iii) S_3 is the system of all polynomials whose degree is *exactly* 3, together with the operations defined for P_3 in example 7.
   (iv) S_4 is the system of all 3×5 real matrices A whose entries are all nonnegative, together with the usual operations of ${}_3R_5$ (see example 3).
   (v) S_5 is the system of all 3-tuples (x_1, x_2, x_3) such that $x_1x_2x_3 = 0$, together with the operations of R_3.

   Complete the following table, placing a "Y" in the jth column of the ith row if system S_i satisfies the jth axiom of definition 2.4 and an "N" if it does not.

	AXIOM									
System	VS 1	VS 2	VS 3	VS 4	VS 5	VS 6	VS 7	VS 8	VS 9	VS 10
S_1		Y		N			Y			
S_2				Y		N				
S_3	N				Y				Y	
S_4		Y			N					
S_5				Y		Y				

5. Verify that the system $C[0, 1]$ defined in example 6 is a vector space. Note that axioms VS 2, VS 3, VS 7, VS 8, VS 9, and VS 10 are all inherited by $C[0, 1]$ from $F[0, 1]$, the vector space of all functions (continuous or not) defined on [0, 1]. (See example 5.)
6. Verify that the system $\{R^+; \oplus; \odot\}$ defined in example 9 is a vector space.
   Hints:
   (for VS 4) Find a number t such that $x + t = xt = x$ for each element x of R.
   (for VS 5) The additive inverse of 3 is $\frac{1}{3}$.
7. Complete the proof of theorem 2.4.
   Selected hints:
   (for part iii) First use axiom VS 8 to show that $(0\alpha) + (0\alpha) = 0\alpha$.
   (for part iv) Use VS 9 and part iii to show that $s\theta = s(0\theta) = 0\theta$.
   (for part vi) Show that $(\alpha + \beta) + [(-\alpha) + (-\beta)] = \theta$ and then use VS 5.
   (for part vii) If $s \neq 0$, use VS 9 to show that $\alpha = \theta$.
   (for part viii) Let $\gamma = \alpha - \beta$ and consider the equation $s\gamma = \theta$.
   (for part ix) If $\alpha \neq \theta$, let $t = s - r$ and consider the equation $t\alpha = \theta$.

8. Let $\alpha_1, \ldots, \alpha_k$ be elements of the vector space V, and let $\beta = r_1\alpha_1 + \ldots + r_k\alpha_k$ and $\gamma = s_1\alpha_1 + \ldots + s_k\alpha_k$, where the r_j and s_j are real numbers. Use the axioms of definition 2.4 together with mathematical induction to prove the following statements:

   (i) β is contained in V.
   (ii) $\beta + \gamma = (r_1 + s_1)\alpha_1 + \ldots + (r_k + s_k)\alpha_k$
   (iii) $t\beta = (tr_1)\alpha_1 + \ldots + (tr_k)\alpha_k$

9. In a vector space V, it is known that vectors β_1 and β_2 can be expressed in terms of α_1 and α_2 as follows:

$$2\alpha_1 - \alpha_2 = \beta_1$$
$$-5\alpha_1 + 2\alpha_2 = \beta_2$$

Solve this system to express α_1 and α_2 in terms of β_1 and β_2. Justify each algebraic step by referring to either the axioms of definition 2.4 or to the results listed in theorem 2.4.

10. Determine which of the ten axioms of definition 2.4 are satisfied by the system $\{I; \oplus; \odot\}$, where I is the set of all integers and the operations $\oplus$ and $\odot$ are defined as follows:

   If α, β are integers, then $\alpha \oplus \beta = \begin{cases} 1 & \text{if } \alpha + \beta \text{ is odd} \\ 0 & \text{if } \alpha + \beta \text{ is even.} \end{cases}$

   If α is an integer and t is a real number, then $t \odot \alpha = \alpha$.

*11. Let S be a collection of ordered 2-tuples which forms a vector space with respect to the operations of R_2. Assume that S contains more than the zero vector $(0, 0)$.

   (i) If (x_1, y_1) is a non-zero element of S, show that S must also contain each 2-tuple (x, y) whose components satisfy $y_1x - x_1y = 0$. Geometrically, this is a line through the origin.

   (ii) If S contains two elements (x_1, y_1) and (x_2, y_2) such that $x_1y_2 - y_1x_2 \neq 0$, show that it must contain every 2-tuple.
   Hint: Show that the system

$$x_1u + x_2v = x$$
$$y_1u + y_2v = y$$

   is always consistent.

   (iii) Deduce from parts (i) and (ii) that S is either a line through the origin or all of R_2.

2.4   SUBSPACES

In general, the only way to determine whether or not a given mathematical system S is a vector space is to see if S satisfies the ten axiomatic conditions of definition 2.4. However, if S happens to be part of a larger system V that is known to be a vector space, much of this tedious verification is unnecessary. To be specific, note that axioms VS 2, VS 3, VS 7, VS 8, VS 9, and VS 10 all describe manipulative rules that must be satisfied by all vectors in V, in particular, for those vectors in S. Therefore, the question of whether or not S is itself a vector space hinges on axioms VS 1, VS 4, VS 5, and VS 6, and in the theorem which follows, we show that only the closure axioms are crucial.

Theorem 2.5.   A non-empty subset S of the vector space V forms a vector space with respect to the same operations used in V if and only if
   (i) $\alpha + \beta$ is contained in S for all vectors α, β of S.
   (ii) $t\alpha$ is contained in S for each scalar t and vector α of S.

Proof: Since (i) and (ii) are just restatements of axioms VS 1 and VS 6 of definition 2.4, they must certainly be valid if S is a vector space.

Conversely, suppose (i) and (ii) are satisfied by the vectors of S. Since these are the closure axioms and since axioms VS 2, VS 3, VS 7, VS 8, VS 9, and VS 10 are all inherited by S from V, we see that S is a vector space if it meets the requirements of axioms VS 4, and VS 5. To verify axiom VS 5, note that if α is a vector of S, then (ii) guarantees that the scalar multiple $(-1)\alpha = -\alpha$ is also contained in S. Finally, since S contains both α and $-\alpha$, condition (i) tells us that it also contains the sum $\alpha + (-\alpha) = \theta$, and this verifies VS 4. Thus, S satisfies all ten axioms of definition 2.4, and it must be a vector space, as claimed. ∎

If a subset S of the vector space V forms a vector space in its own right with respect to the operations used in V, we shall say that S is a *subspace* of V. Thus, theorem 2.5 may be rephrased as follows:

> A non-empty subset S of the vector space V is a subspace of V if and only if it is closed with respect to the operations of V.

If S coincides with V, it is said to be an *improper subspace*, while any other subspace of V is referred to as *proper*.

Many of the vector spaces described in the last section are subspaces of larger systems. For instance, $C[a, b]$ is a subspace of $F[a, b]$, and it is not hard to see that the polynomial space P_m is a subspace of P_n if $n \geq m$. On the other hand, the system $\{R^+; \oplus, \odot\}$ described in example 9 of that section is *not* a subspace of R_1, even though R^+ is a subset of the set of all real numbers, because the operations $\oplus$ and $\odot$ are not the same as the operations of R_1. The following are examples of important subspaces which recur throughout our work.

EXAMPLE 1.   The set $D[a, b]$ of all functions differentiable on the interval $[a, b]$ is a subspace of $C[a, b]$.

This observation may be proved by means of theorem 2.5 and the following results from elementary calculus:

(i) If f is differentiable at a point c, it is also continuous at c.

(ii) If f, g are functions which are differentiable on $[a, b]$ and t is a real number, then $f + g$ and tf are also differentiable on $[a, b]$.

EXAMPLE 2. The set of all functions f which are defined on the interval $[a, b]$ and satisfy the differential equation $2f'' - 3f' + f = 0$ is a subspace of $F[a, b]$.

To prove this statement, let f_1 and f_2 be solutions of the differential equation and let t be a real number. We find that $f_1 + f_2$ and tf_1 are also solutions since

$$2(f_1 + f_2)'' - 3(f_1 + f_2)' + (f_1 + f_2)$$
$$= [2f_1'' - 3f_1' + f_1] + [2f_2'' - 3f_2' + f_2] = 0 + 0 = 0$$

and

$$2(tf_1)'' - 3(tf_1)' + (tf_1) = t[2f_1'' - 3f_1' + f_1] = t \cdot 0 = 0$$

Therefore, according to theorem 2.5, the solution set is a subspace of $F[a, b]$, as claimed.

In general, the set of all solutions of the homogeneous nth order differential equation

$$a_n y^{(n)} + a_{n-1} y^{(n-1)} + \ldots + a_1 y' + a_0 y = 0$$

is a subspace of $F[a, b]$, and we shall refer to this system as the *solution space* of the differential equation in question.

EXAMPLE 3. The set of all solutions of the homogeneous 3×3 linear system

$$2x_1 - 3x_2 + x_3 = 0$$
$$x_1 + x_2 - 5x_3 = 0$$

is a subspace of R_3.

Note that the solution set of this system can be regarded as the collection of all 3-tuples $X = (x_j)$ such that $AX = O$, where

$$A = \begin{bmatrix} 2 & -3 & 1 \\ 1 & 1 & -5 \end{bmatrix}$$

Thus, if X_1 and X_2 are solutions and t is a real number, we find that $X_1 + X_2$ and tX_1 are also solutions since

$$A(X_1 + X_2) = (AX_1) + (AX_2) = O + O = O$$

and

$$A(tX_1) = t(AX_1) = t \cdot O = O$$

Theorem 2.5 can then be invoked to verify that the solution set is a subspace of R_3.

This proof can be generalized to show that the set of all solutions of any $m \times n$ homogeneous linear system is a subspace of R_n. We shall refer to this set as the *solution space* of the system of equations.

EXAMPLE 4. If S and T are subspaces of the vector space V, then $S \cap T$, the collection of all vectors which belong to both S and T, is also a subspace of V.

To prove this statement, let α and β be vectors in $S \cap T$ and let t be a scalar. Since α and β belong to the subspaces S and T, the "only if" part of theorem 2.5 tells us that $\alpha + \beta$ and $t\alpha$ also belong to S and to T. Therefore, $S \cap T$ contains both $\alpha + \beta$ and $t\alpha$, and the "if" part of theorem 2.5 enables us to conclude that $S \cap T$ is a subspace of V.

At this point, it is natural to conjecture that the set $S \cup T$ of all vectors contained in *either* S or T is subspace of V, but this is true only under very special circumstances (see exercise 12). However, the set $S + T$ of all vectors γ of the form $\gamma = \sigma + \tau$, where σ and τ are elements of S and T, respectively, is a subspace of V (see exercise 13). We shall refer to $S + T$ as the *sum* of S and T. The sets $S \cap T$, $S \cup T$, and $S + T$ are depicted in figure 2.14 in the case where S and T are both planes containing the origin O:

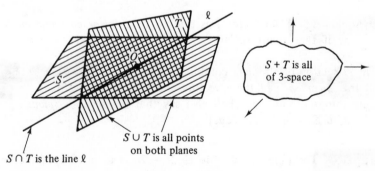

FIGURE 2.14

In developing the main ideas of linear algebra, it is possible to make extensive use of subspace sums $(S + T)$ and intersections $(S \cap T)$. For the most part, we shall proceed along different lines, but the terminology introduced in this example will appear in selected examples and exercises throughout this text.

The systems analyzed above are only a few of the interesting subspaces we shall encounter, and we invite the reader to test his understanding of theorem 2.5 by working exercises 1–4 and 11. The following solved problem is typical of those which appear in the exercises.

PROBLEM 1. Let S denote the collection of all 3-tuples (x, y, z) such that $2x - 3y + z = 0$. Determine whether or not S is a subspace of R_3.

Solution: If $\alpha_1 = (x_1, y_1, z_1)$ and $\alpha_2 = (x_2, y_2, z_2)$ are typical members of S and r is a scalar, we have

$$2(x_1 + x_2) - 3(y_1 + y_2) + (z_1 + z_2)$$
$$= (2x_1 - 3y_1 + z_1) + (2x_2 - 3y_2 + z_2) = 0 + 0 = 0$$

and

$$2(rx_1) - 3(ry_1) + (rz_1) = r(2x_1 - 3y_1 + z_1) = r \cdot 0 = 0$$

Therefore, $\alpha_1 + \alpha_2$ and $r\alpha_1$ are contained in S, and theorem 2.5 assures us that S is a subspace of R_3.

By combining the closure axioms of definition 2.4, it is easy to show that if $\alpha_1, \alpha_2, \ldots, \alpha_k$ are vectors in V, then V also contains each expression of the form $s_1\alpha_1 + s_2\alpha_2 + \ldots + s_k\alpha_k$ (see example 8, section 2.3). Such expressions occur so frequently in linear algebra that it is convenient to give them a name.

Definition 2.5. The vector γ is said to be a *linear combination* of α_1, $\alpha_2, \ldots, \alpha_k$ if there exist scalars $s_1, s_2, \ldots, s_k$ such that $\gamma = s_1\alpha_1 + s_2\alpha_2 + \ldots + s_k\alpha_k$.

The collection of all linear combinations of the vectors $\alpha_1, \alpha_2, \ldots, \alpha_k$ is called the *linear span* of these vectors and is denoted by $\mathrm{Sp}\{\alpha_1, \ldots, \alpha_k\}$. Turning things around, we shall also say that the set S is *spanned* by α_1, $\alpha_2, \ldots, \alpha_k$ if $S = \mathrm{Sp}\{\alpha_1, \ldots, \alpha_k\}$.

EXAMPLE 5. The polynomial $-4t^2 + 2t + 26$ is a linear combination of $t^2 - 3$ and $2t^2 + t + 1$ since

$$-4t^2 + 2t + 26 = -8(t^2 - 3) + 2(2t^2 + t + 1)$$

However, the 3-tuple $\gamma = (-1, 0, 1)$ cannot be expressed as a linear combination of $\alpha_1 = (-1, 3, 5)$ and $\alpha_2 = (4, -1, 2)$. To prove this statement, note that the vector equation $\gamma = s_1\alpha_1 + s_2\alpha_2$ is equivalent to the inconsistent

3×2 linear system

$$-s_1 + 4s_2 = -1$$
$$3s_1 - s_2 = 0$$
$$5s_1 + 2s_2 = 1$$

PROBLEM 2. Show that the linear span of $\alpha_1 = (1, 1, 0)$ and $\alpha_2 = (0, 3, -1)$ is the plane whose equation is $x - y - 3z = 0$.

Solution: The vector equation $s_1(1, 1, 0) + s_2(0, 3, -1) = (x, y, z)$ can be solved if and only if the scalar system

$$s_1 + 0s_2 = x$$
$$s_1 + 3s_2 = y$$
$$0s_1 - s_2 = z$$

is consistent, and the reduction

$$\begin{bmatrix} 1 & 0 & \vdots & x \\ 1 & 3 & \vdots & y \\ 0 & -1 & \vdots & z \end{bmatrix} \longrightarrow \begin{bmatrix} 1 & 0 & \vdots & x \\ 0 & 1 & \vdots & \frac{1}{3}(y - x) \\ 0 & 0 & \vdots & x - y - 3z \end{bmatrix}$$

tells us that this is the case whenever $x - y - 3z = 0$.

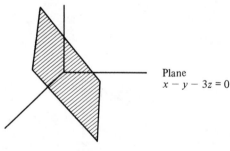

Plane
$x - y - 3z = 0$

FIGURE 2.15

From the closure axioms, it is clear that a given vector space V must contain the linear span of any finite collection of its vectors. Indeed, our next theorem shows that each such linear span must actually be a subspace of V.

Theorem 2.6. If $\alpha_1, \ldots, \alpha_k$ are elements of the vector space V, then $\mathrm{Sp}\{\alpha_1, \ldots, \alpha_k\}$ is a subspace of V.

Proof: Our proof will be based on the criterion established in theorem 2.5. To this end, let β and γ be elements of Sp $\{\alpha_1, \ldots, \alpha_k\}$ and let t be a scalar. According to definition 2.5, there exist scalars r_i and s_i such that $\beta = r_1\alpha_1 + \ldots + r_k\alpha_k$ and $\gamma = s_1\alpha_1 + \ldots + s_k\alpha_k$, and we can use the result of exercise 8 of section 2.3 to show that

$$\beta + \gamma = (r_1 + s_1)\alpha_1 + \ldots + (r_k + s_k)\alpha_k$$

and

$$t\beta = (tr_1)\alpha_1 + \ldots + (tr_k)\alpha_k$$

Since $\beta + \gamma$ and $t\beta$ are both linear combinations of the α's, theorem 2.5 enables us to conclude that Sp $\{\alpha_1, \ldots, \alpha_k\}$ is a subspace of V, as desired. ∎

Earlier in this chapter, we showed that if O is a fixed reference point in space, then each point P can be identified with a unique vector $\overrightarrow{OP}$. The collection of all such vectors forms a vector space V with respect to the operations defined in section 2.1, and each collection of vectors in V is represented by a geometric configuration. In particular, the results of section 2.2 enable us to make the following observations:

1. The linear span of a non-zero vector α_1 is represented by a line through O.
2. If α_2 is not a scalar multiple of α_1, the linear span of α_1 and α_2 is represented by a plane containing O (e.g., see problem 2, above.)
3. If α_3 cannot be expressed as a linear combination of α_1 and α_2, the linear span of α_1, α_2, and α_3 is all of 3-space.

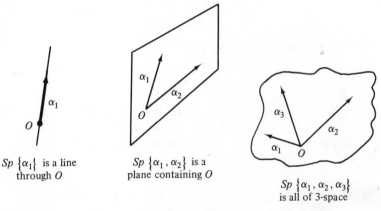

Sp $\{\alpha_1\}$ is a line
through O

Sp $\{\alpha_1, \alpha_2\}$ is a
plane containing O

Sp $\{\alpha_1, \alpha_2, \alpha_3\}$
is all of 3-space

FIGURE 2.16

Using these observations in conjunction with theorem 2.6, it can be shown that each subspace S of 3-space must fall in one of the following categories (see exercise 8):

1. S contains only O.
2. S is a line through O.
3. S is a plane containing O.   (2.7)
4. S is all of 3-space.

If each vector of the vector space V can be expressed as a linear combination of the elements of a fixed collection of vectors S, we shall say that S *spans* V, or equivalently, that V is *generated* by S. In this text, we shall be especially interested in those vector spaces which can be spanned by a finite collection of vectors, and since such systems are so important, we give them a name of their own.

Definition 2.6.   The vector space V is said to be *finitely generated* if there exists a finite collection of vectors $\alpha_1, \alpha_2, \ldots, \alpha_k$ such that $V = \text{Sp} \{\alpha_1, \ldots, \alpha_k\}$.

As we see in the following examples, some of the vector spaces we have discussed are finitely generated and some are not.

EXAMPLE 6. R_3 is finitely generated since each 3-tuple (x, y, z) can be expressed as a linear combination of the vectors $\epsilon_1 = (1, 0, 0)$, $\epsilon_2 = (0, 1, 0)$, and $\epsilon_3 = (0, 0, 1)$. Specifically,

$$(x, y, z) = x(1, 0, 0) + y(0, 1, 0) + z(0, 0, 1) = x\epsilon_1 + y\epsilon_2 + z\epsilon_3$$

More generally, R_n is finitely generated, as is the vector space $_mR_n$ of all $m \times n$ matrices (see exercise 9.)

EXAMPLE 7. P_n is finitely generated since each polynomial p of degree n or less can be expressed as a linear combination of the "power" polynomials, $t^n, t^{n-1}, \ldots, t, 1$. For instance, a typical member of P_2 can be expressed in the form

$$p(t) = a_2 t^2 + a_1 t + a_0 1$$

which is a linear combination of $t^2, t, 1$.

EXAMPLE 8. P_∞, the vector space of all polynomials of all degrees, is *not* finitely generated. It is spanned by the infinite collection $1, t, t^2, \ldots, t^k, \ldots$ but by no finite collection of its elements (see exercise 10).

Other examples of vector spaces which are not finitely generated include the function spaces $F[a, b]$, $C[a, b]$, and $D[a, b]$.

It is easy to show that each 2-tuple (x, y) can be expressed as a unique linear combination of $(1, 0)$ and $(0, 1)$. The vectors $(1, 1)$, $(3, 1)$, $(0, 2)$ also span R_2, but this time, each 2-tuple can be expressed as a linear combination of the given vectors in many different ways. For instance, we have

$$(4, 4) = (1, 1) + (3, 1) + (0, 2)$$
$$= -8(1, 1) + 4(3, 1) + 4(0, 2)$$

In general, if V is spanned by the vectors $\alpha_1, \ldots, \alpha_k$, then each vector γ can be expressed in the form $\gamma = s_1\alpha_1 + \ldots + s_k\alpha_k$, and in some cases, the coefficients which appear in this representation are uniquely determined, while in others, they are not. In the next section, we shall learn how to distinguish between these two cases.

Exercise Set 2.4

1. Which of the following collections of 3-tuples are subspaces of R_3?
   (i) The set of all (x, y, z) such that $3x + y - 5z = 0$.
   (ii) The set of all (x, y, z) such that $xyz = 0$.
   (iii) The set of all (x, y, z) such that $x/7 = z/3$.
   (iv) The set of all (x, y, z) such that $x/-2 = (y + 5)/8 = z$.

2. Let S denote the collection of all polynomials of the form $p(t) = (2a - b)t^2 + (3c - b)t + (a - c)$, where a, b, c are real numbers. Determine whether or not S is a subspace of P_2.

3. Let S denote the set of all 2-tuples (x, y) whose components x and y satisfy either $2x - y = 0$ or $3x + 5y = 0$. Show that S is *not* a subspace of R_2.

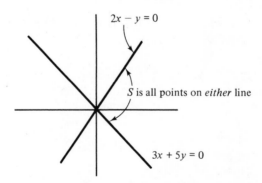

4. Let S denote the set of all 3-tuples (x_1, x_2, x_3) whose components satisfy the non-homogeneous linear system

$$2x_1 - 3x_2 + x_3 = 30$$
$$x_1 + x_2 - 5x_3 = -8$$

(i) Show that S is *not* a subspace of R_3.

(ii) Describe the geometric configuration which corresponds to S.

5. (i) Show that the linear span of the vector $\alpha_1 = (-7, 8, 5)$ is the line whose equation is

$$\frac{x}{-7} = \frac{y}{8} = \frac{z}{5}$$

(ii) Describe the geometric configuration which corresponds to the linear span of the vectors $\alpha_2 = (-1, 1, 3)$ and $\alpha_3 = (2, -1, 4)$.

(iii) Describe Sp $\{\alpha_1, \alpha_2, \alpha_3\}$.

6. Show that $\beta = (13, -7, 2)$ cannot be expressed as a linear combination of $\alpha_1 = (1, 7, -2)$, $\alpha_2 = (-3, 1, 4)$, and $\alpha_3 = (-7, 17, 8)$. What about $\gamma = (13, 3, -18)$?

Hint: Review the method illustrated in example 5.

7. Let α_1, α_2, and α_3 be elements of the vector space V, and let $\beta_1 = 2\alpha_1 - \alpha_2 + 2\alpha_3$ and $\beta_2 = -2\alpha_1 + \alpha_2 - 5\alpha_3$.

(i) Express $\gamma = 3\beta_1 + 7\beta_2$ as a linear combination of the α's.

(ii) Show that each linear combination $s_1\beta_1 + s_2\beta_2$ can be expressed as a linear combination of the α's.

(iii) Find a linear combination of the α's which is *not* contained in Sp $\{\beta_1, \beta_2\}$.

8. Justify the following statements:

(i) Each of the four geometric configurations listed in (2.7) is a subspace of 3-space.

(ii) A subspace of 3-space must fall in one of the four categories listed in (2.7).

9. (i) Show that each 2×2 real matrix can be expressed as a linear combination of the matrices

$$E_{11} = \begin{bmatrix} 1 & 0 \\ 0 & 0 \end{bmatrix}, \quad E_{12} = \begin{bmatrix} 0 & 1 \\ 0 & 0 \end{bmatrix}, \quad E_{21} = \begin{bmatrix} 0 & 0 \\ 1 & 0 \end{bmatrix}, \quad E_{22} = \begin{bmatrix} 0 & 0 \\ 0 & 1 \end{bmatrix}$$

Why does this prove that $_2R_2$ is finitely generated?

(ii) Show that $_mR_n$ is finitely generated.

*10. (i) Let $p_1, p_2, \ldots, p_m$ be a finite collection of polynomials, and let q be a polynomial whose degree is greater than that of any of the p_k. Show that q cannot be expressed as a linear combination of $p_1, \ldots, p_m$.

(ii) Show that P_∞ is not finitely generated.

*11. In each of the following cases, determine whether or not the given collection of matrices is a subspace of $_nR_n$.

(i) The set of all $n \times n$ symmetric matrices (i.e., $A = A^t$).

(ii) The set of all $n \times n$ skew-symmetric matrices ($A = -A^t$).

(iii) The set of all $n \times n$ matrices with trace 0 (see exercise 5, section 1.4).

(iv) The set of all $n \times n$ non-singular matrices.

*12. Let S and T be subspaces of the vector space V.

(i) If α is an element of S but not of T, and β is contained in T but not S, show that $\alpha + \beta$ is contained in neither S nor T (see exercise 3).

(ii) Prove that $S \cup T$ is a subspace of V if and only if S is contained in T or vice versa (see example 4).

*13. If S and T are subspaces of the vector space V, show that $S + T$ is also a subspace (see example 4).

Hint: Let $\alpha_1 = \sigma_1 + \tau_1$ and $\alpha_2 = \sigma_2 + \tau_2$ be typical elements of $S + T$ and show that the sum $\alpha_1 + \alpha_2$ and the scalar multiple $r\alpha_1$ are both vectors of the form $\sigma + \tau$. You will need the "only if" part of theorem 2.5.

*14. Let S denote the collection of all 3-tuples (x, y, z) such that $2x - y + z = 0$, and let T denote the collection of all (x, y, z) such that $x + 3y = 0$.

  (i) Show that S and T are subspaces of R_3 and describe the geometric configurations to which they correspond.

  (ii) Describe the geometric configurations which correspond to $S \cap T$, $S \cup T$, and $S + T$.

15. Show that the solution set of the non-homogeneous $m \times n$ linear system $AX = \beta$ is not a subspace of R_n.

2.5   THE CONSTRUCTION OF A BASIS

In section 2.2, we showed that whenever the non-zero vectors α and β are not parallel, then each planar vector γ can be expressed in the form $\gamma = s\alpha + t\beta$ in exactly one way, and for this reason, we found it appropriate to refer to the ordered pair $\{\alpha, \beta\}$ as a basis for the plane. Such a representation has many desirable features, and the object of this section is to examine the more general notion of basis described in the following definition.

Definition 2.7.   The ordered set $S = \{\alpha_1, \alpha_2, \ldots, \alpha_m\}$ is said to be a *basis* for the vector space V if each vector γ of V can be expressed in the form $\gamma = s_1\alpha_1 + s_2\alpha_2 + \ldots + s_m\alpha_m$ in exactly one way. The trivial vector space which contains only the zero vector is said to have no basis.

It is important to remember that the set S has a specified ordering, and different arrangements of the same vectors are regarded as different bases. For instance, from the standpoint of the definition, the bases $\{(-2, 1), (3, 5)\}$ and $\{(3, 5), (-2, 1)\}$ for R_2 are not the same, even though they involve the same vectors. It may appear that we're nit-picking, but the reason for emphasizing the ordering of vectors in S will become clear in section 2.7 when we use a fixed ordered basis to coordinatize V in much the same way that coordinates were introduced to the plane in section 2.2.

The following examples describe important bases which appear frequently in our work.

EXAMPLE 1.   We have already noted that the ordered pair $\{\alpha, \beta\}$ is a basis for the plane if α and β are not parallel. Similarly, $\{\alpha, \beta, \gamma\}$ is an ordered basis for 3-space if α, β, and γ do not all lie in the same plane. We leave it to the reader to verify that these geometric definitions are equivalent to definition 2.7.

EXAMPLE 2.   The set $\{(1, 0, 0), (0, 1, 0), (0, 0, 1)\}$ is a basis for R_3. More generally, the set $\{\epsilon_1, \epsilon_2, \ldots, \epsilon_n\}$ is a basis for R_n where $\epsilon_j = (0, \ldots, 0, 1, 0, \ldots, 0)$ is the n-tuple with a 1 in the jth component position and zeroes elsewhere. This set is often called the *standard basis* for R_n, and ϵ_j is referred to as the jth standard basis vector.

EXAMPLE 3.   Each polynomial of degree n or less can be expressed as a unique linear combination of the power polynomials $t^n, t^{n-1}, \ldots, t, 1$. We shall refer to the set $\{t^n, t^{n-1}, \ldots, t, 1\}$ as the *natural basis* for the polynomial space P_n.

EXAMPLE 4.   The *standard basis* for $_mR_n$ is the set $\{E_{11}, \ldots, E_{1n}, E_{21}, \ldots, E_{2n}, \ldots, E_{m1}, \ldots, E_{mn}\}$, where E_{ij} is an $m \times n$ matrix with a 1 in the (i, j) position and 0's elsewhere.

As a specific example, the standard basis for $_2R_3$ is comprised of the matrices

$$E_{11} = \begin{bmatrix} 1 & 0 & 0 \\ 0 & 0 & 0 \end{bmatrix}, \quad E_{12} = \begin{bmatrix} 0 & 1 & 0 \\ 0 & 0 & 0 \end{bmatrix}, \quad E_{13} = \begin{bmatrix} 0 & 0 & 1 \\ 0 & 0 & 0 \end{bmatrix}$$

$$E_{21} = \begin{bmatrix} 0 & 0 & 0 \\ 1 & 0 & 0 \end{bmatrix}, \quad E_{22} = \begin{bmatrix} 0 & 0 & 0 \\ 0 & 1 & 0 \end{bmatrix}, \quad E_{23} = \begin{bmatrix} 0 & 0 & 0 \\ 0 & 0 & 1 \end{bmatrix}$$

The reader should have no difficulty in showing that each 2×3 matrix can be expressed as a linear combination of these six basis matrices in exactly one way; for instance,

$$\begin{bmatrix} -7 & 0 & 3 \\ 2 & 4 & -8 \end{bmatrix} = -7E_{11} + 0E_{12} + 3E_{13} + 2E_{21} + 4E_{22} - 8E_{23}$$

With its present wording, definition 2.7 only applies to finitely generated vector spaces (why?), but it is possible to provide alternate definitions of the concept of basis which may be applied to any vector space. For instance, we could say that the infinite ordered set $\{\alpha_1, \alpha_2, \ldots, \alpha_n, \ldots\}$ is a basis for V if each vector of V can be expressed as a unique linear combination of finitely many α's. If V is finitely generated, this definition is the same as definition 2.7, but it can also be used for such purposes as showing that the infinite ordered set $S = \{1, t, t^2, \ldots, t^n, \ldots\}$ is a basis for P_∞. Incidentally, using such a definition, it can be shown that every vector space has a basis, but the general proof of this statement involves nuances of mathematical logic beyond the scope of this text (e.g., see *Linear Algebra*, by R. Stoll and E. Wong, Academic Press, New York, 1968, pp. 28–30).

Our next goal is to show that each finitely generated vector space V has a basis. According to definition 2.6, there exists a finite collection of vectors $S = \{\alpha_1, \ldots, \alpha_k\}$ which spans V, but we cannot assert that S is a basis for V without first proving that each vector γ of V can be expressed as a linear combination of the α's in only one way. The key to determining whether or not a given linear representation of vectors is unique is contained in the following concept, which serves as a generalization of the principle of planar linear independence (recall theorem 2.2).

Definition 2.8. Let $\alpha_1, \alpha_2, \ldots, \alpha_k$ be distinct elements of the vector space V. If the vector equation

$$s_1\alpha_1 + s_2\alpha_2 + \ldots + s_k\alpha_k = \theta \tag{2.8}$$

has only the trivial solution $s_1 = s_2 = \ldots = s_k = 0$, the α's are said to be *linearly independent*. On the other hand, if there exist scalars s_j *not all zero* which satisfy (2.8), the α's are *linearly dependent*.

EXAMPLE 5. The vectors $\alpha_1 = (-1, 2, 4)$, $\alpha_2 = (1, 3, -2)$, and $\alpha_3 = (-3, -4, 8)$ are linearly dependent since $\alpha_1 - 2\alpha_2 - \alpha_3 = \theta$.

EXAMPLE 6. According to theorem 2.2, two planar vectors are linearly independent if they are not parallel. On the other hand, it can be shown that any three vectors which lie in the same plane must be linearly dependent (see exercise 7). Similarly, in 3-space the vectors α, β, γ are linearly independent if they do not all lie in the same plane, while any four vectors must be linearly dependent.

PROBLEM 1. Show that the polynomials $p_1(t) = t^2 - 3t + 1$, $p_2(t) = 4t^2 + 7t + 1$, and $p_3(t) = t - 2$ are linearly independent.

Solution: Let s_1, s_2, s_3 be scalars such that

$$s_1(t^2 - 3t + 1) + s_2(4t^2 + 7t + 1) + s_3(t - 2) = 0t^2 + 0t + 0$$

Equating coefficients in this polynomial equation, we obtain the homogeneous 3×3 linear system

$$s_1 + 4s_2 + 0s_3 = 0$$

$$-3s_1 + 7s_2 + s_3 = 0 \tag{2.9}$$

$$s_1 + s_2 - 2s_3 = 0$$

which may be solved by performing the reduction

$$
\begin{bmatrix}
1 & 4 & 0 \\
-3 & 7 & 1 \\
1 & 1 & -2
\end{bmatrix}
\longrightarrow
\begin{bmatrix}
1 & 0 & 0 \\
0 & 1 & 0 \\
0 & 0 & 1
\end{bmatrix}
$$

Thus, the system displayed in (2.9) has only the trivial solution $s_1 = s_2 = s_3 = 0$ (why?), and it follows that p_1, p_2, p_3 are linearly independent.

In our introduction to definition 2.8, we claimed that the concept of linear independence is closely related to the problem of determining whether or not a collection of vectors which spans the vector space V is a basis for V. This relationship is made explicit in our next theorem.

Theorem 2.7. The ordered set $S = \{\alpha_1, \alpha_2, \ldots, \alpha_m\}$ is a basis for the vector space V if and only if the α's span V and are linearly independent.

Proof: Suppose the α's span V and are linearly independent. We know that each vector γ of V can be expressed as a linear combination of the α's in at least one way (why?). Assume there are two such representations, say

$$\gamma = s_1\alpha_1 + \ldots + s_m\alpha_m \quad \text{and} \quad \gamma = t_1\alpha_1 + \ldots + t_m\alpha_m \qquad (2.10)$$

Since vector equation (2.10) can be rewritten in the form

$$(s_1 - t_1)\alpha_1 + (s_2 - t_2)\alpha_2 + \ldots + (s_m - t_m)\alpha_m = \theta$$

the linear independence of the α's enables us to conclude that $s_j - t_j = 0$, $j = 1, 2, \ldots, m$. In other words, $s_j = t_j$ for each index j, and the two "different" linear representations displayed in (2.10) are really the same. Therefore, each element of V can be expressed as a unique linear combination of the α's, and S is a basis for V. The proof of the converse is outlined in exercise 8. ∎

Suppose the vectors $\alpha_1, \alpha_2, \ldots, \alpha_m$ span V. According to theorem 2.7, the ordered set $S = \{\alpha_1, \ldots, \alpha_m\}$ is a basis for V only if the α's are linearly independent. However, we shall soon see that even when the α's are dependent, it is always possible to "pare off" some of the members of S to form a subset which is a basis for V. In describing the construction of such a subset, we shall make use of the terminology introduced in the following definition.

Definition 2.9. The vector α_j is said to be a *dependent* member of the ordered set $S = \{\alpha_1, \alpha_2, \ldots, \alpha_m\}$ if it can be expressed in the form $\alpha_j = s_1\alpha_1 + \ldots + s_{j-1}\alpha_{j-1}$. Otherwise, α_j is called an *independent member* of S.

In other words, α_j is a dependent member of S if it can be expressed as a linear combination of vectors that precede it in the given ordering. Independent members of S cannot be represented in such a fashion. It is easy to show that the vectors $\alpha_1, \alpha_2, \ldots, \alpha_k$ are linearly independent if and only if $S = \{\alpha_1, \alpha_2, \ldots, \alpha_k\}$ contains no dependent members (see exercise 9). Moreover, if α_j is a dependent member of S, each linear combination of the α's which involves α_j can be rewritten as one which does not. For example, if $\alpha_3 = 2\alpha_1 - \alpha_2$, the expression $5\alpha_1 + 3\alpha_2 - 3\alpha_3 + \alpha_4$ can be rewritten as

$$5\alpha_1 + 3\alpha_2 - 3(2\alpha_1 - \alpha_2) + \alpha_4 = -\alpha_1 + 6\alpha_2 + \alpha_4 \qquad (2.11)$$

Consequently, the linear span of $\alpha_1, \alpha_2, \ldots, \alpha_k$ is the same as the linear span of the independent members of S (see exercise 10), These observations may be used to prove the following important theorem.

Theorem 2.8. If the vectors $\alpha_1, \ldots, \alpha_m$ span the vector space V, then the subset formed by eliminating the dependent members of the ordered set $S = \{\alpha_1, \ldots, \alpha_m\}$ is a basis for V.

EXAMPLE 7. Let $\alpha_1 = (-1, 3, 7)$, $\alpha_2 = (5, -2, 1)$, $\alpha_3 = (7, 5, 23)$, and $\alpha_4 = (0, 1, 1)$. It can be shown that $\alpha_3 = 3\alpha_1 + 2\alpha_2$ and that $\alpha_1, \alpha_2,$ and α_4 are independent members of $S = \{\alpha_1, \alpha_2, \alpha_3, \alpha_4\}$. Therefore, $T = \{\alpha_1, \alpha_2, \alpha_4\}$ is a basis for $\mathrm{Sp}\{\alpha_1, \alpha_2, \alpha_3, \alpha_4\}$.

Thanks to theorem 2.8, we now know that each finite collection of vectors whose members span the vector space V can be pared down to a basis for V. Next, we shall use theorem 2.8 to show how it is also possible to form a basis for V by adding on to a given collection of independent vectors.

Theorem 2.9. If $\alpha_1, \ldots, \alpha_k$ are linearly independent vectors in the vector space V, there exists a basis $\{\alpha_1, \ldots, \alpha_k, \alpha_{k+1}, \ldots, \alpha_n\}$ for V whose first k members are the given vectors. In other words, a linearly independent set can always be extended to form a basis.

Proof: Let $\gamma_1, \ldots, \gamma_m$ be a collection of vectors which spans V. Then the members of the ordered set $S = \{\alpha_1, \ldots, \alpha_k, \gamma_1, \ldots, \gamma_m\}$ also span V (why?), and theorem 2.8 assures us that the ordered subset T comprised of the independent members of S must be a basis for V. Since the α's are linearly independent and occur first in the ordering of S, they must be the first k vectors in T, and the proof is complete. ∎

EXAMPLE 8. The vectors $\alpha_1 = (3, -5, 2)$ and $\alpha_2 = (0, 2, 1)$ are linearly independent. To extend the ordered set $\{\alpha_1, \alpha_2\}$ to a basis for R_3, we must eliminate the dependent members of $S = \{\alpha_1, \alpha_2, \epsilon_1, \epsilon_2, \epsilon_3\}$, where the ϵ's are the standard basis vectors of R_3 (see example 2). We find that α_1, α_2, and ϵ_1 are independent and that $\epsilon_2 = -\frac{1}{9}\alpha_1 + \frac{2}{9}\alpha_2 + \frac{1}{3}\epsilon_1$ and $\epsilon_3 = \frac{2}{9}\alpha_1 + \frac{5}{9}\alpha_2 - \frac{2}{3}\epsilon_1$. Therefore, $\{\alpha_1, \alpha_2, \epsilon_1\}$ is a basis for R_3.

The situations analyzed in examples 7 and 8 are fairly easy to handle, but to be honest, it is usually quite difficult to locate the dependent members of a given set of vectors. However, in the important special case where the vectors in question are n-tuples, there is a simple computational technique for performing this task. This technique, which we call the *dependency relationship algorithm*, is discussed in the supplementary section which follows.

SUPPLEMENTARY SECTION: THE DEPENDENCY RELATIONSHIP ALGORITHM

In linear algebra, there are many different situations in which it is necessary to determine all scalars $s_1, s_2, \ldots, s_k$ such that $s_1\alpha_1 + \ldots + s_k\alpha_k = \theta$, and the object of this supplementary section is to develop an algorithm for finding such scalars in the important special case where the α's are n-tuples. The principal ideas behind this algorithm are illustrated in the following solved problem.

PROBLEM 1.   Given the vectors: $\alpha_1 = (-1, 1, 2, 5)$, $\alpha_2 = (3, -2, 0, 1)$, $\alpha_3 = (-5, 4, 4, 9)$, $\alpha_4 = (4, 0, -3, 2)$, and $\alpha_5 = (18, -7, -8, 2)$.

(i) Identify the independent members of $S = \{\alpha_1, \ldots, \alpha_5\}$.
(ii) Express each dependent member of S as a linear combination of vectors which precede it.

Solution: Equating components on both sides of the vector equation

$$s_1\alpha_1 + s_2\alpha_2 + s_3\alpha_3 + s_4\alpha_4 + s_5\alpha_5 = \theta \tag{2.12}$$

we find that this equation is equivalent to the 4×5 homogeneous linear system

$$-s_1 + 3s_2 - 5s_3 + 4s_4 + 18s_5 = 0$$

$$s_1 - 2s_2 + 4s_3 + 0s_4 - 7s_5 = 0$$

$$2s_1 + 0s_2 + 4s_3 - 3s_4 - 8s_5 = 0$$

$$5s_1 + s_2 + 9s_3 + 2s_4 + 2s_5 = 0$$

which may be solved by performing the reduction

$$
\begin{array}{ccccc}
\alpha_1 & \alpha_2 & \alpha_3 & \alpha_4 & \alpha_5
\end{array}
\qquad \downarrow \;\; \downarrow \qquad\quad \downarrow
$$

$$
\begin{bmatrix}
-1 & 3 & -5 & 4 & 18 \\
1 & -2 & 4 & 0 & -7 \\
2 & 0 & 4 & -3 & -8 \\
5 & 1 & 9 & 2 & 2
\end{bmatrix}
\longrightarrow
\begin{bmatrix}
1 & 0 & 2 & 0 & -1 \\
0 & 1 & -1 & 0 & 3 \\
0 & 0 & 0 & 1 & 2 \\
0 & 0 & 0 & 0 & 0
\end{bmatrix}
\tag{2.13}
$$

We find that the scalars s_j satisfy vector equation (2.12) if and only if

$$
\begin{aligned}
s_1 &= -2t_1 + t_2 \\
s_2 &= t_1 - 3t_2 \\
s_3 &= t_1 \\
s_4 &= -2t_2 \\
s_5 &= t_2
\end{aligned}
\tag{2.14}
$$

where t_1 and t_2 are parameters. These parametric equations can be used to answer practically any question regarding linear relationships among the α's. In particular, the vectors α_1, α_2, and α_4 are independent members of S, for if s_1, s_2, and s_4 satisfy $s_1\alpha_1 + s_2\alpha_2 + s_4\alpha_4 = \theta$, they must also satisfy (2.12) in the special case where $s_3 = s_5 = 0$. Checking (2.14), we see that $s_3 = s_5 = 0$ if and only if $t_1 = t_2 = 0$ and this in turn forces $s_1 = s_2 = s_4 = 0$. Moreover, since

$$
-2\alpha_1 + \alpha_2 + \alpha_3 = \theta \qquad [t_1 = 1, t_2 = 0]
$$

and

$$
\alpha_1 - 3\alpha_2 - 2\alpha_4 + \alpha_5 = \theta \qquad [t_1 = 0, t_2 = 1]
$$

we find that

$$
\begin{aligned}
\alpha_3 &= 2\alpha_1 - \alpha_2 \\
\alpha_5 &= -\alpha_1 + 3\alpha_2 + 2\alpha_4
\end{aligned}
\tag{2.15}
$$

Thanks to (2.15), we know that α_3 and α_5 are dependent members of S, and we also have the dependency relationships requested in part (ii).

In retrospect, we see that all our conclusions regarding the α's could have been reached by simply giving the proper interpretation to certain features of the reduced row echelon matrix on the right of reduction (2.13). To be specific, note that the independent vectors $\alpha_1, \alpha_2, \alpha_4$ correspond to the basic columns of the echelon matrix, while the entries in the non-basic columns appear as coefficients in the vector equations displayed in (2.15).

These observations suggest that any problem of the same general type as problem 1 may be solved as follows:

1. Given the ordered set $S = \{\alpha_1, \ldots, \alpha_m\}$ of n-tuples, let A denote the $n \times m$ matrix whose jth column vector is α_j. Use elementary row operations to place A in reduced row echelon form E.
2. The vector α_j is an independent member of S if the jth column of E is a basic column.
3. If the kth column of E is not a basic column, the vector α_k is a dependent member of S, and the entries of the kth column appear as coefficients when α_k is expressed as a linear combination of the independent members of the subset $\{\alpha_1, \ldots, \alpha_{k-1}\}$.

For example, if E is the matrix

$$\begin{array}{c} \text{basic columns} \\ \text{correspond to independent members of } S \end{array}$$

$$E = \begin{bmatrix} 1 & 0 & -3 & 0 & 0 & -8 & 0 \\ 0 & 1 & 4 & 0 & 0 & 1 & 0 \\ 0 & 0 & 0 & 1 & 0 & 3 & 0 \\ 0 & 0 & 0 & 0 & 1 & 0 & 1 \\ 0 & 0 & 0 & 0 & 0 & 0 & 0 \end{bmatrix}$$

$$\text{non-basic columns}$$

we can draw the following conclusions about the corresponding collection $S = \{\alpha_1, \ldots, \alpha_7\}$ of 5-tuples:

1. $\alpha_1, \alpha_2, \alpha_4$, and α_5 are the independent members of S
2. $\alpha_3 = -3\alpha_1 + 4\alpha_2$ \qquad\qquad (third column entries)
   $\alpha_6 = -8\alpha_1 + \alpha_2 + 3\alpha_4 + 0\alpha_5$ \quad (sixth column entries
   $\alpha_7 = 0\alpha_1 + 0\alpha_2 + 0\alpha_4 + \alpha_5$ \quad (seventh column entries)

We shall refer to the process described in the preceding paragraph as the *dependency relationship algorithm*. This algorithm is extremely useful, and we shall employ it for many different purposes. The following solved problem shows how the algorithm can be used to perform several different tasks at once.

PROBLEM 2. Let V denote the linear span of the vectors $\alpha_1 = (-1, 3, 0, 2)$, $\alpha_2 = (5, 1, 1, -3)$, $\alpha_3 = (3, 7, 1, 1)$, $\alpha_4 = (0, 1, -2, 4)$, $\alpha_5 = (7, 12, 0, 4)$.

(i) Find a basis for V.

(ii) Determine whether or not the vector $\beta = (-17, 7, -11, 29)$ belongs to V. What about $\gamma = (1, 1, 0, 1)$?

Solution: We first perform the reduction

$$
\begin{array}{ccccccc}
\alpha_1 & \alpha_2 & \alpha_3 & \alpha_4 & \alpha_5 & \beta & \gamma \\
\end{array}
$$

$$
\begin{bmatrix}
-1 & 5 & 3 & 0 & 7 & -17 & 1 \\
3 & 1 & 7 & 1 & 12 & 7 & 1 \\
0 & 1 & 1 & -2 & 0 & -11 & 0 \\
2 & -3 & 1 & 4 & 4 & 29 & 1
\end{bmatrix}
\longrightarrow
\begin{bmatrix}
1 & 0 & 2 & 0 & 3 & 2 & 0 \\
0 & 1 & 1 & 0 & 2 & -3 & 0 \\
0 & 0 & 0 & 1 & 1 & 4 & 0 \\
0 & 0 & 0 & 0 & 0 & 0 & ①
\end{bmatrix}
$$

"bad" entry

As indicated by the arrows, the basic columns are those which correspond to $\alpha_1, \alpha_2, \alpha_4$, and γ. It follows that α_1, α_2, and α_4 are the only independent members of the ordered set $S = \{\alpha_1, \ldots, \alpha_5\}$, and thus, $\{\alpha_1, \alpha_2, \alpha_4\}$ is a basis for V. Moreover, since the sixth column is not basic, we conclude that β is contained in the linear span of the α's, and the entries of this column tell us that $\beta = 2\alpha_1 - 3\alpha_2 + 4\alpha_4$. Finally, since the seventh column is basic, the vectors $\alpha_1, \alpha_2, \alpha_4$, and γ are linearly independent, and γ cannot be contained in V (why not?).

In our final solved problem, we show how the dependency relationship algorithm may be used to extend a linearly independent collection of n-tuples to a basis for all of R_n.

PROBLEM 3.   Find a basis for R_4 which includes the independent members of the ordered set $S = \{\alpha_1, \alpha_2, \alpha_3, \alpha_4\}$, where $\alpha_1 = (-5, 1, -3, 1)$, $\alpha_2 = (2, 4, 5, -7)$, $\alpha_3 = (-11, 11, 1, -11)$, and $\alpha_4 = (4, -5, -2, 6)$.

Solution: The vectors $\alpha_1, \alpha_2, \alpha_3, \alpha_4, \epsilon_1, \epsilon_2, \epsilon_3, \epsilon_4$ certainly span R_4 (recall that the ϵ_j are the standard basis vectors of R_4). Therefore, the reduction

$$
\begin{array}{cccccccc}
\alpha_1 & \alpha_2 & \alpha_3 & \alpha_4 & \epsilon_1 & \epsilon_2 & \epsilon_3 & \epsilon_4
\end{array}
$$

$$
\begin{bmatrix}
-5 & 2 & -11 & 4 & 1 & 0 & 0 & 0 \\
1 & 4 & 11 & -5 & 0 & 1 & 0 & 0 \\
-3 & 5 & 1 & -2 & 0 & 0 & 1 & 0 \\
1 & -7 & -11 & 6 & 0 & 0 & 0 & 1
\end{bmatrix}
$$

$$
\longrightarrow
\begin{bmatrix}
1 & 0 & 3 & 0 & 1 & 0 & -\frac{40}{16} & -\frac{24}{16} \\
0 & 1 & 2 & 0 & 1 & 0 & -\frac{34}{16} & -\frac{22}{16} \\
0 & 0 & 0 & 1 & 1 & 0 & -\frac{33}{16} & -\frac{19}{16} \\
0 & 0 & 0 & 0 & 0 & 1 & \frac{11}{16} & \frac{17}{16}
\end{bmatrix}
$$

enables us to conclude that $\alpha_1, \alpha_2, \alpha_4$ are the independent members of S and that $\{\alpha_1, \alpha_2, \alpha_4, \epsilon_2\}$ is a basis for R_4. Note that $\epsilon_1 = \alpha_1 + \alpha_2 + \alpha_4$, as indicated by the entries in the fifth column of the reduced row echelon matrix.

We invite the reader to test his understanding of the dependency relationship algorithm by using it to solve exercise 6 of section 2.4 and exercises 1–4 and 13 in the exercise set which follows.

Exercise Set 2.5

1. In each of the following cases, determine whether or not the given collection of vectors is linearly independent.
   (i) $\alpha_1 = (-7, 1, 3), \alpha_2 = (4, 1, -2), \alpha_3 = (1, 3, -1)$
   (ii) $p_1(t) = -7t^2 + t + 3, p_2(t) = 4t^2 + t - 2, p_3(t) = t^2 + 3t - 1$
   (iii) $\alpha_1 = (1, 1, 0, 1), \alpha_2 = (1, 0, 1, 0), \alpha_3 = (5, -3, 2, 7),$
   $\alpha_4 = (-2, 5, -1, -5)$

2. Which of the following are bases for R_3?
   (i) $\{\alpha_1, \alpha_2, \alpha_3\}$ where $\alpha_1 = (1, -1, 4), \alpha_2 = (1, 0, 1), \alpha_3 = (2, 1, 0)$
   (ii) $\{\beta_1, \beta_2\}$ where $\beta_1 = (5, 3, -2), \beta_2 = (7, 1, 1)$
   (iii) $\{\gamma_1, \gamma_2, \gamma_3, \gamma_4\}$ where $\gamma_1 = (-3, 1, 5), \gamma_2 = (5, -2, -4),$
   $\gamma_3 = (3, -2, 8),$ and $\gamma_4 = (1, 0, 1)$

3. Let V denote the linear span of the vectors $\alpha_1 = (-3, 1, 0, -4), \alpha_2 = (4, -3, 5, 2), \alpha_3 = (1, -7, 20, -12),$ and $\alpha_4 = (11, 3, -2, 8)$.
   (i) Find a basis for V.
   (ii) Express the vector $\beta = (8, 9, -17, 14)$ as a linear combination of the α's.

4. Find a basis for R_4 which includes the vectors $\alpha_1 = (-2, 3, 2, -1)$ and $\alpha_2 = (2, -8, -6, 3)$.

5. If the vectors $\alpha_1, \alpha_2, \alpha_3$ are linearly independent, find scalars s_1, s_2 which satisfy

$$s_1(\alpha_1 - 2\alpha_2 + 3\alpha_3) - 3s_2(-2\alpha_1 + \alpha_2 - 2\alpha_3) = 8\alpha_1 - 7\alpha_2 + 12\alpha_3$$

6. Prove the following statements:
   (i) Any collection of vectors which includes θ is linearly dependent. Thus, θ cannot be contained in a basis.
   (ii) If the vectors $\alpha_1, \alpha_2, \ldots, \alpha_k$ are contained in a linearly independent collection, they themselves must be linearly independent.
   (iii) If the vectors $\beta_1, \beta_2, \ldots, \beta_m$ are linearly dependent, then any collection of vectors which contains the β's must also be linearly dependent.

7. (i) Show that two planar vectors α and β are linearly independent if and only if they are not parallel.
   (ii) Show that three vectors α, β, γ which lie in the same plane must be linearly dependent.

8. Let $S = \{\alpha_1, \alpha_2, \ldots, \alpha_m\}$ be a basis for the vector space V. Complete the proof of theorem 2.7 by showing that the α's must span V and be linearly independent.
   Hint: First note that the condition for linear independence is the same as saying that θ can be expressed as a unique linear combination of the α's.

9. (i) If the vectors $\alpha_1, \alpha_2, \ldots, \alpha_m$ are linearly dependent, show that there exists an index j such that $\alpha_j = s_1\alpha_1 + \ldots + s_{j-1}\alpha_{j-1}$.
   (ii) Show that the α's are linearly independent if and only if the ordered set $S = \{\alpha_1, \ldots, \alpha_m\}$ contains no dependent members.

10. Let $S = \{\alpha_1, \alpha_2, \alpha_3, \alpha_4, \alpha_5\}$ be a sequence of vectors in the vector space V. Suppose α_1, α_2, and α_4 are independent members of S and that $\alpha_3 = 3\alpha_1 - 5\alpha_2$ and $\alpha_5 = 2\alpha_1 + \alpha_2 - 4\alpha_4$.
   (i) Rewrite the vector expression

$$\gamma = 4\alpha_1 - 7\alpha_2 + 2\alpha_3 + \alpha_4 - 8\alpha_5$$

   as a linear combination of α_1, α_2, and α_4.
   (ii) Generalize part (i) by showing that each linear combination of the α's is contained in Sp $\{\alpha_1, \alpha_2, \alpha_4\}$. Why does this mean that Sp $\{\alpha_1, \alpha_2, \alpha_4\} =$ Sp $\{\alpha_1, \alpha_2, \alpha_3, \alpha_4, \alpha_5\}$?

11. If the vectors $\alpha_1, \alpha_2, \ldots, \alpha_k$ are linearly independent, show that β is contained in Sp $\{\alpha_1, \ldots, \alpha_k\}$, if and only if $\alpha_1, \alpha_2, \ldots, \alpha_k, \beta$ are linearly dependent.

*12. Show that the functions $f_1(x) = e^{2x}$ and $f_2(x) = e^{-3x}$ are linearly independent by completing the steps in the following outline:
   (i) If $s_1 e^{2x} + s_2 e^{-3x} = 0$ for all x, show that the equation $2s_1 e^{2x} - 3s_2 e^{-3x} = 0$ must also be satisfied.
   Hint: Use differentiation.
   (ii) In particular, when $x = 0$ show that the two equations which appear in part (i) have the unique solution $s_1 = s_2 = 0$.

13. For a certain collection of 5-tuples $\{\alpha_1, \ldots, \alpha_6\}$, it is known that α_1, α_4, and α_6 are linearly independent and that $\alpha_4 = 4\alpha_1 + 2\alpha_2 + \alpha_3$ and $\alpha_5 = 2\alpha_1 - 3\alpha_2 + \alpha_6$. It is known that if $E = (e_{ij})$ is the reduced row echelon form of the 5×6 matrix whose jth column vector is α_j, then $e_{14} = -4, e_{22} = 0$, and $e_{23} = 1$, as indicated below:

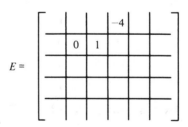

$$E =$$

   (i) Use this information to show that α_1, α_3, and α_5 are the independent members of the ordered set $\{\alpha_1, \ldots, \alpha_6\}$. (Note that even though α_1, α_4, and α_6 are *linearly independent*, α_4 and α_6 are *dependent* members of the ordered set of α's. Why is this not a contradiction?)
   (ii) Fill in the remaining 27 entries of E.

14. Let $\{\alpha_1, \ldots, \alpha_n\}$ be a basis for the vector space V.
   (i) If $S = $ Sp $\{\alpha_1, \ldots, \alpha_k\}$ and $T = $ Sp $\{\alpha_{k+1}, \ldots, \alpha_n\}$, show that $S \cap T$ is the zero vector.

(ii) Show that $S + T$ is V.

(iii) If γ is a vector in V, show that there exist *unique* vectors σ in S and τ in T for which $\gamma = \sigma + \tau$.

2.6 DIMENSION

Each finitely generated vector space has many different bases, and the object of this section is to show that all such bases have at least one thing in common —the number of vectors they contain. This property, which is often called the invariance of dimensionality, is one of the most important features of vector spaces, and after proving that the property is generally valid, we shall examine some of its many implications.

The key to proving the invariance of dimensionality property lies in a simple observation: Each collection of vectors which contains more elements than a basis must be linearly dependent. A more precise statement of this principle and its proof are given in the following lemma.

Lemma 2.1. Let $\alpha_1, \ldots, \alpha_n$ be a basis for the vector space V, and let $\beta_1, \ldots, \beta_m$ be vectors in V. If $m > n$, the β's must be linearly dependent.

Proof: We shall provide a proof for the special case where $n = 2$ and $m = 3$. The proof for more general m and n is completely analogous and is left to the reader.

Specifically, let $S = \{\alpha_1, \alpha_2\}$ be a basis for V and let $\beta_1, \beta_2, \beta_3$ be vectors in V. Since S is a basis for V, there exist unique scalars a_{ij} such that

$$\beta_1 = a_{11}\alpha_1 + a_{21}\alpha_2$$

$$\beta_2 = a_{12}\alpha_1 + a_{22}\alpha_2$$

$$\beta_3 = a_{13}\alpha_1 + a_{23}\alpha_2$$

Therefore, if s_1, s_2, s_3 are scalars such that $s_1\beta_1 + s_2\beta_2 + s_3\beta_3 = \theta$, we have

$$s_1(a_{11}\alpha_1 + a_{21}\alpha_2) + s_2(a_{12}\alpha_1 + a_{22}\alpha_2) + s_3(a_{13}\alpha_1 + a_{23}\alpha_2) = \theta$$

Regrouping terms, we obtain the equation

$$(s_1a_{11} + s_2a_{12} + s_3a_{13})\alpha_1 + (s_1a_{21} + s_2a_{22} + s_3a_{23})\alpha_2 = \theta$$

and since the α's are linearly independent, it follows that

$$a_{11}s_1 + a_{12}s_2 + a_{13}s_3 = 0$$

$$a_{21}s_1 + a_{22}s_2 + a_{23}s_3 = 0$$

(2.16)

When (2.16) is regarded as a 2×3 homogeneous linear system in the variables s_1, s_2, s_3, it becomes apparent that there exist non-trivial solutions, since the system has more variables than equations. In other words, there are scalars s_1, s_2, s_3 not all zero which satisfy (2.16) and hence satisfy $s_1\beta_1 + s_2\beta_2 + s_3\beta_3 = \theta$. Consequently, the β's must be linearly dependent, as claimed. ∎

If the vector space V is known to have an ordered basis with n elements and if the vectors $\beta_1, \beta_2, \ldots, \beta_m$ are linearly independent, then according to lemma 2.1, we must have $m \leq n$. Using this form of the lemma, it is now an easy matter to prove the invariance of dimensionality property.

Theorem 2.10.   All the bases of a vector space V contain the same number of elements.

Proof: Let $S = \{\alpha_1, \ldots, \alpha_n\}$ and $T = \{\beta_1, \ldots, \beta_m\}$ be bases for V. Since S is a basis and $\beta_1, \ldots, \beta_m$ are linearly independent vectors, the lemma tells us that $m \leq n$. Turning things around, since T is a basis and the vectors $\alpha_1, \ldots, \alpha_n$ are linearly independent, we also have $n \leq m$. Combining these inequalities, we find that $m = n$, which tells us that each pair of bases for V have the same number of elements, and this is logically equivalent to saying that all such bases have this property. ∎

Definition 2.10.   A vector space V is said to have *dimension n* if all its bases contain n elements. The trivial vector space which contains only θ is said to have dimension 0.

If V has dimension n, we shall say that it is n-dimensional and shall write dim $V = n$. In certain contexts, we shall also find it convenient to associate V with its dimension by writing V_n. Henceforth, we shall refer to any vector space that has a finite basis as being *finite-dimensional*, while those that have no such basis will be called *infinite-dimensional*. Note that V is finite-dimensional if and only if it is finitely generated (why?).

The dimension of a given vector space can be determined by simply counting the elements in any one of its bases, and this is especially easy when the vector space has a "standard" or "natural" basis, as in the following examples.

EXAMPLE 1.   The vector space R_n is n-dimensional, since its standard basis $\{\epsilon_1, \ldots, \epsilon_n\}$ has n elements (see example 2, section 2.5).

EXAMPLE 2.   Since the natural basis for the polynomial space P_n is $\{t^n, t^{n-1}, \ldots, t, 1\}$, it follows that dim $P_n = n + 1$.

EXAMPLE 3.   In $_mR_n$, the vector space of all $m \times n$ real matrices, the standard basis contains mn elements (see example 4, section 2.5). Therefore, $_mR_n$ has dimension mn. For instance, $_2R_3$ is a 6-dimensional vector space.

EXAMPLE 4.   As might be expected, 3-space has dimension 3. A 2-dimensional subspace of 3-space is a plane containing the origin O, while a 1-dimensional subspace is a line through O.

EXAMPLE 5.   The polynomial space P_∞ and the function spaces $F[0, 1]$, $C[0, 1]$, and $D[0, 1]$ are not finitely generated and are thus infinite-dimensional.

EXAMPLE 6.   Using theoretical tools usually developed in advanced calculus, it can be shown that the solution space of the homogeneous differential equation

$$a_n y^{(n)} + a_{n-1} y^{(n-1)} + \ldots + a_2 y'' + a_1 y' + a_0 y = 0$$

must have dimension n. (Also, see supplementary exercise 16 of chapter 4).

For instance, the solution space of the 3rd order differential equation $y''' - 4y'' - 3y' + 18y = 0$ has dimension 3. The reader may wish to convince himself of this fact by verifying that each solution of the given equation can be expressed as a unique linear combination of the functions $y_1 = e^{-2x}$, $y_2 = e^{3x}$, and $y_3 = xe^{3x}$.

Ordinarily, to show that a given vector sequence S is a basis for the vector space V, we must verify that its members are linearly independent *and* that they span V. However, if V has dimension n and S contains n members, our next theorem shows that it is enough to verify *either* of these two conditions.

Theorem 2.11.   Let $S = \{\alpha_1, \ldots, \alpha_n\}$ be a collection of n vectors in the n-dimensional vector space V. Then

(i) S is a basis if the α's span V.

or

(ii) S is a basis if the α's are linearly independent.

Proof: We shall prove (i) here and outline the proof of (ii) in exercise 6.

If the α's span V, then according to theorem 2.8, the subset formed by eliminating the dependent members of S is a basis for V. However, since dim $V = n$, this subset must contain n elements, and it follows that none of the α's is dependent. Therefore, the α's are linearly independent and since we already know that they span V, we conclude that S must be a basis. ∎

When the dimension of V is known, the criterion established in part (ii) of theorem 2.11 can sometimes be used to simplify the process of extending a linearly independent collection of vectors to a basis for V. Consider the following example.

EXAMPLE 7.   The polynomials $p_1(t) = t^2 - 1$ and $p_2(t) = t^2 + 3t$ are clearly linearly independent. Therefore, since P_2 has dimension 3, we can obtain a basis for P_2 which includes p_1 and p_2 by simply locating another polynomial which cannot be expressed as a linear combination of p_1 and p_2. By inspection, we see that $p_3(t) = t$, among others, satisfies this requirement.

Often it is quite difficult to find the dimension of a given vector space V. However, in the special case where V is spanned by a finite collection of n-tuples, its dimension can be determined by the dependency relationship algorithm, as illustrated in the following solved problem.

PROBLEM 1.   Find the dimension of the vector space V spanned by the 4-tuples $\alpha_1 = (1, 1, 0, 1)$, $\alpha_2 = (-3, 7, 3, 1)$, $\alpha_3 = (-1, 9, 3, 3)$, and $\alpha_4 = (-5, 5, 3, -1)$.

Solution: As in the dependency relationship algorithm, we begin by applying elementary row operations to the 4×4 matrix whose jth column vector is α_j. The downward phase of the reduction yields

$$
\begin{array}{cccc}
\alpha_1 & \alpha_2 & \alpha_3 & \alpha_4 \\
\end{array}
\qquad \downarrow \qquad \downarrow
$$

$$
\begin{bmatrix}
1 & -3 & -1 & -5 \\
1 & 7 & 9 & 5 \\
0 & 3 & 3 & 3 \\
1 & 1 & 3 & -1
\end{bmatrix}
\longrightarrow
\begin{bmatrix}
1 & -3 & -1 & -5 \\
0 & 1 & 1 & 1 \\
0 & 0 & 0 & 0 \\
0 & 0 & 0 & 0
\end{bmatrix}
$$

From the form of the matrix on the right, it is clear that $\{\alpha_1, \alpha_2\}$ is a basis for V (why?), and we conclude that V has dimension two.

In our next solved problem, we illustrate a useful general procedure for finding a basis for the solution space of a homogeneous $m \times n$ linear system.

PROBLEM 2.   Find a basis for the solution space of the 4×6 linear system

$$-x_1 + 5x_2 + 3x_3 \qquad + 7x_5 - 17x_6 = 0$$

$$3x_1 + x_2 + 7x_3 + x_4 + 12x_5 + 7x_6 = 0$$

$$x_2 + x_3 - 2x_4 \qquad - 11x_6 = 0 \tag{2.17}$$

$$2x_1 - 3x_2 + x_3 + 4x_4 + 4x_5 + 29x_6 = 0$$

Solution: The reduction

$$
\begin{bmatrix}
-1 & 5 & 3 & 0 & 7 & -17 \\
3 & 1 & 7 & 1 & 12 & 7 \\
0 & 1 & 1 & -2 & 0 & -11 \\
2 & -3 & 1 & 4 & 4 & 29
\end{bmatrix}
\longrightarrow
\begin{bmatrix}
1 & 0 & 2 & 0 & 3 & 2 \\
0 & 1 & 1 & 0 & 2 & -3 \\
0 & 0 & 0 & 1 & 1 & 4 \\
0 & 0 & 0 & 0 & 0 & 0
\end{bmatrix}
\quad(2.18)
$$

tells us that the given system has the following parametric solution:

$$x_1 = -2s_1 - 3s_2 - 2s_3$$

$$x_2 = -s_1 - 2s_2 + 3s_3$$

$$x_3 = s_1$$

$$x_4 = -s_2 - 4s_3$$

$$x_5 = s_2$$

$$x_6 = s_3$$

In other words, the column 6-tuple $X = (x_j)$ is a solution of (2.17) if and only if there exist real numbers s_1, s_2, s_3 such that

$$
X =
\begin{bmatrix}
x_1 \\ x_2 \\ x_3 \\ x_4 \\ x_5 \\ x_6
\end{bmatrix}
= s_1
\begin{bmatrix}
-2 \\ -1 \\ 1 \\ 0 \\ 0 \\ 0
\end{bmatrix}
+ s_2
\begin{bmatrix}
-3 \\ -2 \\ 0 \\ -1 \\ 1 \\ 0
\end{bmatrix}
+ s_3
\begin{bmatrix}
-2 \\ 3 \\ 0 \\ -4 \\ 0 \\ 1
\end{bmatrix}
$$

Thus, the vectors $\eta_1 = (-2, -1, 1, 0, 0, 0)$, $\eta_2 = (-3, -2, 0, -1, 1, 0)$, and $\eta_3 = (-2, 3, 0, -4, 0, 1)$ span the solution space of (2.17), and since these vectors are clearly linearly independent (why?), the desired basis is $\{\eta_1, \eta_2, \eta_3\}$.   THM, 2.7

In general, a basis for the solution space V of a given $m \times n$ homogeneous linear system may be found in this fashion. If r is the number of basic columns in the reduced row echelon form of the coefficient matrix of the system, there will be $p = n - r$ solutions analogous to η_1, η_2, η_3 of problem 2. Since these vectors form a basis for V, we have dim $V = n - r$. Note that the given system has a unique solution if and only if $r = n$ (why?). For future reference, the n-tuple η_k obtained by this method will be called the kth *fundamental solution* of the system, and the ordered set $\{\eta_1, \eta_2, \ldots, \eta_p\}$ will be called the *fundamental basis* for the solution space. In the special case where $p = 0$, the solution space is just the zero vector, which has no basis.

The set of all solutions of a non-homogeneous linear system is not a vector space (recall exercise 15, section 2.4), but it can still be characterized by the methods of this chapter. Such a characterization is provided by our next theorem. Incidentally, in stating this result, we find it convenient to say that two $m \times n$ linear systems are *related* if they have the same coefficient matrix. For example, the systems

$$2x_1 - 3x_2 + x_3 = 1 \quad \text{and} \quad 2x_1 - 3x_2 + x_3 = 0$$

$$-4x_1 + x_2 + 7x_3 = -8 \qquad\qquad -4x_1 + x_2 + 7x_3 = 0$$

are related.

Theorem 2.12. If a given $m \times n$ non-homogeneous linear system has at least one solution ρ_0, then the n-tuple ρ is a solution if and only if $\rho = \rho_0 + \eta$, where η is a solution of the related homogeneous system.

Proof: The non-homogeneous system can be written in the form $AX = \beta$, where A is the coefficient matrix, β is the column m-tuple of right side constants, and X is an n-tuple of variables. Suppose ρ_0 satisfies $A\rho_0 = \beta$. If η satisfies the homogeneous equation $A\eta = O$ and if $\rho = \rho_0 + \eta$, we find that

$$A\rho = A(\rho_0 + \eta) = A\rho_0 + A\eta = \beta + O = \beta$$

and it follows that β is also a solution of the non-homogeneous system. Conversely, if ρ is a solution, we have

$$A(\rho - \rho_0) = A\rho - A\rho_0 = \beta - \beta = O$$

and $\rho - \rho_0$ must be a solution of the related homogeneous system. ∎

If S is a subset of the vector space V and ρ is a fixed vector, we write $\rho + S$ to denote the collection of all vectors γ which can be expressed as $\gamma = \rho + \sigma$, where σ is an element of S. According to theorem 2.12, if T is the solution set of a consistent non-homogeneous $m \times n$ linear system, then $T = \rho_0 + S$, where ρ_0 is a fixed solution and S is the solution space of the related homogeneous system. Even though T is not a subspace of R_n, its geometrical structure is still quite similar to that of S. In fact, T can be regarded as an exact copy of S which has been translated away from the origin, as indicated in figure 2.17.

In section 2.5, we found that the geometric configuration which corresponds to a proper subspace S of 3-space must be either a single point (the

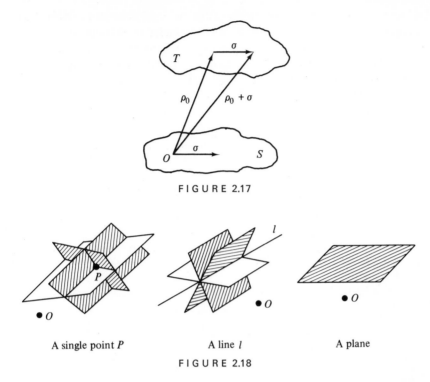

FIGURE 2.17

A single point P A line l A plane

FIGURE 2.18

origin O), a line through O, or a plane containing O. Therefore, as a result of the observations made in the preceding paragraph, we see that the solution set of a consistent non-homogeneous linear system in three unknowns must correspond to either a point, a line, or a plane which has been translated away from O (see figure 2.18). It is also interesting to examine the geometric configurations associated with inconsistent systems. For instance, an inconsistent 3×3 linear system must fall into one of the categories depicted in figure 2.19:

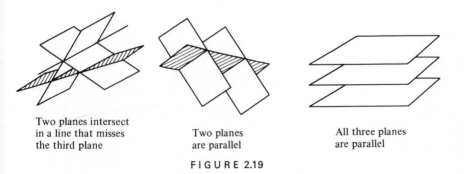

Two planes intersect
in a line that misses
the third plane

Two planes
are parallel

All three planes
are parallel

FIGURE 2.19

So far, we have found that most of the concepts discussed earlier in connection with the plane and 3-space can be extended to abstract vector spaces. In the next section, we shall continue this development by showing that it is possible to coordinatize all vector spaces in much the same way that coordinates were introduced into the plane and 3-space in section 2.2.

Exercise Set 2.6

1. In each of the following cases, find the dimension of the linear span of the given vectors.

   (i) $\alpha_1 = (3, -2, 5)$, $\alpha_2 = (-4, 1, 3)$, $\alpha_3 = (2, -3, 13)$

   (ii) $\beta_1 = (1, 0, 1, 1)$, $\beta_2 = (0, 1, 0, 0)$, $\beta_3 = (-3, 1, 1, 4)$,
   $\beta_4 = (11, 0, 3, -3)$

   (iii) $\gamma_1 = (-3, 5, -1, 2)$, $\gamma_2 = (4, 4, 1, -3)$, $\gamma_3 = (10, -6, 3, -7)$,
   $\gamma_4 = (3, -5, 1, -2)$

2. In each of the following cases, either find the fundamental solutions of the given linear system or show that the system has a unique solution.

   (i) $\begin{aligned} x_1 - 3x_2 + x_3 &= 0 \\ 2x_1 + x_2 + 3x_3 &= 0 \\ x_1 + 4x_2 + 2x_3 &= 0 \\ x_1 + 18x_2 + 4x_3 &= 0 \end{aligned}$ (ii) $\begin{aligned} 2x_1 - 5x_2 + 3x_3 + 7x_4 &= 0 \\ -x_1 + 3x_2 + x_3 + 2x_4 &= 0 \end{aligned}$

   (iii) $\begin{aligned} 2x_1 + x_2 \quad\quad - 4x_4 + 3x_5 &= 0 \\ -x_1 + 3x_2 + x_3 + 2x_4 - 5x_5 &= 0 \\ 4x_1 + 9x_2 + 2x_3 - 8x_4 - x_5 &= 0 \\ 6x_1 + 17x_2 + 4x_3 - 12x_4 - 5x_5 &= 0 \end{aligned}$

3. In each of the following cases, find a basis for the solution space of the given system and describe the geometric configuration to which it corresponds.

   (i) $3x_1 - 5x_2 + x_3 = 0$ (ii) $\dfrac{x_1}{7} = \dfrac{x_2}{-3} = \dfrac{x_3}{4}$

   (iii) $\begin{aligned} 2x_1 + x_2 - x_3 &= 0 \\ 7x_1 + 2x_2 + 4x_3 &= 0 \end{aligned}$ (iv) $\begin{aligned} -2x_1 + x_2 + x_3 &= 0 \\ 3x_1 + x_2 - 6x_3 &= 0 \\ x_1 - 3x_2 + 4x_3 &= 0 \end{aligned}$

4. The coefficient matrix of a certain 4×5 homogeneous linear system is row equivalent to

$$\begin{bmatrix} 1 & -3 & 0 & -5 & 2 \\ 0 & 0 & 1 & 7 & -1 \\ 0 & 0 & 0 & 0 & 0 \\ 0 & 0 & 0 & 0 & 0 \end{bmatrix}$$

   (i) Find the fundamental solutions of the system and determine the dimension of its solution space.

   (ii) Determine whether or not $\eta = (3, 2, 6, -1, -1)$ is also a solution of the system.

   (iii) A non-homogeneous system related to the system under consideration is known to have the solution $p_0 = (4, -1, 14, 6, -5)$. Is the vector $p = (-3, 3, 15, -2, 1)$ also a solution of this system?

5. Describe the six different geometric configurations which can correspond to a given 3×3 non-homogeneous linear system and provide examples of specific systems which correspond to each configuration (see figures 2.18 and 2.19).

6. Let $\alpha_1, \alpha_2, \ldots, \alpha_n$ be a collection of n linearly independent vectors in the n-dimensional vector space V.
   (i) If β is a vector in V, show that the vectors $\alpha_1, \alpha_2, \ldots, \alpha_n, \beta$ are linearly dependent.
   Hint: Use lemma 2.1.
   (ii) Complete the proof of theorem 2.11 by showing that $S = \{\alpha_1, \ldots, \alpha_n\}$ is a basis for V.
   Hint: Use theorem 2.9.

7. Let V denote the subspace of P_2 which is spanned by the polynomials $p_1(t) = t^2 + 1$, $p_2(t) = -3t^2 + 2t + 1$, $p_3(t) = t + 2$, and $p_4(t) = t^2 + t + 3$.
   (i) Show that $s_1 p_1(t) + s_2 p_2(t) + s_3 p_3(t) + s_4 p_4(t) = 0$ for all t if and only if

$$s_1 - 3s_2 + 0s_3 + s_4 = 0$$

$$0s_1 + 2s_2 + s_3 + s_4 = 0$$

$$s_1 + s_2 + 2s_3 + 3s_4 = 0$$

   (ii) Use part (i) to locate the dependent members of the ordered set $\{p_1, p_2, p_3, p_4\}$.
   (iii) What is the dimension of V?
   (iv) In what way is this problem like that of finding the dimension of the linear span of the 3-tuples $(1, 0, 1)$, $(-3, 2, 1)$, $(0, 1, 2)$, and $(1, 1, 3)$?

8. Find the dimension of the subspace of $_2R_2$ which is spanned by the matrices

$$A_1 = \begin{bmatrix} 4 & 1 \\ 5 & 14 \end{bmatrix}, \quad A_2 = \begin{bmatrix} -1 & 0 \\ -1 & -3 \end{bmatrix}, \quad A_3 = \begin{bmatrix} 0 & 3 \\ 3 & 6 \end{bmatrix}, \quad A_4 = \begin{bmatrix} -5 & 7 \\ 2 & -1 \end{bmatrix}$$

   Hint: Modify the approach outlined in problem 7. In what sense is this problem the same as finding the dimension of the linear span of $(4, 1, 5, 14)$, $(-1, 0, -1, -3)$, $(0, 3, 3, 6)$, and $(-5, 7, 2, -1)$?

9. A certain 5×6 homogeneous linear system has the fundamental solutions $\eta_1 = (-7, 3, 1, 0, 0, 0)$, $\eta_2 = (5, -1, 0, 1, 0, 0)$, and $\eta_3 = (1, -3, 0, 0, -2, 1)$. Use this information to determine the entries of the reduced row echelon matrix which is row equivalent to the coefficient matrix of the system.

*10. Let A be a fixed 3×3 matrix, and let S denote the ordered set of matrices $\{I_3, A, A^2, \ldots, A^9\}$.
   (i) Show that S contains a dependent member.
   Hint: Recall that $_3R_3$ has dimension 9 and use lemma 2.1.
   (ii) Use part (i) to show that there exists an integer $k \le 9$ such that $A^k = s_0 I_3 + s_1 A + \ldots + s_{k-1}A^{k-1}$, but the matrices $I_3, A, \ldots, A^{k-1}$ are linearly independent.
   (iii) Generalize this argument to show that each $n \times n$ matrix A must satisfy an equation of the form $A^m + s_{m-1}A^{m-1} + \ldots + s_1 A + s_0 I_n = 0$,

where $m \leqslant n^2$. Incidentally, using more sophisticated methods, it can be shown that the smallest such m never exceeds n.

*11. If S and T are subspaces of the vector space V, it can be shown that

$$\dim S + \dim T = \dim (S + T) + \dim (S \cap T)$$

(see supplementary exercise 8 at the end of this chapter). Use this formula to justify the following statements about subspaces of 3-space:

  (i) If S and T are two distinct planes which contain the origin O, then $S + T$ must be all of 3-space.

  (ii) If S is a line through O and T is a plane which does not contain S, then $S + T$ is all of 3-space.

  (iii) If S and T are two distinct lines through O, then $S + T$ is a plane.

2.7   COORDINATES

In section 2.2, we found that if $S = \{\alpha_1, \alpha_2\}$ is a basis for the plane, then each planar vector γ can be dealt with as if it were the ordered pair (s_1, s_2), where the s_j are the uniquely determined coefficients that appear in the linear representation $\gamma = s_1 \alpha_1 + s_2 \alpha_2$. The resulting coordinatization of the plane is extremely useful, for it enables us to manipulate vector expressions and solve vector equations by means of ordinary arithmetic and algebra. The object of this section is to explore the use of coordinates in vector spaces other than the plane, and it is appropriate to begin with the following definition.

Definition 2.11.   Let $S = \{\alpha_1, \alpha_2, \ldots, \alpha_n\}$ be a basis for the vector space V. Then the vector γ of V is said to have S-coordinates $(s_1, s_2, \ldots, s_n)$ if $\gamma = s_1 \alpha_1 + s_2 \alpha_2 + \ldots + s_n \alpha_n$. We denote this relationship by writing $[\gamma]_S = (s_1, s_2, \ldots, s_n)$.

The vector spaces R_n, P_n, and $_m R_n$ all have standard or natural bases, and we shall refer to the coordinates associated with these bases as the *standard coordinates* of their respective spaces.

Example 1.   It should come as no great surprise to learn that the vector $\gamma = (-1, 3, 5)$ has standard coordinates $(-1, 3, 5)$. However, the reader should guard against thinking that this is the only possible way to represent γ by an ordered 3-tuple. For instance, if $\alpha_1 = (7, -1, 4)$, $\alpha_2 = (4, 0, 1)$, and $\alpha_3 = (-3, 5, 0)$, the set $S = \{\alpha_1, \alpha_2, \alpha_3\}$ is a basis for R_3, and since $\gamma = 2\alpha_1 - 3\alpha_2 + \alpha_3$ (verify this), it follows that $[\gamma]_S = (2, -3, 1)$.

Example 2.   In P_2, the polynomial $p(t) = at^2 + bt + c$ has standard coordinates (a, b, c). It is important to remember that a change in the order of vectors in a basis will change the coordinates of all vectors. Thus, the polynomial

$13t^2 - 7t + 5$ has standard coordinates $(13, -7, 5)$, but with respect to the basis $S_1 = \{1, t, t^2\}$, its coordinates are $(5, -7, 13)$. The polynomial is the same in both cases, but its representation changes.

EXAMPLE 3.   The standard basis for $_2R_3$ is described in example 4 of section 2.5. We find that the matrix

$$A = \begin{bmatrix} 8 & -3 & 0 \\ 2 & 5 & -7 \end{bmatrix}$$

has standard coordinates $(8, -3, 0, 2, 5, -7)$.

The following solved problem illustrates how the dependency relationship algorithm can be used to find the coordinates of a given n-tuple.

PROBLEM 1.   If $\alpha_1 = (-3, 2, 1)$, $\alpha_2 = (7, 4, 3)$, $\alpha_3 = (1, 0, -2)$, then $S = \{\alpha_1, \alpha_2, \alpha_3\}$ is a basis for R_3.

(i) Find a vector γ with S-coordinates $(-4, 1, 5)$.
(ii) Find the S-coordinates of $\beta = (-5, 18, 19)$.

Solution: (i) According to definition 2.13, we have

$$\gamma = -4\alpha_1 + \alpha_2 + 5\alpha_5 = -4(-3, 2, 1) + (7, 4, 3) + 5(1, 0, -2)$$

Therefore, γ is the 3-tuple $(24, -4, -11)$.
(ii) Applying the dependency relationship algorithm to the ordered set $\{\alpha_1, \alpha_2, \alpha_3, \beta\}$, we perform the reduction

$$\begin{array}{cccc} \alpha_1 & \alpha_2 & \alpha_3 & \beta \end{array}$$
$$\begin{bmatrix} -3 & 7 & 1 & -5 \\ 2 & 4 & 0 & 18 \\ 1 & 3 & -2 & 19 \end{bmatrix} \longrightarrow \begin{bmatrix} 1 & 0 & 0 & 5 \\ 0 & 1 & 0 & 2 \\ 0 & 0 & 1 & -4 \end{bmatrix}$$

and conclude that $\beta = 5\alpha_1 + 2\alpha_2 - 4\alpha_3$. Therefore, $[\beta]_S = (5, 2, -4)$.

If $S = \{\alpha_1, \alpha_2, \ldots, \alpha_n\}$ is a basis for the vector space V and $\alpha = a_1\alpha_1 + \ldots + a_n\alpha_n$ and $\beta = b_1\alpha_1 + \ldots + b_n\alpha_n$ are typical members of V, we know that

1. $\alpha = \beta$ if and only if $a_j = b_j$ for $j = 1, 2, \ldots, n$
2. $\alpha + \beta = (a_1\alpha_1 + \ldots + a_n\alpha_n) + (b_1\alpha_1 + \ldots + b_n\alpha_n)$
   $= (a_1 + b_1)\alpha_1 + \ldots + (a_n + b_n)\alpha_n$
3. $t\alpha = t(a_1\alpha_1 + \ldots + a_n\alpha_n) = (ta_1)\alpha_1 + \ldots + (ta_n)\alpha_n$

These vector identities show that the S-coordinates of α and β must satisfy the following rules:

1. $[\alpha]_S = [\beta]_S$ if and only if $\alpha = \beta$
2. $[\alpha + \beta]_S = [\alpha]_S + [\beta]_S$ (2.19)
3. $[t\alpha]_S = t[\alpha]_S$

More generally, if $\gamma = s_1\gamma_1 + s_2\gamma_2 + \ldots + s_m\gamma_m$ is a linear combination of vectors in V, then the n-tuple $[\gamma]_S$ can be expressed as the corresponding linear combination

$$[\gamma]_S = s_1[\gamma_1]_S + s_2[\gamma_2]_S + \ldots + s_m[\gamma_m]_S$$

This observation enables us to make the following general statement:

The Fundamental Isomorphism Principle

If S is a basis for the n-dimensional vector space V, then any problem which deals with linear combinations of vectors in V can be transformed into an equivalent problem in R_n which involves the S-coordinates of those vectors.

Two mathematical systems are *said to be isomorphic* if they differ only in such superficial matters as the "names" used to designate their elements and operations. Using a precise definition of isomorphism (see exercise 9), it can be shown that any two n-dimensional vector spaces are isomorphic, and the fundamental principle stated above is a consequence of this fact. The concept of isomorphism is of great importance in advanced work in all areas of abstract algebra, and the reader who wishes to know more about this concept should begin by examining the pertinent material in a good introductory text, such as *Topics in Algebra*, by I. N. Herstein, Ginn-Blaisdell, Waltham, Massachusetts, 1964.

In addition to its theoretical importance, the fundamental isomorphism principle is an extremely valuable computational tool, for it enables us to deal with abstract vectors as if they were n-tuples and to use techniques such as the dependency relationship algorithm to solve problems in vector spaces other than R_n. To be specific, in order to solve a problem in the vector space V, we simply coordinatize V with respect to a convenient basis (the standard basis, if V has one), solve the corresponding problem in R_n, and then translate the solution back into the same context as the original problem. This approach is illustrated in the following solved problems.

PROBLEM 2.   Let V denote the linear span of the polynomials $p_1(t) = t^3 - 3t^2 + t + 7$, $p_2(t) = t^3 - 2t + 4$, $p_3(t) = -t^3 - 3t^2 + 5t - 1$, and

$p_4(t) = 2t^3 + t^2 - 5t + 11$. Find a basis for V, and determine whether or not V contains the polynomial $q(t) = 4t^3 - 5t^2 + 17t - 3$.

Solution: The standard coordinates of the given polynomials are, respectively, $\alpha_1 = (1, -3, 1, 7)$, $\alpha_2 = (1, 0, -2, 4)$, $\alpha_3 = (-1, -3, 5, -1)$, and $\alpha_4 = (2, 1, -5, 11)$. Let V' denote the linear span of the α's and let $\beta = (4, -5, 17, -3)$. Then the reduction

$$
\begin{array}{ccccc}
\alpha_1 & \alpha_2 & \alpha_3 & \alpha_4 & \beta
\end{array}
$$

$$
\left[
\begin{array}{cccc|c}
1 & 1 & -1 & 2 & 4 \\
-3 & 0 & -3 & 1 & -5 \\
1 & -2 & 5 & -5 & 17 \\
7 & 4 & -1 & 11 & -3
\end{array}
\right]
\longrightarrow
\left[
\begin{array}{cccc|c}
1 & 0 & 1 & 0 & 0 \\
0 & 1 & -2 & 0 & 0 \\
0 & 0 & 0 & 1 & 0 \\
0 & 0 & 0 & 0 & \textcircled{1}
\end{array}
\right]
$$

enables us to conclude that $\{\alpha_1, \alpha_2, \alpha_4\}$ is a basis for V' and that β is not contained in V'. Consequently, $\{p_1, p_2, p_4\}$ is a basis for V, and q is not contained in V.

PROBLEM 3. Let V denote the linear span of the 2×2 matrices

$$
A_1 = \begin{bmatrix} 3 & -1 \\ 7 & 5 \end{bmatrix}, \quad
A_2 = \begin{bmatrix} 5 & 1 \\ 9 & 2 \end{bmatrix}, \quad
A_3 = \begin{bmatrix} -1 & -5 \\ 3 & 11 \end{bmatrix}, \quad
A_4 = \begin{bmatrix} 7 & 4 \\ -1 & 2 \end{bmatrix}
$$

(i) Find a basis S for V.
(ii) Find the S-coordinates of the matrix

$$
B = \begin{bmatrix} 19 & 18 \\ -13 & -10 \end{bmatrix}
$$

Solution: The 4-tuples $\alpha_1 = (3, -1, 7, 5)$, $\alpha_2 = (5, 1, 9, 2)$, $\alpha_3 = (-1, -5, 3, 11)$, $\alpha_4 = (7, 4, -1, 2)$, and $\beta = (19, 18, -13, -10)$, are the standard coordinate vectors of A_1, A_2, A_3, A_4, and B, respectively. The reduction

$$
\begin{array}{ccccc}
\alpha_1 & \alpha_2 & \alpha_3 & \alpha_4 & \beta
\end{array}
$$

$$
\left[
\begin{array}{cccc|c}
3 & 5 & -1 & 7 & 19 \\
-1 & 1 & -5 & 4 & 18 \\
7 & 9 & 3 & -1 & -13 \\
5 & 2 & 11 & 2 & -10
\end{array}
\right]
\longrightarrow
\left[
\begin{array}{cccc|c}
1 & 0 & 3 & 0 & -4 \\
0 & 1 & -2 & 0 & 2 \\
0 & 0 & 0 & 1 & 3 \\
0 & 0 & 0 & 0 & 0
\end{array}
\right]
$$

tells us that α_3 is the only dependent member of $\{\alpha_1, \alpha_2, \alpha_3, \alpha_4\}$ and that $\beta = -4\alpha_1 + 2\alpha_2 + 3\alpha_4$. Therefore, we conclude that $S = \{A_1, A_2, A_4\}$ is a basis for V and that $[B]_S = (-4, 2, 3)$.

The final goal of this section is to investigate the effect of a change in basis on the coordinatized representation of a given vector. For example, let $S = \{\alpha_1, \alpha_2, \alpha_3\}$ and $T = \{\beta_1, \beta_2, \beta_3\}$ be bases for the vector space V, and suppose that

$$\alpha_1 = -2\beta_1 + 3\beta_2 + 7\beta_3$$

$$\alpha_2 = 3\beta_1 - 4\beta_2 + \beta_3 \qquad\qquad (2.20)$$

$$\alpha_3 = 9\beta_1 + 2\beta_2 + 5\beta_3$$

If $\gamma = a_1\alpha_1 + a_2\alpha_2 + a_3\alpha_3$ is a typical member of V, we have

$$\gamma = a_1(-2\beta_1 + 3\beta_2 + 7\beta_3) + a_2(3\beta_1 - 4\beta_2 + \beta_3) + a_3(9\beta_1 + 2\beta_2 + 5\beta_3)$$
$$= (-2a_1 + 3a_2 + 9a_3)\beta_1 + (3a_1 - 4a_2 + 2a_3)\beta_2 + (7a_1 + a_2 + 5a_3)\beta_3$$

and it follows that γ has T-coordinates (b_1, b_2, b_3), where

$$b_1 = -2a_1 + 3a_2 + 9a_3$$

$$b_2 = 3a_1 - 4a_2 + 2a_3 \qquad\qquad (2.21)$$

$$b_3 = 7a_1 + a_2 + 5a_3$$

Rewriting (2.21) in vector form, we obtain the equation

$$[\gamma]_T = \begin{bmatrix} b_1 \\ b_2 \\ b_3 \end{bmatrix} = \begin{bmatrix} -2 & 3 & 9 \\ 3 & -4 & 2 \\ 7 & 1 & 5 \end{bmatrix} \begin{bmatrix} a_1 \\ a_2 \\ a_3 \end{bmatrix} = C[\gamma]_S \qquad (2.22)$$

where C is the *transpose* of the coefficient matrix of .vector system (2.20). Since we know the entries of the matrix C, formula (2.22) can be used to transform the S-coordinates of any vector in V into T-coordinates, and for this reason, it is natural to call C the S to T change of basis matrix. This notion is given precise form in the following definition.

Definition 2.12. If $S = \{\alpha_1, \ldots, \alpha_n\}$ and $T = \{\beta_1, \ldots, \beta_n\}$ are bases for the vector space V and the scalars c_{ij} satisfy

$$\alpha_1 = c_{11}\beta_1 + c_{21}\beta_2 + \ldots + c_{n1}\beta_n$$

$$\alpha_2 = c_{12}\beta_1 + c_{22}\beta_2 + \ldots + c_{n2}\beta_n$$
$$\begin{matrix} \cdot & \cdot & \cdot & \cdot \\ \cdot & \cdot & \cdot & \cdot \\ \cdot & \cdot & \cdot & \cdot \end{matrix} \qquad\qquad (2.23)$$

$$\alpha_n = c_{1n}\beta_1 + c_{2n}\beta_2 + \ldots + c_{nn}\beta_n$$

then the $n \times n$ matrix

$$C = \begin{bmatrix} c_{11} & c_{12} & \cdots & c_{1n} \\ c_{21} & c_{22} & & c_{2n} \\ \cdot & & & \cdot \\ \cdot & & & \cdot \\ \cdot & & & \cdot \\ c_{n1} & c_{n2} & \cdots & c_{nn} \end{bmatrix}$$

is called the *S to T change of basis matrix.*

In our next theorem, we generalize the result obtained in our example by deriving a formula which represents the transformation of coordinates from the basis S to the basis T.

Theorem 2.13. Let S and T be bases for the n-dimensional vector space V, and let C denote the S to T change of basis matrix. Then we have

$$[\gamma]_T = C[\gamma]_S \tag{2.24}$$

for each vector γ in V.

Proof: Suppose that $S = \{\alpha_1, \ldots, \alpha_n\}$ and $T = \{\beta_1, \ldots, \beta_n\}$, and let $(a_1, \ldots, a_n)$ and $(b_1, \ldots, b_n)$ be the S- and T-coordinates of γ, respectively. Since $\alpha_j = \sum_{i=1}^{n} c_{ij}\beta_i$ for $j = 1, 2, \ldots, n$, it follows that

$$\gamma = \sum_{j=1}^{n} a_j\alpha_j = \sum_{j=1}^{n} a_j\left(\sum_{i=1}^{n} c_{ij}\beta_i\right) = \sum_{i=1}^{n} \left(\sum_{j=1}^{n} a_j c_{ij}\right)\beta_i$$

Therefore, for each index i, we have $b_i = \sum_{j=1}^{n} c_{ij}a_j$, and we conclude that

$$[\gamma]_T = \begin{bmatrix} b_1 \\ b_2 \\ \cdot \\ \cdot \\ \cdot \\ b_n \end{bmatrix} = \begin{bmatrix} c_{11} & \cdots & c_{1n} \\ c_{21} & \cdots & c_{2n} \\ \cdot & & \cdot \\ \cdot & & \cdot \\ \cdot & & \cdot \\ c_{n1} & \cdots & c_{nn} \end{bmatrix}\begin{bmatrix} a_1 \\ a_2 \\ \cdot \\ \cdot \\ \cdot \\ a_n \end{bmatrix} = C[\gamma]_S$$

as desired. ∎

It is important to remember that C is the *transpose* of the coefficient matrix of system (2.23) and not the coefficient matrix itself. Moreover, since the entries of the jth column of C are the T-coordinates of α_j (why is this true?), it follows that the entries of C can be found by applying the method illustrated in problem 1 n different times, once for each member of S. This promises to be tedious, but fortunately, the whole process can be condensed into a single reduction, as illustrated in our next solved problem.

PROBLEM 4.   Let $\alpha_1 = (-1, 8, 5)$, $\alpha_2 = (9, -1, -10)$, $\alpha_3 = (1, 13, 4)$, and $\beta_1 = (-3, 1, 4)$, $\beta_2 = (5, 7, -2)$, $\beta_3 = (0, 1, 1)$.

Given that $S = \{\alpha_1, \alpha_2, \alpha_3\}$ and $T = \{\beta_1, \beta_2, \beta_3\}$ are bases for R_3, find the S to T change of basis matrix.

Solution: To find the desired change of basis matrix C, we must determine the T-coordinates of α_1, α_2, and α_3. Modifying the procedure established in problem 1 to handle three related problems at once, we first perform the reduction

$$
\begin{array}{cccccc}
\beta_1 & \beta_2 & \beta_3 & \alpha_1 & \alpha_2 & \alpha_3
\end{array}
\qquad\qquad I_3 \qquad\qquad C
$$

$$
\begin{bmatrix}
-3 & 5 & 0 & -1 & 9 & 1 \\
1 & 7 & 1 & 8 & -1 & 13 \\
4 & -2 & 1 & 5 & -10 & 4
\end{bmatrix}
\longrightarrow
\begin{bmatrix}
1 & 0 & 0 & 2 & -3 & 3 \\
0 & 1 & 0 & 1 & 0 & 2 \\
0 & 0 & 1 & -1 & 2 & -4
\end{bmatrix}
\qquad (2.25)
$$

Then, from the column vectors of the reduced row echelon matrix on the right of (2.25), we obtain the vector system

$$
\alpha_1 = 2\beta_1 + \beta_2 - \beta_3
$$
$$
\alpha_2 = -3\beta_1 \qquad + 2\beta_3
$$
$$
\alpha_3 = 3\beta_1 + 2\beta_2 - 4\beta_3
$$

and conclude that

$$
C = \begin{bmatrix}
2 & -3 & 3 \\
1 & 0 & 2 \\
-1 & 2 & -4
\end{bmatrix}
$$

As indicated in (2.25), the matrix C is the 3×3 block of entries located to the right of the dashed line in the reduced row echelon matrix. We invite the reader to test his understanding of the computational methods of this chapter by verifying that a change of basis matrix can always be found in this manner.

By reversing the roles of S and T in problem 4, it can be shown that the T to S change of basis matrix is

$$
C' = \frac{1}{8}\begin{bmatrix}
4 & 6 & 6 \\
-2 & 5 & 1 \\
-2 & 1 & -3
\end{bmatrix}
$$

which happens to be the multiplicative inverse of C—that is, $C'C = I_3$. This observation is a consequence of the following general result, which plays an important role in our discussion of matrix similarity in chapter 4.

Theorem 2.14.   A change of basis matrix is always non-singular. More-over, if S and T are bases for the vector space V, the S to T and T to S change of basis matrices are multiplicative inverses of each other.

Proof: Outlined in exercise 8.

We now have a fairly complete picture of what goes on inside a vector space. In the next section, we shall conclude the work of this chapter by reexamining some of the results of chapter 1 in the light of our new found knowledge of vector spaces.

Exercise Set 2.7

1. Let $\alpha_1 = (5, 1, -3)$, $\alpha_2 = (7, -2, 4)$, $\alpha_3 = (1, 0, 1)$, and $\beta_1 = (4, -3, 1)$, $\beta_2 = (7, -5, 2)$, $\beta_3 = (1, 1, 1)$. Given that $S = \{\alpha_1, \alpha_2, \alpha_3\}$ and $T = \{\beta_1, \beta_2, \beta_3\}$ are both bases for R_3, perform the following computations:
   (i) Find the vector γ whose S-coordinates are $(-5, 3, 1)$.
   (ii) Find the T-coordinates of the vector $p = (2, -4, 15)$.
   (iii) Determine the S to T change of basis matrix.
   (iv) Find the S-coordinates of $p = (2, -4, 15)$ and use formula (2.24) to find its T-coordinates. Compare your answer with the coordinates found in part (ii).

2. Let $S = \{\alpha_1, \alpha_2, \alpha_3\}$ be a basis for the vector space V, and let $\beta_1, \beta_2, \beta_3$ be vectors such that $[\beta_1]_S = (4, -2, 1)$, $[\beta_2]_S = (1, 3, -3)$, and $[\beta_3]_S = (5, -13, 11)$.
   (i) Find the S-coordinates of the vectors $\gamma_1 = 2\alpha_1 + \alpha_3$, $\gamma_2 = \beta_1 - \beta_2 + 3\beta_3$, and $\gamma_3 = \alpha_1 - 2\alpha_2 + 5\beta_1 - \beta_2$.
   (ii) Find scalars s_1, s_2, s_3 such that $s_1\beta_1 + s_2\beta_2 + s_3\beta_3 = \theta$.
   (iii) What is the dimension of the linear span of β_1, β_2, and β_3? Explain your reasoning.

3. Let V denote the linear span of the polynomials $p_1(t) = 3 - t + 5t^2 + 2t^3$, $p_2(t) = 4 - 3t + 9t^2 + 7t^3$, $p_3(t) = 5 + 6t^2 - t^3$, and $p_4(t) = 7 + 6t - 17t^3$.
   (i) Find a basis S for V.
   (ii) Determine which of the following polynomials is contained in V and find the S-coordinates of those that are.

$$q_1(t) = 6 - 7t + 17t^2 + 17t^3, \qquad q_3(t) = 5 - 9t + 4t^3$$

$$q_2(t) = 2t - 11t^2, \qquad q_4(t) = 2 + t + t^2 - 3t^3$$

4. Find a basis for the linear span of the matrices

$$A_1 = \begin{bmatrix} 3 & -1 \\ 5 & 2 \end{bmatrix}, \quad A_2 = \begin{bmatrix} 4 & -3 \\ 9 & 7 \end{bmatrix}, \quad A_3 = \begin{bmatrix} 5 & 0 \\ 6 & -1 \end{bmatrix}, \quad A_4 = \begin{bmatrix} 7 & 6 \\ 0 & -17 \end{bmatrix}$$

In what way is this exercise like exercise 3?

5. Let S denote the fundamental basis of the solution space of the linear system

$$3x_1 - 7x_2 + 2x_3 + x_4 = 0$$

$$x_1 + 5x_2 - 3x_3 + 4x_4 = 0$$

$$-5x_1 + 19x_2 - 7x_3 + 2x_4 = 0$$

$$13x_1 - x_2 - 6x_3 + 19x_4 = 0$$

(i) Find the S-coordinates of the solutions $\beta_1 = (-1, 0, 1, 1)$ and $\beta_2 = (12, 7, 9, -5)$.

(ii) Note that $T = \{\beta_1, \beta_2\}$ is also a basis for the solution space of the given system (why?). Compute both the T to S and S to T change of basis matrices. *Hint:* One of these follows easily from the result of part (i).

6. Let V denote the solution space of the homogeneous differential equation $y''' - 3y'' + y' - 3y = 0$. It can be shown that $S = \{e^{3x}, \cos x, \sin x\}$ is a basis for V.

(i) Verify that $f_1(x) = 2e^{3x}$, $f_2(x) = \cos x + \sin x$, and $f_3(x) = e^{3x} - \sin x$ are all contained in V, and find the S-coordinates of each of these functions.

(ii) Show that $T = \{f_1, f_2, f_3\}$ is a basis for V.

(iii) Find the S to T change of basis matrix, and use it to find the T-coordinates of the function $g(x) = 5e^{3x} - 2\sin x + 4\cos x$.

7. Let $S = \{\alpha_1, \alpha_2, \alpha_3\}$ and $T = \{\beta_1, \beta_2, \beta_3\}$ be bases for R_3. It is known that $\alpha_1 = (-3, 5, 2)$, $\alpha_2 = (4, 1, 1)$, and $\beta_2 = (4, 0, -7)$ and that the T to S change of basis matrix is

$$C = \begin{bmatrix} -3 & 1 & 5 \\ 1 & -1 & 2 \\ 2 & -1 & -1 \end{bmatrix}$$

Use this information to find α_3, β_1, and β_3.

*8. Let S and T be ordered bases for the n-dimensional vector space V, and let C and C' be the S to T and T to S change of basis matrices, respectively.

(i) Use formula (2.24) to show that

$$[\gamma]_S = C'[\gamma]_T = C'C[\gamma]_S$$

for each vector γ of V.

(ii) Let $B = C'C$. Show that $B\epsilon_j = \epsilon_j$ for each standard basis n-tuple ϵ_j.

(iii) Show that $BI_n = I_n$, and conclude that $C'C = I_n$. Use this fact to prove theorem 2.14.

*9. Let U and V be vector spaces. A function F from U to V is said to be an *isomorphism* if it has the following properties:

1. If v is a vector in V, then $F(u) = v$ for exactly one vector u in U.
2. $F(u_1 + u_2) = F(u_1) + F(u_2)$ for all u_1, u_2 in U.
3. $F(ru) = rF(u)$ for each scalar r and vector u in U.

If such a function exists, we say that U is isomorphic to V.

(i) Show that U is isomorphic to V if and only if V is isomorphic to U.
Hint: Consider the function F^{-1}.

(ii) Show that if U is isomorphic to V and V to W, then U is also isomorphic to W.

(iii) If U has dimension n, show that it is isomorphic to R_n.
Hint: Let F be the function from U to R_n defined as follows:

$$F(u) = [u]_S \quad \text{for each vector } u \text{ in } U$$

Then use the properties displayed in (2.19).

(iv) Show that any two n-dimensional vector spaces are isomorphic to each other.
Hint: Combine parts (i), (ii), and (iii).

2.8 THE RANK OF A MATRIX

The main purpose of this final section is to show how vector methods can be used to analyze the solution set of a system of linear equations. In the process, we shall discuss a few new ideas and develop more elegant proofs for several results of chapter 1. We begin by introducing some new terminology.

Definition 2.13.   Let $A = (a_{ij})$ be a fixed $m \times n$ matrix. Then:
(i) The *row space* of A is the subspace of R_n which is spanned by the m row vectors of A. The dimension of this space is called the *row rank* of A.

(ii) The *column space* of A is the linear span of its n column vectors, and the dimension of this subspace of R_m is called the *column rank* of A.

EXAMPLE 1.   Let

$$A = \begin{bmatrix} -1 & 3 & 7 & -5 \\ 4 & -2 & 1 & 3 \\ 2 & 4 & 15 & -7 \end{bmatrix}$$

The row vectors of A are $\alpha_1 = (-1, 3, 7, -5)$, $\alpha_2 = (4, -2, 1, 3)$, and $\alpha_3 = (2, 4, 15, -7)$. The row space is $\mathrm{Sp}\,\{\alpha_1, \alpha_2, \alpha_3\}$, and it can be shown that the row rank is 2 (specifically, $\alpha_3 = 2\alpha_1 + \alpha_2$ is the only dependent member of $\{\alpha_1, \alpha_2, \alpha_3\}$).

The column vectors of A are

$$\beta_1 = \begin{bmatrix} -1 \\ 4 \\ 2 \end{bmatrix}, \quad \beta_2 = \begin{bmatrix} 3 \\ -2 \\ 4 \end{bmatrix}, \quad \beta_3 = \begin{bmatrix} 7 \\ 1 \\ 15 \end{bmatrix}, \quad \beta_4 = \begin{bmatrix} -5 \\ 3 \\ -7 \end{bmatrix}$$

and the column space is Sp $\{\beta_1, \beta_2, \beta_3, \beta_4\}$. We find that β_1 and β_2 are linearly independent, but $\beta_3 = \frac{17}{10}\beta_1 + \frac{29}{10}\beta_2$ and $\beta_4 = -\frac{1}{10}\beta_1 - \frac{17}{10}\beta_2$. Thus, the column rank is 2.

Note that the column rank and row rank are equal. Later in this section, we shall prove that this is always the case.

Questions concerning the consistency of a given non-homogeneous linear system can be answered by examining the column space of its coefficient matrix. For example, note that x_1, x_2, x_3, x_4 are real numbers which satisfy the linear system

$$2x_1 - 3x_2 - x_3 + 5x_4 = 5$$

$$3x_1 + 2x_2 + 5x_3 + 14x_4 = 7 \qquad (2.26)$$

$$x_1 - 8x_2 - 7x_3 - 4x_4 = 2$$

whenever these numbers satisfy the vector equation

$$x_1 \begin{bmatrix} 2 \\ 3 \\ 1 \end{bmatrix} + x_2 \begin{bmatrix} -3 \\ 2 \\ -8 \end{bmatrix} + x_3 \begin{bmatrix} -1 \\ 5 \\ -7 \end{bmatrix} + x_4 \begin{bmatrix} 5 \\ 14 \\ -4 \end{bmatrix} = \begin{bmatrix} 5 \\ 7 \\ 2 \end{bmatrix}$$

Therefore, system (2.26) is consistent if and only if it is possible to express the vector

$$\beta = \begin{bmatrix} 5 \\ 7 \\ 2 \end{bmatrix}$$

as a linear combination of the column vectors of the coefficient matrix A of the system, and this is the same as saying that β must be contained in the column space of A. We invite the reader to generalize this argument and by so doing to prove the following theorem.

Theorem 2.15.   The linear system $AX = \beta$ has a solution if and only if β is contained in the column space of the coefficient matrix A.

Recall that the augmented matrix $(A \,\vdots\, \beta)$ of the $m \times n$ linear system $AX = \beta$ is the $m \times (n + 1)$ matrix whose first n columns coincide with those of A and whose $(n + 1)$st column is β. For example, the augmented matrix of system (2.26) is

$$(A \,\vdots\, \beta) = \begin{bmatrix} 2 & -3 & -1 & 5 & \vdots & 5 \\ 3 & 2 & 5 & 14 & \vdots & 7 \\ 1 & -8 & -7 & -4 & \vdots & 2 \end{bmatrix}$$

In chapter 1, we observed that a linear system is consistent if and only if the reduced row echelon form of its augmented matrix has no "bad" entries. As an alternative to this rather awkward consistency criterion, we have the following theorem.

Theorem 2.16. The $m \times n$ linear system $AX = \beta$ has a solution if and only if the column rank of A is the same as that of the augmented matrix $(A \mathbin{\vdots} \beta)$.

Proof: Let $\gamma_1, \gamma_2, \ldots, \gamma_n$ denote the column vectors of A. If the system is consistent, β can be expressed as a linear combination of the γ's and must therefore be a dependent member of $S' = \{\gamma_1, \ldots, \gamma_n, \beta\}$. Consequently, S' and $S = \{\gamma_1, \ldots, \gamma_n\}$ have the same independent members, and the column rank of A is the same as that of $(A \mathbin{\vdots} \beta)$.

If the system is inconsistent, β is an independent member of S' (why?), and it follows that the column rank of $(A \mathbin{\vdots} \beta)$ is greater (by 1) than that of A. ∎

For theoretical purposes, this new consistency criterion is quite useful. However, despite its more elegant appearance, theorem 2.16 is not really very different from the old "bad" entry criterion established in chapter 1. For example, suppose $\gamma_1, \ldots, \gamma_4$ are the column vectors of the coefficient matrix A of system (2.26) and β is the column vector of right side constants. Then the reduction

$$
\begin{array}{ccccc}
\gamma_1 & \gamma_2 & \gamma_3 & \gamma_4 & \beta
\end{array}
$$

$$
\begin{bmatrix}
2 & -3 & -1 & 5 & \vdots & 5 \\
3 & 2 & 5 & 14 & \vdots & 7 \\
1 & -8 & -7 & -4 & \vdots & 2
\end{bmatrix}
\longrightarrow
\begin{bmatrix}
1 & 0 & 1 & 4 & \vdots & 0 \\
0 & 1 & 1 & 1 & \vdots & 0 \\
0 & 0 & 0 & 0 & \vdots & \boxed{1}
\end{bmatrix}
$$

"bad" entry

tells us that $\gamma_3 = \gamma_1 + \gamma_2$ and $\gamma_4 = 4\gamma_1 + \gamma_2$ are the only dependent members of $\{\gamma_1, \gamma_2, \gamma_3, \gamma_4, \beta\}$. Therefore, the coefficient matrix A has column rank 2, and the augmented matrix $(A \mathbin{\vdots} \beta)$ has column rank 3. Note that the presence of the "bad" entry in column 5 is what guarantees that β cannot be expressed as a linear combination of the γ's.

Next, we turn our attention to the study of the row space of a given $m \times n$ matrix A. For definiteness, we shall consider the matrix

$$
A = \begin{bmatrix}
-3 & 4 & 9 & 2 \\
7 & 5 & 3 & 0 \\
1 & -8 & 1 & 5
\end{bmatrix}
\begin{array}{l}
\leftarrow \rho_1 \\
\leftarrow \rho_2 \\
\leftarrow \rho_3
\end{array}
$$

but our remarks will be just as valid for any other matrix. Let A_1 denote the matrix obtained by performing a single elementary row operation on A—say, the matrix

$$A_1 = \begin{bmatrix} -3 & 4 & 9 & 2 \\ 7 & 5 & 3 & 0 \\ -14 & 12 & 46 & 15 \end{bmatrix} \begin{matrix} \longleftarrow \rho_1' = \rho_1 \\ \longleftarrow \rho_2' = \rho_2 \\ \longleftarrow \rho_3' = 5\rho_1 + \rho_3 \end{matrix}$$

obtained by adding 5 times the first row of A to the third row. If we let ρ_1, ρ_2, ρ_3 denote the row vectors of A, then the row vectors of A_1 are $\rho_1' = \rho_1$, $\rho_2' = \rho_2$, and $\rho_3' = 5\rho_1 + \rho_3$, and it follows that any linear combination of the row vectors of A_1 can be expressed as a linear combination of those of A. For example,

$$2\rho_1' - 3\rho_2' + 4\rho_3' = 2\rho_1 - 3\rho_2 + 4(5\rho_1 + \rho_3) = 22\rho_1 - 3\rho_2 + 4\rho_3$$

Since $\rho_1 = \rho_1'$, $\rho_2 = \rho_2'$, and $\rho_3 = \rho_3' - 5\rho_1'$, we can also show that each linear combination of the row vectors of A may be expressed as a linear combination of those of A_1.

The same kind of reasoning may be used in the case where A is a general $m \times n$ matrix and A_1 is derived from A by one of the other elementary row operations, and we can make the following statement:

> If the matrix A_1 is derived from A by performing a single elementary row operation, then A and A_1 have the same row space.

Thus, if A_2 is derived from A by performing two elementary row operations, our observation allows us to conclude that A_2 has the same row space as A_1 (why?) and hence, the same row space as A. By formalizing the logic in this illustration, we can prove the following theorem.

Theorem 2.17.   Row equivalent matrices have the same row space and hence, the same row rank.

Theorem 2.17 provides a simple mechanism for finding a basis for the row space of a given matrix A. To be specific, if E is the reduced row echelon form of A, the theorem tells us that A and E have the same row space. However, thanks to the staggered positioning of the leading 1's in E, it is clear that none of its non-zero rows can be expressed as a linear combination of the others. Consequently, these non-zero rows form basis for the row space of E and hence, for the row space of A also. These remarks constitute a proof of our next theorem.

Theorem 2.18. Let A be an $m \times n$ matrix whose reduced row echelon form has exactly r non-zero rows, $\bar{\rho}_1, \bar{\rho}_2, \ldots, \bar{\rho}_r$. Then A has row rank r and $\{\bar{\rho}_1, \bar{\rho}_2, \ldots, \bar{\rho}_r\}$ is a basis for its row space.

PROBLEM 1. Find a basis for the row space of the matrix

$$A = \begin{bmatrix} 2 & 7 & -1 & 2 & 5 \\ 1 & 4 & 0 & 3 & 2 \\ 0 & -1 & -1 & -4 & 1 \\ 6 & 23 & -1 & 14 & 13 \end{bmatrix}$$

Solution: Applying the reduction algorithm to A, we obtain

$$\begin{bmatrix} 2 & 7 & -1 & 2 & 5 \\ 1 & 4 & 0 & 3 & 2 \\ 0 & -1 & -1 & -4 & 1 \\ 6 & 23 & -1 & 14 & 13 \end{bmatrix} \longrightarrow \begin{bmatrix} 1 & 0 & -4 & -13 & 6 \\ 0 & 1 & 1 & 4 & -1 \\ 0 & 0 & 0 & 0 & 0 \\ 0 & 0 & 0 & 0 & 0 \end{bmatrix}$$

Therefore, A has row rank 2, and a basis for its row space is $\{\bar{\rho}_1, \bar{\rho}_2\}$, where $\bar{\rho}_1 = (1, 0, -4, -13, 6)$, and $\bar{\rho}_2 = (0, 1, 1, 4, -1)$.

Thus, we have arrived at a point where we know how to find the row rank of a given matrix, but we need to know the column rank for certain interesting applications. Fortunately, the following theorem provides a way out of this dilemma.

Theorem 2.19. The row rank and column rank of a matrix are always equal.

Proof: Let $A = (a_{ij})$ be an $m \times n$ matrix, and denote its row and column vectors by $\alpha_1, \ldots, \alpha_m$ and $\beta_1, \ldots, \beta_n$, respectively. Suppose that A has row rank r and column rank s. The argument used in the proof is as follows:

(i) Show that $s \leq r$ by using a basis $\{\rho_1, \rho_2, \ldots, \rho_r\}$ for the row space of A to construct a collection of r vectors $\gamma_1, \ldots, \gamma_r$ which span the column space.

(ii) Reverse the roles of row space and column space in step (i) to conclude that $r \leq s$ and hence, that $r = s$.

This may seem rather simple, and it is. However, to describe the construction which takes place in step (i), we must enter a veritable quagmire of indexing and labeling difficulties which serve only to obscure the real issues. Instead, we shall indicate the general lines of the construction by providing a detailed analysis of a typical case.

Accordingly, consider the 3×4 matrix

$$
A = \begin{array}{c} \\ \end{array}
\begin{array}{cccc}
\beta_1 & \beta_2 & \beta_3 & \beta_4 \\
\downarrow & \downarrow & \downarrow & \downarrow \\
\end{array}
$$

$$
A = \begin{bmatrix}
a_{11} & a_{12} & a_{13} & a_{14} \\
a_{21} & a_{22} & a_{23} & a_{24} \\
a_{31} & a_{32} & a_{33} & a_{34}
\end{bmatrix}
\begin{array}{l}
\longleftarrow \alpha_1 \\
\longleftarrow \alpha_2 \\
\longleftarrow \alpha_3
\end{array}
$$

and assume that the row rank r is 2. Let $\{\rho_1, \rho_2\}$ be a basis for the row space of A, where $\rho_1 = (r_1, r_2, r_3, r_4)$ and $\rho_2 = (s_1, s_2, s_3, s_4)$. There exist scalars t_{ij} such that

$$
\alpha_1 = t_{11}\rho_1 + t_{12}\rho_2
$$

$$
\alpha_2 = t_{21}\rho_1 + t_{22}\rho_2 \tag{2.27}
$$

$$
\alpha_3 = t_{31}\rho_1 + t_{32}\rho_2
$$

and by equating components in these three vector equations, we obtain four separate scalar systems, one for each component. For instance, the systems associated with the 1st and 4th components are the following:

$$
\begin{array}{ll}
\text{1st component} & \text{4th component} \\
a_{11} = t_{11}(r_1) + t_{12}(s_1) & a_{14} = t_{11}(r_4) + t_{12}(s_4) \\
a_{21} = t_{21}(r_1) + t_{22}(s_1) & a_{24} = t_{21}(r_4) + t_{22}(s_4) \\
a_{31} = t_{31}(r_1) + t_{32}(s_1) & a_{34} = t_{31}(r_4) + t_{32}(s_4)
\end{array} \tag{2.28}
$$

Rewriting the linear systems displayed in (2.28) in vector form, we find that

$$
\begin{bmatrix} a_{11} \\ a_{21} \\ a_{31} \end{bmatrix} = r_1 \begin{bmatrix} t_{11} \\ t_{21} \\ t_{31} \end{bmatrix} + s_1 \begin{bmatrix} t_{12} \\ t_{22} \\ t_{32} \end{bmatrix} \quad \text{and} \quad \begin{bmatrix} a_{14} \\ a_{24} \\ a_{34} \end{bmatrix} = r_4 \begin{bmatrix} t_{11} \\ t_{21} \\ t_{31} \end{bmatrix} + s_4 \begin{bmatrix} t_{12} \\ t_{22} \\ t_{32} \end{bmatrix}
$$

Similar representations can be found for the systems associated with the 2nd and 3rd components in (2.27), and we see that

$$
\beta_1 = r_1\gamma_1 + s_1\gamma_2, \qquad \beta_2 = r_2\gamma_1 + s_2\gamma_2
$$

$$
\beta_3 = r_3\gamma_1 + s_3\gamma_2, \qquad \beta_4 = r_4\gamma_1 + s_4\gamma_2
$$

where the β_j are the column vectors of A and

$$\gamma_1 = \begin{bmatrix} t_{11} \\ t_{21} \\ t_{31} \end{bmatrix} \quad \text{and} \quad \gamma_2 = \begin{bmatrix} t_{12} \\ t_{22} \\ t_{32} \end{bmatrix}$$

Since each column vector of A can be expressed as a linear combination of γ_1 and γ_2, it follows that the column space is spanned by these two vectors. Therefore, the column rank of A is at most 2 (why?), as claimed.

We invite the reader to test his understanding of the construction of the γ's by working the numerical problem outlined in exercise 12 at the end of this section. ∎

Thanks to theorem 2.19, we no longer need to distinguish between the row rank and column rank of a matrix A. Henceforth, either of these (equal) numbers will be referred to as simply the *rank* of A.

In section 2.6, we observed that the $m \times n$ homogeneous linear system $AX = O$ has $n - r$ fundamental solutions, where r is the number of non-zero rows in the reduced row echelon form of the coefficient matrix A. In the terminology of this section, r is the (row) rank of A, and the aforementioned result can be restated as follows.

Theorem 2.20.   The homogeneous $m \times n$ linear system $AX = O$ has a unique solution if its coefficient matrix has rank n. If the rank r is less than n, the solution space of the system has dimension $n - r$.

We conclude the work of this section by listing four important consequences of theorem 2.20. In each case, the proof is outlined in the exercises at the end of the section.

Corollary 2.1.   A consistent $m \times n$ non-homogeneous linear system $AX = \beta$ has a unique solution if the coefficient matrix A has rank n, and infinitely many solutions if the rank is less than n.

Corollary 2.2.   If $m < n$, the $m \times n$ homogeneous linear system $AX = O$ has infinitely many solutions.

Corollary 2.3.   If A has rank m, the $m \times n$ non-homogeneous linear system $AX = \beta$ must be consistent.

Corollary 2.4.   The $n \times n$ matrix A is non-singular if and only if it has rank n.

The concept of rank plays an important role in linear algebra. We shall encounter this concept in another form in chapter 4 and again in chapter 5.

Exercise Set 2.8

1. Find the rank of each of the following matrices.

$$A_1 = \begin{bmatrix} -3 & 1 & 5 \\ 2 & 7 & 1 \\ 1 & -3 & 2 \end{bmatrix}, \quad A_2 = \begin{bmatrix} 8 & -3 & 1 & 2 \\ -5 & 1 & 2 & 7 \\ 1 & -3 & 8 & 25 \end{bmatrix}$$

$$A_3 = \begin{bmatrix} -7 & 9 & 4 & 5 & 3 \\ 5 & 1 & 0 & -2 & 1 \\ 3 & 11 & 4 & 1 & 5 \\ -5 & -1 & 0 & 2 & -1 \end{bmatrix}$$

2. Find a basis for the row space of each of the following matrices.

$$A_1 = \begin{bmatrix} -2 & 5 & 1 & 7 \\ 1 & -3 & 5 & 2 \\ -1 & 1 & 17 & 20 \end{bmatrix}, \quad A_2 = \begin{bmatrix} 1 & 2 & 3 & 1 \\ 0 & 1 & 4 & -1 \\ 3 & 0 & -15 & 9 \\ 5 & 8 & 7 & 7 \end{bmatrix}$$

3. Recall that the transpose A^t of the $m \times n$ matrix $A = (a_{ij})$ is the $n \times m$ matrix whose (i, j) entry is a_{ji} (see exercise 9, section 1.4).
   (i) Show that the column space of A is the same as the row space of A^t.
   (ii) Show that A and A^t always have the same rank.
   (iii) If A has rank r, what is the dimension of the solution space of the $n \times m$ linear system $A^t X = O$?
4. Show that the $m \times n$ matrix A has rank r if and only if it has r linearly independent rows and r linearly independent columns.
5. If the $m \times n$ matrix A has rank r, show that $r \leq m$ and $r \leq n$.
6. Find a basis for the column space of the matrix

$$\begin{bmatrix} -5 & 2 & -1 & 1 & -9 \\ 7 & 3 & 13 & 3 & 17 \\ 1 & -5 & -9 & 8 & 10 \\ 3 & 8 & 19 & 4 & 10 \end{bmatrix}$$

7. (i) If the non-homogeneous $m \times n$ linear system $AX = \beta$ is consistent, show that it has as many solutions as the related homogeneous system $AX = O$.
   Hint: See theorem 2.12.
   (ii) Use theorem 2.20 and part (i) to prove corollary 2.1.

8. If $m < n$, show that an $m \times n$ matrix A cannot have rank n. Use this fact to prove corollary 2.2.

9. If the $m \times n$ matrix A has rank m, show that the augmented matrix of the linear system $AX = \beta$ also has rank m. Use this fact to prove corollary 2.3.

10. Show that an $n \times n$ matrix A has rank n if and only if each linear system of the form $AX = \beta$ has exactly one solution.

Hint: For the "if" part, use corollary 2.1, and for the "only if" part, use corollaries 2.1 and 2.3.

11. Prove corollary 2.4 by justifying the following statements:

(i) If the $n \times n$ matrix A has rank n, there exist (unique) n-tuples $\gamma_1, \ldots, \gamma_n$ such that $A\gamma_j = \epsilon_j$, for $j = 1, 2, \ldots, n$. As usual, ϵ_j denotes the jth standard basis n-tuple.

(ii) If C is the $m \times n$ matrix whose jth column vector is γ_j, then $AC = I_n$. Thus, A is non-singular and $A^{-1} = C$.

(iii) Conversely, if A is non-singular, each linear system of the form $AX = \beta$ has the unique solution $X = A^{-1}\beta$. Therefore, A has rank n.

12. Let A be the 3×4 matrix

$$
\begin{array}{cccc}
\beta_1 & \beta_2 & \beta_3 & \beta_4 \\
\downarrow & \downarrow & \downarrow & \downarrow
\end{array}
$$
$$
A = \begin{bmatrix} 4 & -20 & -1 & -13 \\ 2 & -10 & 7 & 1 \\ 3 & -15 & -2 & -11 \end{bmatrix} \begin{array}{l} \longleftarrow \alpha_1 \\ \longleftarrow \alpha_2 \\ \longleftarrow \alpha_3 \end{array}
$$

and let $\alpha_1, \alpha_2, \alpha_3$ and $\beta_1, \beta_2, \beta_3, \beta_4$ denote the row vectors and column vectors of A, as indicated.

(i) Show that A has row rank 2 and find a basis $\{\rho_1, \rho_2\}$ for the row space.

(ii) Find scalars t_{ij} such that $\alpha_i = t_{i1}\rho_1 + t_{i2}\rho_2$ for $i = 1, 2, 3$.

(iii) Express each column vector β_j as a linear combination of the vectors

$$
\gamma_1 = \begin{bmatrix} t_{11} \\ t_{21} \\ t_{31} \end{bmatrix} \quad \text{and} \quad \gamma_2 = \begin{bmatrix} t_{12} \\ t_{22} \\ t_{32} \end{bmatrix}
$$

(iv) Show that the vector

$$
\beta = \begin{bmatrix} 2 \\ 4 \\ 1 \end{bmatrix}
$$

can be expressed as a linear combination of γ_1 and γ_2. Does the linear system $AX = \beta$ have a solution? Explain.

13. A certain 4×5 matrix A is known to have rank 3. Furthermore, if $\alpha_1, \alpha_2, \alpha_3, \alpha_4$ are the row vectors of A, it is known that there exist vectors ρ_1, ρ_2, ρ_3

such that

$$\alpha_1 = -7\rho_1 + 2\rho_2 - 5\rho_3$$

$$\alpha_2 = \ \ 4\rho_1 + 0\rho_2 + \ \ \rho_3$$

$$\alpha_3 = \ \ 9\rho_1 + \ \ \rho_2 - 2\rho_3$$

$$\alpha_4 = -2\rho_1 + \ \ \rho_2 + 3\rho_3$$

Use this information for the following purposes.

(i) Show that $\{\rho_1, \rho_2, \rho_3\}$ is a basis for the row space of A.

(ii) Find three vectors $\gamma_1, \gamma_2, \gamma_3$ which span the column space of A.

(iii) Determine whether or not the linear system $AX = \beta$ has a solution, where

$$\beta = \begin{bmatrix} 0 \\ 3 \\ 13 \\ -4 \end{bmatrix}$$

SUPPLEMENTARY EXERCISES

1. In problem 2 of section 2.1, we used vector methods to prove that the medians of a triangle meet at a point (called the *center* of the triangle) located two-thirds the distance from each vertex to the midpoint of the opposite side. Generalize this result by completing the steps in the following outline:

   (i) Let P_1, P_2, P_3, P_4 denote the vertices of a tetrahedron in 3-space, and let $\alpha_1 = \overrightarrow{P_1P_2}, \alpha_2 = \overrightarrow{P_1P_3}, \alpha_3 = \overrightarrow{P_1P_4}$. Note that $S = \{\alpha_1, \alpha_2, \alpha_3\}$ is a basis for 3-space. Find the S-coordinates of C_1, the center of the triangular face whose vertices are $P_2, P_3,$ and P_4.

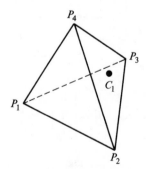

   (ii) Find the coordinates of the point P located three-fourths the distance from P_1 to C_1.

(iii) For $j = 2, 3, 4$, show that P is located three-fourths the distance from P_j to the center C_j of the opposite triangular face. Appropriately, P is called the *center* or *centroid* of the tetrahedron.

2. In the figure below, triangle ABC is equilateral and the points M, N, O are located so that

$$|AM| = \tfrac{1}{3}|AB|, \qquad |BN| = \tfrac{1}{3}|BC|, \qquad \text{and} \quad |CO| = \tfrac{1}{3}|CA|$$

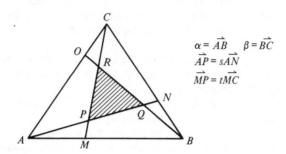

By symmetry, triangle PQR is also equilateral and, $|PM| = |QN| = |RO|$ and $|AP| = |BQ| = |CR|$.

(i) Use the vector methods developed in section 2.1 to show that $\overrightarrow{PQ} = \tfrac{3}{7}\overrightarrow{AN}$.

(ii) Show that $|AN|^2 = \tfrac{7}{9}|AB|^2$.
Hint: Use the law of cosines.

(iii) Verify that the area of triangle PQR is one-seventh that of triangle ABC. Do you think the same result would hold if triangle ABC were not equilateral? Explain.

3. In each of the following cases, determine whether or not the given collection of functions is a subspace of F, the vector space of all real-valued functions of a real variable.

(i) All functions which are integrable on $[0, 1]$.

(ii) All solutions of the differential equation $y'' - (\sin x)y' + e^x y = 0$.

(iii) All functions f such that $f(x) \geq f(y)$ if $x \geq y$.

(iv) All even functions, that is, $f(x) = f(-x)$.

(v) All functions f such that $\lim_{x \to 1} f(x) = 0$.

(vi) All functions f such that $f(x + 1) = f(x)$ for all real numbers x.

(vii) All functions f which are integrable on $[0, 1]$ and satisfy $\int_0^1 f(t) \sin t \, dt = 0$.

(viii) All functions f which satisfy $f(x) \geq 0$ for all real numbers x.

4. Let

$$A = \begin{bmatrix} 0 & 1 & 4 \\ 6 & 1 & -8 \\ -9 & 3 & 15 \end{bmatrix}$$

(i) Show that the collection V_1 of all 3-tuples X such that $AX = 3X$ is a 2-dimensional subspace of R_3. Find a basis $S_1 = \{\rho_1, \rho_2\}$ for V_1.

(ii) Show that V_2, the collection of all X such that $AX = 10X$, is a 1-dimensional subspace of R_3. Find a vector ρ_3 which spans V_2.

(iii) Show that $\{\rho_1, \rho_2, \rho_3\}$ is a basis for R_3.

(iv) Let P denote the 3×3 matrix whose jth column vector is ρ_j. Verify that
$$P^{-1}AP = \text{diag}(3, 3, 10)$$
This exercise previews a few ideas we shall develop in detail in chapters 4 and 6.

5. Let A be an $n \times n$ matrix, and let $\alpha_1, \ldots, \alpha_k$ be a collection of linearly independent n-tuples. Let $\beta_1, \ldots, \beta_k$ be vectors which satisfy $\beta_j = A\alpha_j$ for $j = 1, 2, \ldots, k$. Show that the β's must be linearly independent if A is non-singular.

6. Let A be an $n \times n$ matrix, and let γ be a fixed non-zero n-tuple.

(i) Show that the vectors $\gamma, A\gamma, A^2\gamma, \ldots, A^n\gamma$ are linearly dependent.
   Hint: Use lemma 2.1.

(ii) Show that the subspace $\text{Sp}\{\gamma, A\gamma, \ldots, A^n\gamma\}$ has dimension 1 if and only if there exists a scalar c such that $A\gamma = c\gamma$.

7. The vector space V is said to be the *direct sum* of the subspaces S and T if $S + T = V$ and if $S \cap T$ is the zero vector, in which case, we write $V = S \oplus T$.

(i) If $V = S \oplus T$ and if $\{\alpha_1, \ldots, \alpha_k\}$ and $\{\alpha_{k+1}, \ldots, \alpha_n\}$ are bases for S and T, respectively, show that $\{\alpha_1, \ldots, \alpha_n\}$ is a basis for V.

(ii) Let S and T denote the planes $2x + y + z = 0$ and $3x - y - z + 0$, respectively. Show that $S + T$ is all of 3-space. If $\{\alpha_1, \alpha_2\}$ and $\{\alpha_3, \alpha_4\}$ are bases for S and T, respectively, is $\{\alpha_1, \alpha_2, \alpha_3, \alpha_4\}$ a basis for R_3? Explain why this does not contradict the result of part (i).

8. If S and T are subspaces of the vector space V, it can be shown that
$$\dim S + \dim T = \dim(S + T) + \dim(S \cap T)$$

Verify this formula in the case where
$$\dim S = 3, \qquad \dim T = 4, \qquad \dim(S \cap T) = 2$$

by completing the steps in the following outline.

(i) Let $\{\alpha_1, \alpha_2\}$ be a basis for $S \cap T$. Explain why it is possible to find vectors β_3 and γ_3, γ_4 such that $\{\alpha_1, \alpha_2, \beta_3\}$ and $\{\alpha_1, \alpha_2, \gamma_3, \gamma_4\}$ are bases for S and T, respectively.

(ii) If $\rho = \sigma + \tau$, where σ and τ are elements of S and T, respectively, show that ρ can be expressed as a linear combination of the vectors $\alpha_1, \alpha_2, \beta_3, \gamma_3, \gamma_4$.

(iii) Suppose $r_1\alpha_1 + r_2\alpha_2 + s_3\beta_3 + t_3\gamma_3 + t_4\gamma_4 = \theta$. Show that the vector $\gamma = t_3\gamma_3 + t_4\gamma_4$ is contained in T and S. Since γ is an element of $S \cap T$, there exist scalars c_1 and c_2 such that $\gamma = c_1\alpha_1 + c_2\alpha_2$ (why?). Since $\{\alpha_1, \alpha_2, \gamma_3, \gamma_4\}$ is a basis for T, it follows that $c_1 = c_2 = t_3 = t_4 = 0$ (why?). Use a similar argument to show that $s_3 = 0$ and hence that $r_1 = r_2 = 0$.

(iv) Combine parts (ii) and (iii) to show that $\{\alpha_1, \alpha_2, \beta_3, \gamma_3, \gamma_4\}$ is a basis for $S + T$, and conclude the proof by noting that $\dim(S + T) = 5 = 3 + 4 - 2 = \dim S + \dim T - \dim(S \cap T)$.

The ambitious reader may wish to construct a general proof along the lines suggested here. The only difficulty occurs in part (iii), where the labeling of vectors becomes very messy.

9. An $n \times n$ matrix N is said to be *nilpotent of index* k if $N^k = O$ but N^{k-1} contains at least one non-zero entry.

   (i) Suppose N is nilpotent of index 2. If γ is an n-tuple such that $N\gamma \neq O$, show that γ and $N\gamma$ are linearly independent.
   Hint: If $s_0\gamma + s_1N\gamma = O$, note that $s_0N\gamma + s_1N^2\gamma = 0$. Since $N^2 = O$, it follows that $s_0N\gamma = O$.

   (ii) Generalize the result of part (i) by showing that if N is nilpotent of index k and γ is an n-tuple such that $N^{k-1}\gamma \neq O$, then the vectors γ, $N\gamma$, $\ldots$, $N^{k-1}\gamma$ are linearly independent.

   (iii) Show that the index of nilpotency k can be at most n.
   Hint: Use lemma 2.1 and the result of part (ii).

   (iv) Verify that

$$N = \begin{bmatrix} -2 & -3 & -7 \\ 1 & -1 & -4 \\ 0 & 1 & 3 \end{bmatrix}$$

   is nilpotent of index 3. Let

$$\gamma = \begin{bmatrix} 1 \\ 1 \\ 1 \end{bmatrix}$$

   First use part (ii) to show that $S = \{\gamma, N\gamma, N^2\gamma\}$ is a basis for R_3, and then find the S-coordinates of each of the standard basis 3-tuples.

10. Let $\{\alpha_1, \ldots, \alpha_n\}$ be a basis for R_n. Show that the $n \times n$ matrix P is non-singular if and only if the vectors $P\alpha_1, P\alpha_2, \ldots, P\alpha_n$ are linearly independent. (See exercise 5).

*11. If A is an $m \times n$ matrix with rank r, there exist non-singular matrices P and Q such that

$$QAP = J_r = \begin{bmatrix} I_r & O \\ O & O \end{bmatrix}$$

In other words, J_r is an $m \times n$ matrix with 1's in the first r diagonal positions and 0's elsewhere.
Prove this statement by completing the steps outlined below:

   (i) If E is the reduced row echelon form of A, explain why there exists a non-singular matrix Q such that $QA = E$.

   (ii) Show that the reduced row echelon form of E^t is the matrix J_r.
   Hint: Note that the form of E guarantees that the first r column vectors of E^t are linearly independent and all other columns contain only zeroes.

   (iii) Let P be a non-singular matrix which satisfies $P^tE^t = J_r$. (Why does such a matrix exist?). Complete the proof by showing that $QAP = J_r$.

*12. The change of coordinates represented by the following equations plays an important role in the special theory of relativity and is known as the *Lorentz transformation*:

$$X = k_v(x - vt)$$

$$T = k_v\left(\frac{-vx}{c^2} + t\right)$$

where $|v|$ is the speed of a moving object, c is the speed of light (a constant), and $k_v = c(c^2 - v^2)^{-1/2}$.

  (i) Describe the matrix $A(v)$ which represents the transformation of coordinates from the basis $\{X, T\}$ to the basis $\{x, t\}$.

 (ii) Show that $A(v)$ is non-singular whenever $|v| < c$. Describe $A^{-1}(v)$ and show that it also represents a Lorentz transformation.

(iii) If v_1, v_2 are two different speeds, show that there exists a third speed v_3 such that $A(v_1)A(v_2) = A(v_3)$.

Note: Using the result obtained in part (iii), it can be shown that the effect of performing a finite sequence of Lorentz transformations may be duplicated by performing a single, well-chosen transformation of the same type.

CHAPTER

3

Real Inner Product Spaces

3.1 METRIC PROPERTIES OF VECTORS IN 3-SPACE

The object of this section is to show how vector methods can be used in the plane and 3-space for measuring such things as the distance between two points, the length of a line segment, the angle between two rays, and the area of a triangle.

Throughout this section, we shall assume that O is the center of a fixed Cartesian coordinate system for 3-space—that is, a system in which the three coordinate axes are mutually perpendicular. As usual, if the point P has coordinates (a_1, a_2, a_3) and α is the vector $\overrightarrow{OP}$, we shall write $\alpha = (a_1, a_2, a_3)$. Our first objective is to obtain a formula for $\|\alpha\|$, the length of α. To this end, note that since the coordinate axes are mutually perpendicular, the point $P(a_1, a_2, a_3)$ is located $|a_1|$ units from the YZ-plane, $|a_2|$ units from the XZ-plane, and $|a_3|$ units from the XY-plane. This situation is depicted in figure 3.1 for the special case where a_1, a_2, a_3 are all positive. In figure 3.2, the point labeled Q lies directly "below" P in the XY-plane, while R is the point in the XZ-plane which is closest to Q. Applying the pythagorean theorem first to right triangle ORQ and then to OQP, we find that $l = |OQ| = \sqrt{a_1^2 + a_2^2}$ and $|OP| = \sqrt{l^2 + a_3^2} = \sqrt{a_1^2 + a_2^2 + a_3^2}$, and it is easy

149

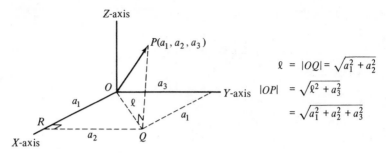

FIGURE 3.1

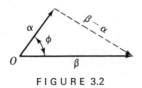

FIGURE 3.2

to see that the same relationships hold even when a_1, a_2, a_3 are not all positive. Since $\alpha = \overrightarrow{OP}$, it follows that $\|\alpha\| = |OP|$, and we can make the following general statement.

Lemma 3.1. If $\alpha = (a_1, a_2, a_3)$ is a vector in 3-space, then $\|\alpha\| = \sqrt{a_1^2 + a_2^2 + a_3^2}$.

EXAMPLE 1. If $\alpha = (-3, 4, 12)$, then

$$\|\alpha\| = \sqrt{(-3)^2 + (4)^2 + (12)^2} = 13$$

Next, suppose $\alpha = (a_1, a_2, a_3)$ and $\beta = (b_1, b_2, b_3)$ are vectors in 3-space, and let ϕ denote the angle between them, as indicated in figure 3.2. The distance from α to β, which we shall denote by $d(\alpha, \beta)$, is just the length of the vector $\beta - \alpha = (b_1 - a_1, b_2 - a_2, b_3 - a_3)$, and it follows that

$$d(\alpha, \beta) = \|\beta - \alpha\| = \sqrt{(b_1 - a_1)^2 + (b_2 - a_2)^2 + (b_3 - a_3)^2}$$

As the first step in determining the angle ϕ, we obtain the following equation by applying the law of cosines to the sides of triangle OP_1P_2:

$$\|\beta - \alpha\|^2 = \|\alpha\|^2 + \|\beta\|^2 - 2\|\alpha\|\|\beta\|\cos\phi$$

Solving for $\cos \phi$ and substituting coordinates, we find that

$$\cos \phi = \frac{\|\alpha\|^2 + \|\beta\|^2 - \|\beta - \alpha\|^2}{2\|\alpha\|\|\beta\|}$$

$$= \frac{(a_1^2 + a_2^2 + a_3^2) + (b_1^2 + b_2^2 + b_3^2)}{2\sqrt{a_1^2 + a_2^2 + a_3^2}\sqrt{b_1^2 + b_2^2 + b_3^2}}$$

$$- \frac{[(b_1 - a_1)^2 + (b_2 - a_2)^2 + (b_3 - a_3)^2]}{2\sqrt{a_1^2 + a_2^2 + a_3^2}\sqrt{b_1^2 + b_2^2 + b_3^2}}$$

$$= \frac{a_1b_1 + a_2b_2 + a_3b_3}{\sqrt{a_1^2 + a_2^2 + a_3^2}\sqrt{b_1^2 + b_2^2 + b_3^2}}$$

(3.1)

We can compute $\cos \phi$ by substituting in formula (3.1), but many different angles have the same cosine. Therefore, to guarantee that each expression of the form displayed in (3.1) is the cosine of exactly one angle ϕ, we insist that $0 \leq \phi \leq \pi$, where ϕ is measured in radians.

EXAMPLE 2.   If $\alpha = (1, 0, 1)$ and $\beta = (0, 1, -1)$, then

$$\cos \phi = \frac{(1)(0) + (0)(1) + (1)(-1)}{\sqrt{1^2 + 0^2 + 1^2}\sqrt{0^2 + 1^2 + (-1)^2}} = \frac{-1}{\sqrt{2}\sqrt{2}} = \frac{-1}{2}$$

It follows that ϕ must be $2\pi/3$ radians (120°), since this is the only angle between 0 and π whose cosine is $-\frac{1}{2}$.

The formulas we have derived for length, distance, and angle are rather awkward to use in algebraic manipulations but we can eliminate this difficulty by using the notational device defined as follows.

Definition 3.1.   If $\alpha = (a_1, a_2, a_3)$ and $\beta = (b_1, b_2, b_3)$ are vectors in 3-space, the number $a_1b_1 + a_2b_2 + a_3b_3$ is called the *scalar product* of α by β and is denoted by $\alpha \cdot \beta$.

EXAMPLE 3.   If $\alpha = (-1, 3, 2)$ and $\beta = (7, 5, -3)$, we have $\alpha \cdot \beta = (-1)(7) + (3)(5) + (2)(-3) = 2$.

The scalar product notation enables us to express each formula we have derived so far in a more compact form. To be specific, if $\alpha = (a_1, a_2, a_3)$

and $\beta = (b_1, b_2, b_3)$, we have

$$\| \alpha \| = \sqrt{a_1^2 + a_2^2 + a_3^2} = \sqrt{\alpha \cdot \alpha}$$

$$d(\alpha, \beta) = \| \beta - \alpha \| = \sqrt{(\beta - \alpha) \cdot (\beta - \alpha)} \qquad (3.2)$$

$$\cos \phi = \frac{a_1 b_1 + a_2 b_2 + a_3 b_3}{\sqrt{a_1^2 + a_2^2 + a_3^2}\sqrt{b_1^2 + b_2^2 + b_3^2}} = \frac{\alpha \cdot \beta}{\| \alpha \| \| \beta \|}$$

Before we can say much more about these formulas, we need to know more about the scalar product, and our first theorem provides a list of the most important properties of this operation.

Theorem 3.1. If α, β, γ are vectors in 3-space, then

(i) $\alpha \cdot \alpha > 0$ if α is not the zero vector

(ii) $\alpha \cdot \beta = \beta \cdot \alpha$

(iii) $(t\alpha) \cdot \beta = t(\alpha \cdot \beta)$ for each real number t

(iv) $(\alpha + \beta) \cdot \gamma = (\alpha \cdot \gamma) + (\beta \cdot \gamma)$

Proof: We shall leave the first two parts as an exercise. To verify (iii) and (iv), let $\alpha = (a_1, a_2, a_3)$, $\beta = (b_1, b_2, b_3)$, and $\gamma = (c_1, c_2, c_3)$. Then we have

(iii) $(t\alpha) \cdot \beta = (ta_1)b_1 + (ta_2)b_2 + (ta_3)b_3$

$$= t(a_1 b_1 + a_2 b_2 + a_3 b_3) = t(\alpha \cdot \beta)$$

(iv) $(\alpha + \beta) \cdot \gamma = (a_1 + b_1)c_1 + (a_2 + b_2)c_2 + (a_3 + b_3)c_3$

$$= (a_1 c_1 + a_2 c_2 + a_3 c_3) + (b_1 c_1 + b_2 c_2 + b_3 c_3)$$

$$= (\alpha \cdot \gamma) + (\beta \cdot \gamma) \quad \blacksquare$$

By combining the last three properties of theorem 3.1, we find that

$$\alpha \cdot (\beta + \gamma) = (\beta + \gamma) \cdot \alpha = (\beta \cdot \alpha) + (\gamma \cdot \alpha) = \alpha \cdot \beta + \alpha \cdot \gamma$$

and

$$\alpha \cdot (t\beta) = (t\beta) \cdot \alpha = t(\beta \cdot \alpha) = t(\alpha \cdot \beta)$$

These observations yield the following useful result.

Corollary 3.1. If $\rho = s_1 \alpha_1 + s_2 \alpha_2$ and $\sigma = t_1 \beta_1 + t_2 \beta_2$, then

$$\rho \cdot \sigma = s_1 t_1 (\alpha_1 \cdot \beta_1) + s_1 t_2 (\alpha_1 \cdot \beta_2) + s_2 t_1 (\alpha_2 \cdot \beta_1) + s_2 t_2 (\alpha_2 \cdot \beta_2)$$

EXAMPLE 4. Suppose $\rho = 3\alpha_1 - 2\alpha_2$ and $\sigma = 4\beta_1 + \beta_2$. Then $\rho \cdot \sigma = 12(\alpha_1 \cdot \beta_1) + 3(\alpha_1 \cdot \beta_2) - 8(\alpha_2 \cdot \beta_1) - 2(\alpha_2 \cdot \beta_2)$

Note that when corollary 3.1 is written in terms of the summation notation, we have

$$\left(\sum_{i=1}^{2} s_i \alpha_i \right) \cdot \left(\sum_{j=1}^{2} t_j \beta_j \right) = \sum_{i=1}^{2} \sum_{j=1}^{2} s_i t_j (\alpha_i \cdot \beta_j)$$

and a slightly more general version of this result is outlined in exercise 13.

Earlier, we observed that the cosine of the angle ϕ between the vectors
α and β may be expressed as

$$\cos \phi = \frac{\alpha \cdot \beta}{\|\alpha\| \|\beta\|}$$

Turning things around, this same formula can be used to derive the following additional properties of the scalar product.

Theorem 3.2. If α and β are vectors in 3-space, we have
(i) $\alpha \cdot \beta = 0$ if and only if α and β are perpendicular.
(ii) $|\alpha \cdot \beta| \leq \|\alpha\| \|\beta\|$ with equality if and only if $\beta = t\alpha$ for some real number t.

Proof: (i) For α and β to be perpendicular, the angle ϕ between them
must be $\pi/2$ radians (90°). In other words, $\cos \phi$ must be 0, and this is equivalent to the condition $\alpha \cdot \beta = 0$.

(ii) The inequality is an immediate consequence of the fact that $|\cos \phi| \leq$
1. To verify the statement regarding equality, note that $|\alpha \cdot \beta| = \|\alpha\| \|\beta\|$ is
equivalent to $|\cos \phi| = 1$, and this occurs only when ϕ is 0 or π. Therefore,
we have equality if and only if α and β are parallel, that is, whenever β is a
scalar multiple of α. ∎

EXAMPLE 5. The vectors $\alpha = (3, -2, 5)$ and $\beta = (7, 3, -3)$ are perpendicular since

$$\alpha \cdot \beta = (3)(7) + (-2)(3) + (5)(-3) = 0$$

If $\gamma = (0, 3, -4)$ and $\delta = (3, -4, 12)$, we have

$$\gamma \cdot \delta = (0)(3) + (3)(-4) + (-4)(12) = -60$$

$$\|\gamma\| = \sqrt{0^2 + 3^2 + (-4)^2} = 5$$

$$\|\delta\| = \sqrt{3^2 + (-4)^2 + (12)^2} = 13$$

and we find that

$$|-60| \leq (5)(13)$$

$$|\gamma\cdot\delta| \leq \|\gamma\|\|\delta\|$$

as predicted by part (ii) of theorem 3.2.

The properties listed in theorems 3.1 and 3.2 can be used for many different purposes in geometry, several of which are illustrated in the solved problems that follow.

PROBLEM 1.   Show that the triangle whose vertices are $P_1(1, -2, 3)$, $P_2(4, 2, 15)$, and $P_3(-3, 10, 0)$ is an isosceles right triangle with right angle at P_1.

Solution: If α and β denote the vectors $\overrightarrow{P_1P_2}$ and $\overrightarrow{P_1P_3}$, respectively, the problem can be restated as indicated in figure 3.3:

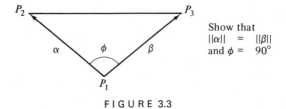

FIGURE 3.3

Since $\alpha = (3, 4, 12)$ and $\beta = (-4, 12, -3)$, we have

$$\|\alpha\| = \sqrt{(3)^2 + (4)^2 + (12)^2} = 13$$

$$\|\beta\| = \sqrt{(-4)^2 + (12)^2 + (-3)^2} = 13$$

and

$$\alpha\cdot\beta = (3)(-4) + (4)(12) + (12)(-3) = 0$$

Therefore, α and β are perpendicular vectors which have the same length, and it follows that $P_1P_2P_3$ is an isosceles right triangle.

PROBLEM 2.   Show that a rectangle is the only kind of parallelogram whose diagonals are of equal length.

Solution: Let P_1, P_2, P_3, P_4 be the vertices of a parallelogram. If we set $\alpha = \overrightarrow{P_1P_2}$ and $\beta = \overrightarrow{P_1P_4}$, then the diagonals are $\overrightarrow{P_1P_3} = \alpha + \beta$ and $\overrightarrow{P_4P_2} =$

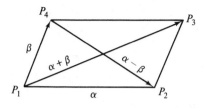

Show that
$||\alpha + \beta|| = ||\alpha - \beta||$
if and only if α is
perpendicular to β

FIGURE 3.4

$\alpha - \beta$, as indicated in figure 3.4. Using properties (ii), (iii), and (iv) of theorem 3.1, we find that

$$||\alpha + \beta||^2 = (\alpha + \beta)\cdot(\alpha + \beta) = (\alpha\cdot\alpha) + 2(\alpha\cdot\beta) + (\beta\cdot\beta)$$

and

$$||\alpha - \beta||^2 = (\alpha - \beta)\cdot(\alpha - \beta) = (\alpha\cdot\alpha) - 2(\alpha\cdot\beta) + (\beta\cdot\beta)$$

Therefore, we have

$$||\alpha + \beta||^2 - ||\alpha - \beta||^2 = 4(\alpha\cdot\beta)$$

and it follows that $||\alpha + \beta|| = ||\alpha - \beta||$ if and only if $\alpha\cdot\beta = 0$. In other words, the two diagonals have the same length whenever α is perpendicular to β, and this occurs only when $P_1 P_2 P_3 P_4$ is a rectangle.

PROBLEM 3.   Show that the distance from the point $P_0(x_0, y_0, z_0)$ to the plane $Ax + By + Cz = D$ is given by the formula

$$d = \left| \frac{Ax_0 + By_0 + Cz_0 - D}{\sqrt{A^2 + B^2 + C^2}} \right|$$

Solution: First, observe that the vector $\alpha = (A, B, C)$ is perpendicular to every vector in the plane. To prove this statement, let $P_1(x_1, y_1, z_1)$ and $P_2(x_2, y_2, z_2)$ be points in the plane, and let $\beta = \overrightarrow{P_1 P_2} = (x_2 - x_1, y_2 - y_1, z_2 - z_1)$. Then we have

$$\alpha\cdot\beta = A(x_2 - x_1) + B(y_2 - y_1) + C(z_2 - z_1)$$

$$= (Ax_2 + By_2 + Cz_2) - (Ax_1 + By_1 + Cz_1) = D - D = 0$$

Next, let $\gamma = \overrightarrow{P_1 P_0} = (x_1 - x_0, y_1 - y_0, z_1 - z_0)$ and consider figure 3.5.

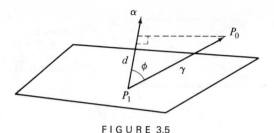

FIGURE 3.5

We find that

$$d = \left| \|\gamma\| \cos \phi \right| = \left| \frac{\alpha \cdot \gamma}{\|\alpha\|} \right|$$

$$= \left| \frac{A(x_1 - x_0) + B(y_1 - y_0) + C(z_1 - z_0)}{\sqrt{A^2 + B^2 + C^2}} \right|$$

$$= \left| \frac{Ax_0 + By_0 + Cz_0 - D}{\sqrt{A^2 + B^2 + C^2}} \right|$$

where the last equation follows from the fact that $Ax_1 + By_1 + Cz_1 = D$ (why is this true?)

We conclude this section by stating two theorems which provide useful information about length and distance.

Theorem 3.3.   If α and β are vectors in 3-space, we have
(i) $\|\alpha\| > 0$ unless α is the zero vector
(ii) $\|t\alpha\| = |t| \|\alpha\|$ for each real number t
(iii) $\|\alpha + \beta\| \leq \|\alpha\| + \|\beta\|$

Proof: Parts (i) and (ii) follow immediately from theorem 3.1. To derive the inequality in part (iii), note that

$$\|\alpha + \beta\|^2 = (\alpha + \beta) \cdot (\alpha + \beta) = (\alpha \cdot \alpha) + 2(\alpha \cdot \beta) + (\beta \cdot \beta)$$

$$= \|\alpha\|^2 + 2(\alpha \cdot \beta) + \|\beta\|^2$$

Since $(\alpha \cdot \beta) \leq \|\alpha\| \|\beta\|$, we have

$$\|\alpha + \beta\|^2 \leq \|\alpha\|^2 + 2\|\alpha\| \|\beta\| + \|\beta\|^2 = (\|\alpha\| + \|\beta\|)^2$$

and it follows that $\|\alpha + \beta\| \leq \|\alpha\| + \|\beta\|$. ∎

Incidentally, the inequality established in part (iii) of theorem 3.3 is often referred to as the *triangle inequality* since it is related to the fact that the sum of the lengths of any two sides of a triangle is greater than the length of the third.

Theorem 3.4. If α, β, γ are vectors in 3-space, then
 (i) $d(\alpha, \alpha) = 0$ and $d(\alpha, \beta) > 0$ if $\alpha \neq \beta$
 (ii) $d(\alpha, \beta) = d(\beta, \alpha)$
 (iii) $d(\alpha, \beta) \leq d(\alpha, \gamma) + d(\gamma, \beta)$

Proof: Parts (i) and (ii) follow from the corresponding parts of theorem 3.3, and the proof of part (iii) is outlined in exercise 12.

Our survey of the metric properties of vectors in 3-space has demonstrated the importance of the scalar product, and in the sections which follow, we shall show how a more general version of this operation may be used to make measurements in abstract vector spaces.

Exercise Set 3.1

1. In each of the following cases, find $\| \alpha \|$, $\| \beta \|$, $\alpha \cdot \beta$, $d(\alpha, \beta)$, and $\cos \phi$, where ϕ is the angle between α and β.
    (i) $\alpha = (12, -5, 0)$ and $\beta = (0, 3, -4)$
    (ii) $\alpha = (1, -1, 3)$ and $\beta = (-5, 1, 2)$
    (iii) $\alpha = (-35, 12, 0)$ and $\beta = (0, 8, 15)$
2. Find the perimeter of the triangle whose vertices are $P_1(1, -2, 5)$, $P_2(4, -6, 17)$, and $P_3(4, 14, -4)$.
3. Show that the triangle whose vertices are $P_1(-3, 0, -4)$, $P_2(0, 0, 0)$, and $P_3(4, 5\sqrt{3}, -3)$ is a right triangle and find its area.
4. If $\alpha = (0, 8, 15)$, find a vector β such that $\| \beta \| = 14$ and $\cos \phi = \frac{1}{17}$, where ϕ is the angle between α and β.
5. (i) If α and β are vectors in 3-space, show that

$$(\alpha + \beta) \cdot (\alpha - \beta) = \| \alpha \|^2 - \| \beta \|^2$$

   (ii) Use part (i) to show that the diagonals of a rhombus are perpendicular to each other. (Recall that a rhombus is a parallelogram whose sides all have the same length.)
6. (i) If α and β are vectors in 3-space, show that

$$\| \alpha + \beta \|^2 + \| \alpha - \beta \|^2 = 2\| \alpha \|^2 + 2\| \beta \|^2$$

   (ii) Use part (i) to state and prove a theorem concerning the diagonals and sides of a parallelogram.
7. Let P_1, P_2, P_3 be points in 3-space, and set $\alpha = \overrightarrow{P_1 P_2}$ and $\beta = \overrightarrow{P_1 P_3}$. Show that triangle $P_1 P_2 P_3$ has a right angle at P_1 if and only if

$$\| \alpha + \beta \|^2 = \| \alpha \|^2 + \| \beta \|^2$$

8. Use the triangle inequality (see theorem 3.3) to show that

$$\| \alpha - \beta \| \geq \left| \|\alpha\| - \|\beta\| \right|$$

for all vectors α, β in 3-space.

9. In the figure below, γ is perpendicular to β, $\alpha \cdot \beta = 14$ and $\|\beta\| = 7$:

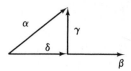

    (i) Find a real number t for which $\delta = t\beta$.

    (ii) Find real numbers s_1, s_2 for which $\gamma = s_1\alpha + s_2\beta$.

10. Find the distance from the point $P_0(-3, 7, 2)$ to the plane $12x - 3y + 4z = 8$.

11. Find the equation of the plane which contains the point $P_0(-3, 7, 4)$ and is perpendicular to the vector $\alpha = (-2, 1, 5)$.

    Hint: If $P(x, y, z)$ is a general point in the plane and $\beta = \overrightarrow{P_0P}$, then $\alpha \cdot \beta$ must be 0.

12. Let α, β, γ be vectors in 3-space.

    (i) Use the triangle inequality to show that

$$\| \beta - \alpha \| = \| \gamma - \alpha + \beta - \gamma \| \leq \| \gamma - \alpha \| + \| \beta - \gamma \|$$

    (ii) Complete the proof of theorem 3.4 by using part (i) to show that

$$d(\alpha, \beta) \leq d(\alpha, \gamma) + d(\gamma, \beta)$$

*13. If $\alpha_1, \ldots, \alpha_m$ and $\beta_1, \ldots, \beta_n$ are vectors in 3-space, use mathematical induction to show that

$$\left(\sum_{i=1}^{m} s_i\alpha_i \right) \cdot \left(\sum_{j=1}^{n} t_j\beta_j \right) = \sum_{j=1}^{n} \sum_{i=1}^{m} s_i t_j (\alpha_i \cdot \beta_j)$$

*14. Show that the triangle determined by the non-zero vectors α and β has area

$$\frac{1}{2} \sqrt{\|\alpha\|^2 \|\beta\|^2 - (\alpha \cdot \beta)^2}$$

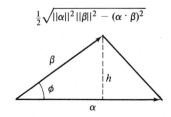

15. Find the area of the triangle whose vertices are $P_1(-3, 1, 0)$, $P_2(0, -1, 4)$, and $P_3(-2, 2, -2)$.

3.2 PROPERTIES OF A REAL INNER PRODUCT

In this section, we shall discuss what is meant by length, distance, and angle in a general vector space. In developing the metric properties of 3-space, we were guided by intuition and familiar results such as the pythagorean theorem and the law of cosines, but this approach cannot be used in other spaces. After all, what kind of intuition do we have for such things as the length of a polynomial or the angle between two matrices? Thus, before we can even begin to discuss measurements in an abstract vector space, we must first provide precise, meaningful definitions of the various metric functions we plan to use. We begin this program by defining a fundamental metric function which has the same basic properties as the scalar product.

Definition 3.2. Let V be a real vector space. A function $(\ \ | \ \ )$ which assigns to each ordered pair $\{\alpha, \beta\}$ of vectors in V a unique real number [denoted by $(\alpha \,|\, \beta)$] is said to be a *real inner product* on V if it also satisfies the following conditions:

(i) $(\alpha \,|\, \alpha) > 0$ if α is a non-zero vector in V
(ii) $(\alpha \,|\, \beta) = (\beta \,|\, \alpha)$ for all α, β in V
(iii) $(t\alpha \,|\, \beta) = t(\alpha \,|\, \beta)$ for α, β in V and each real number t
(iv) $(\alpha + \beta \,|\, \gamma) = (\alpha \,|\, \gamma) + (\beta \,|\, \gamma)$ for all α, β, γ in V

The vector space V together with a specified real inner product function is called a *real inner product space*.

Until now, we have done little to discourage the reader from thinking that our results are just as valid when the scalar field is the set of all complex numbers as when it is the real numbers, and this is true, at least as far as the material covered in the first two chapters is concerned. It is even possible to define the notion of a complex inner product space, but this definition is not entirely analogous to definition 3.2. In this chapter, we shall concentrate on the properties of real inner product spaces, but the corresponding properties of complex inner product spaces are also important and will be developed in the supplementary exercises at the end of the chapter.

According to theorem 3.1, the scalar product is a real inner product on the vector space R_3, and the following examples show how inner product functions can be defined on other familiar spaces.

EXAMPLE 1.   If $\alpha = (a_1, a_2, \ldots, a_n)$ and $\beta = (b_1, b_2, \ldots, b_n)$ are vectors in R_n, the function

$$(\alpha \,|\, \beta) = \sum_{i=1}^{n} a_i b_i$$

is a real inner product (see exercise 3). For future reference, we shall call this function the *standard inner product* on R_n. Note that when $n = 3$, $(\alpha \mid \beta)$ is just the scalar product, $\alpha \cdot \beta$.

EXAMPLE 2.   If p and q are polynomials in P_n, then the function $(\ \mid\ )$ defined by

$$(p \mid q) = \int_0^1 p(t)q(t)\, dt$$

is a real inner product, which we shall refer to as the *integral inner product* on P_n. To verify that this function actually does satisfy the axioms of definition 3.2, we require only the basic properties of integration:

(i) $(p \mid p) = \int_0^1 p^2(t)\, dt > 0$ unless p is the zero polynomial

(ii) $(p_1 \mid p_2) = \int_0^1 p_1(t)p_2(t)\, dt = \int_0^1 p_2(t)p_1(t)\, dt = (p_2 \mid p_1)$

(iii) $(rp_1 \mid p_2) = \int_0^1 [rp_1(t)]p_2(t)\, dt = r\int_0^1 p_1(t)p_2(t)\, dt = r(p_1 \mid p_2)$

(iv) $(p_1 + p_2 \mid p_3) = \int_0^1 [p_1(t) + p_2(t)]p_3(t)\, dt$

$$= \int_0^1 [p_1(t)p_3(t)]\, dt + \int_0^1 [p_2(t)p_3(t)]\, dt = (p_1 \mid p_3) + (p_2 \mid p_3)$$

Incidentally, it can be shown that the function $(f \mid g) = \int_0^1 f(t)g(t)\, dt$ is an inner product on $C[0, 1]$ and its subspaces, notably P_∞ and $D[0, 1]$. The only real difficulty in verifying this fact lies in showing that $\int_0^1 f^2(t)\, dt$ is positive whenever f is a non-zero continuous function, a result which is usually derived in advanced calculus.

EXAMPLE 3.   The function defined by

$$(A \mid B) = \operatorname{tr}(A^tB)\quad \text{for } m \times n \text{ matrices } A,\ B$$

is a real inner product on the vector space ${}_mR_n$ (see exercise 9).
Note: Recall that the trace of the $n \times n$ matrix $C = (c_{ij})$ is given by $\operatorname{tr} C = \sum_{k=1}^n c_{kk}$ (see exercise 5, section 1.4).

It is possible to have more than one inner product function for the same vector space. For instance, the standard inner product $(\ \mid\ )_1$ on R_2 is defined by

$$(\alpha \mid \beta)_1 = a_1b_1 + a_2b_2 \tag{3.3}$$

for all $\alpha = (a_1, a_2)$ and $\beta = (b_1, b_2)$, and it can be shown (see exercise 4) that the function

$$(\alpha \mid \beta)_2 = 4a_1b_1 - a_1b_2 - a_2b_1 + 4a_2b_2 \qquad (3.4)$$

is also an inner product on R_2. This example shows that any reference to "the inner product space R_2" has no meaning until we identify a specific inner product function to be used in the space.

At first glance, our definition of a real inner product function seems to be biased in favor of the left member of the ordered pair $\{\alpha, \beta\}$ but, as in the case of the scalar product, this disparity can be eliminated by observing that

$$(\gamma \mid \alpha + \beta) = (\alpha + \beta \mid \gamma) = (\alpha \mid \gamma) + (\beta \mid \gamma)$$

$$= (\gamma \mid \alpha) + (\gamma \mid \beta)$$

and

$$(\alpha \mid t\beta) = (t\beta \mid \alpha) = t(\beta \mid \alpha) = t(\alpha \mid \beta)$$

If we combine these two observations and generalize by induction, we obtain a formula for the inner product of any two linear combinations.

Theorem 3.5.   If V is a real inner product space, we have

$$\left(\sum_{i=1}^{m} s_i\alpha_i \,\bigg|\, \sum_{j=1}^{n} t_j\beta_j \right) = \sum_{i=1}^{m} s_i\left[\sum_{j=1}^{n} t_j(\alpha_i \mid \beta_j) \right] = \sum_{i=1}^{m} \sum_{j=1}^{n} s_it_j(\alpha_i \mid \beta_j)$$

EXAMPLE 4.   Let $\rho = 3\alpha_1 - \alpha_2 + 4\alpha_3$ and $\sigma = 2\beta_1 - 3\beta_2$. Then we have

$$(\rho \mid \sigma) = 6(\alpha_1 \mid \beta_1) - 9(\alpha_1 \mid \beta_2) - 2(\alpha_2 \mid \beta_1) + 3(\alpha_2 \mid \beta_2)$$

$$+ 8(\alpha_3 \mid \beta_1) - 12(\alpha_3 \mid \beta_2)$$

In the special case where the α's and β's are the same, theorem 3.5 assumes the following form.

Corollary 3.2.   If $\alpha = a_1\alpha_1 + \ldots + a_n\alpha_n$ and $\beta = b_1\alpha_1 + \ldots + b_n\alpha_n$ are vectors in the real inner product space V, then

$$(\alpha \mid \beta) = \sum_{i=1}^{n} a_i\left[\sum_{j=1}^{n} b_j(\alpha_i \mid \alpha_j) \right] = \sum_{i=1}^{n} \sum_{j=1}^{n} a_ib_j(\alpha_i \mid \alpha_j)$$

In the last section, we found that each metric function in 3-space can be expressed in terms of the scalar product. Now, we turn things around by using the inner product function and the formulas displayed in (3.1) to *define* length, distance, and angle in a real inner product space.

Definition 3.3. Let V be a real inner product space with inner product $(\ \mid\ )$. Then

(i) The *length* (or *norm*) of the vector α is $\|\alpha\| = \sqrt{(\alpha\mid\alpha)}$.

(ii) The *distance* from the vector α to β is $d(\alpha,\beta) = \|\alpha - \beta\|$.

(iii) The cosine of the angle ϕ between the non-zero vectors α and β is given by

$$\cos\phi = \frac{(\alpha\mid\beta)}{\|\alpha\|\|\beta\|}$$

EXAMPLE 5. In the real inner product space R_4 with the standard inner product, let $\alpha = (1, 0, 2, -2)$, and $\beta = (-1, 1, 1, -1)$. Then we have

$$\|\alpha\| = \sqrt{(\alpha\mid\alpha)} = \sqrt{(1)^2 + (0)^2 + (2)^2 + (-2)^2} = 3$$

$$\|\beta\| = \sqrt{(-1)^2 + (1)^2 + (1)^2 + (-1)^2} = 2$$

$$(\alpha\mid\beta) = (1)(-1) + (0)(1) + (2)(1) + (-2)(-1) = 3$$

$$d(\alpha,\beta) = \|\alpha - \beta\|$$

$$= \sqrt{(1+1)^2 + (0-1)^2 + (2-1)^2 + (-2+1)^2} = \sqrt{7}$$

$$\cos\phi = \frac{(\alpha\mid\beta)}{\|\alpha\|\|\beta\|} = \frac{3}{(3)(2)} = \frac{1}{2}$$

From the last computation, we see that α and β are separated by an angle of $\pi/3$ radians (60°).

EXAMPLE 6. In the inner product space P_2 with the integral inner product (recall example 2), let $p(t) = 2t - 1$ and $q(t) = t^2 - t + \frac{3}{2}$. Then we have

$$\|p\| = \sqrt{(p\mid p)} = \left[\int_0^1 (2t - 1)^2\, dt\right]^{1/2} = \sqrt{\tfrac{1}{3}}$$

$$\|q\| = \left[\int_0^1 (t^2 - t + \tfrac{3}{2})^2\, dt\right]^{1/2} = \sqrt{\tfrac{107}{60}}$$

$$(p\mid q) = \int_0^1 (2t - 1)(t^2 - t + \tfrac{3}{2})\, dt = 0$$

$$d(p, q) = \|p - q\| = \left[\int_0^1 (-t^2 + 3t - \tfrac{5}{2})^2\, dt\right]^{1/2} = \sqrt{\tfrac{127}{60}}$$

$$\cos\phi = \frac{(p\mid q)}{\|p\|\|q\|} = 0 \quad \text{since } (p\mid q) = 0$$

When two vectors α and β are separated by $90°$ (as are p and q in example 6), it is natural to think of them as being perpendicular, and since this occurs if and only if $(\alpha \,|\, \beta) = 0$, we make the following definition.

Definition 3.4. The vectors α and β are said to be *orthogonal* (that is, perpendicular) with respect to the real inner product $(\quad|\quad)$ if $(\alpha \,|\, \beta) = 0$.

It is important to note that orthogonality is not an innate property associated with a given pair of vectors but depends very much on the nature of the inner product under consideration. For instance, the vectors $\alpha = (1, 2)$ and $\beta = (2, -1)$ are orthogonal with respect to the standard inner product on R_2 since

$$(1)(2) + (2)(-1) = 0$$

but if $(\quad|\quad)_2$ is the inner product defined in (3.4), we find that

$$(\alpha \,|\, \beta)_2 = 4(1)(2) - (1)(-1) - (2)(2) + 4(2)(-1) = -3$$

We shall have a great deal more to say about orthogonal vectors in the next section.

The reader may have already observed that it is meaningless to say that the quotient

$$\frac{(\alpha \,|\, \beta)}{\|\alpha\| \|\beta\|}$$

is the cosine of an angle unless we can show that this expression always lies between 1 and -1. This fact follows immediately from the following generalization of theorem 3.2, which, incidentally, is one of the most important elementary facts in all mathematics.

Theorem 3.6. (Cauchy-Schwarz.) If α and β are vectors in the real inner product space V, then

$$(\alpha \,|\, \beta)^2 \le (\alpha \,|\, \alpha)(\beta \,|\, \beta)$$

with equality if and only if there exists a real number t such that $\beta = t\alpha$.

Proof: Let α and β be arbitrary but fixed vectors in V. If $\alpha = \theta$, the inequality is clearly valid. If α is not the zero vector, consider the (variable) vector expression $x\alpha + \beta$, where x is a real variable. According to the axioms of definition 3.2, we have

$$0 \le (x\alpha + \beta \,|\, x\alpha + \beta) = (\alpha \,|\, \alpha)x^2 + 2(\alpha \,|\, \beta)x + (\beta \,|\, \beta) \qquad (3.5)$$

for all real x, with equality if and only if $x\alpha + \beta = \theta$. Completing the square on the right side of inequality (3.5), we obtain

$$0 \le (\alpha \,|\, \alpha)\left[x + \frac{(\alpha \,|\, \beta)}{(\alpha \,|\, \alpha)} \right]^2 + \left[(\beta \,|\, \beta) - \frac{(\alpha \,|\, \beta)^2}{(\alpha \,|\, \alpha)} \right]$$

for all real x, and in particular, this inequality must be valid when

$$x = -\frac{(\alpha \,|\, \beta)}{(\alpha \,|\, \alpha)}$$

For this value of x, the inequality becomes

$$0 \le (\beta \,|\, \beta) - \frac{(\alpha \,|\, \beta)^2}{(\alpha \,|\, \alpha)}$$

and since $(\alpha \,|\, \alpha) > 0$, it follows that

$$(\alpha \,|\, \alpha)(\beta \,|\, \beta) \ge (\alpha \,|\, \beta)^2,$$

as desired.

The proof of the statement regarding equality is outlined in exercise 11. ∎

EXAMPLE 7.   If $\alpha = (a_1, a_2, \ldots, a_n)$ and $\beta = (b_1, b_2, \ldots, b_n)$ are elements of the real inner product space R_n with the standard inner product, the Cauchy-Schwarz inequality assumes the following form:

$$\left(\sum_{i=1}^{n} a_i b_i \right)^2 \le \left(\sum_{i=1}^{n} a_i^2 \right)\left(\sum_{i=1}^{n} b_i^2 \right)$$

A great deal of time and effort may be saved by using theorem 3.6 to obtain this inequality and others like it (e.g., see exercise 8). The reader who doubts this fact may find it instructive to construct a direct proof using the coordinates of α and β.

If α and β are non-zero vectors in the real inner product space V, the Cauchy-Schwarz inequality can be written as

$$\left[\frac{(\alpha \,|\, \beta)}{\|\alpha\|\|\beta\|} \right]^2 \le 1$$

Therefore, we have

$$-1 \le \frac{(\alpha \,|\, \beta)}{\|\alpha\|\|\beta\|} \le 1$$

which means that our definition of $\cos \varphi$ is reasonable.

Our next two theorems show that the basic properties of length and distance in 3-space are valid in any real inner product space (recall theorems 3.3 and 3.4.)

Theorem 3.7. If V is a real inner product space, then

(i) $||\alpha|| > 0$ for each non-zero vector α

(ii) $||t\alpha|| = |t|\,||\alpha||$ for each real number t and vector α

(iii) $||\alpha + \beta|| \leq ||\alpha|| + ||\beta||$ for all α, β in V

Proof: Exercise. *Hint:* Use the proof of theorem 3.3 as a model.

Theorem 3.8. If α, β, γ are vectors in the real inner product space V, then

(i) $d(\alpha, \beta) > 0$ if $\alpha \neq \beta$

(ii) $d(\alpha, \beta) = d(\beta, \alpha)$

(iii) $d(t\alpha, t\beta) = |t|\,d(\alpha, \beta)$

(iv) $d(\alpha, \beta) \leq d(\alpha, \gamma) + d(\gamma, \beta)$

Proof: Exercise. *Hint:* The proof of (iv) follows the lines suggested in exercise 12, section 3.1.

We now have a fairly complete list of the basic properties of the functions used to carry out measurements in a real inner product space. In the next section, we shall continue our study of inner product spaces by taking a closer look at the concept of orthogonality.

Exercise Set 3.2

1. Let (|) denote the standard inner product on R_4. If $\alpha = (-5, 3, 1, -1)$ and $\beta = (8, 4, 1, 0)$, compute $(\alpha\,|\,\beta)$, $||\alpha||$, $||\beta||$, and $d(\alpha, \beta)$, and find the angle ϕ between α and β.

2. Let (|) denote the integral inner product on P_2. If $p(t) = 2t - 3$ and $q(t) = t^2 - 7t + 1$, find $(p|q)$, $||p||$, and $d(p, q)$.

3. Verify that the standard inner product on R_n actually does satisfy the conditions of definition 3.2.

4. Verify that the function defined by

$$(\alpha\,|\,\beta) = 4a_1b_1 - a_2b_1 - a_1b_2 + 4a_2b_2$$

for $\alpha = (a_1, a_2)$ and $\beta = (b_1, b_2)$ is an inner product on R_2.

5. Each of the following is a function which assigns a unique real number to the ordered pair $\{\alpha, \beta\}$ in R_2, where $\alpha = (a_1, a_2)$ and $\beta = (b_1, b_2)$. In each case, either show that the given function is a real inner product or provide an example to show that it violates a condition of definition 3.2.

(i) $(\alpha\,|\,\beta)_1 = 2a_1b_1 - 3a_1b_2 - 3a_2b_1 + 5a_2b_2$

(ii) $(\alpha\,|\,\beta)_2 = a_1b_1 - 2a_1b_2 - 2a_2b_1 + 3a_2b_2$

(iii) $(\alpha\,|\,\beta)_3 = 4a_1b_1 - a_1b_2 + a_2b_1 + a_2b_2$

(iv) $(\alpha\,|\,\beta)_4 = a_1b_1a_2b_2$

6. Let $\{\sigma_1, \sigma_2, \ldots, \sigma_n\}$ be a basis for the vector space V, and let (|) be the function defined as follows:

$$(\alpha\,|\,\beta) = a_1b_1 + a_2b_2 + \ldots + a_nb_n$$

where $\alpha = a_1\sigma_1 + \ldots + a_n\sigma_n$ and $\beta = b_1\sigma_1 + \ldots + b_n\sigma_n$. Show that $(\ \mid\ )$ is a real inner product.

7. Let V be a real inner product space with inner product $(\ \mid\ )$. If α, β, γ are vectors in V, show that each of the following statements is valid.

   (i) $(\alpha - \beta \mid \gamma) = (\alpha \mid \gamma) - (\beta \mid \gamma)$

   (ii) $(\alpha \mid \theta) = 0$

   (iii) $\|\alpha - \beta\| \geqslant \big|\|\alpha\| - \|\beta\|\big|$

   (iv) (Parallelogram law)

   $$\|\alpha + \beta\|^2 + \|\alpha - \beta\|^2 = 2(\|\alpha\|^2 + \|\beta\|^2)$$

   (v) (Law of cosines)

   $$\|\beta - \alpha\|^2 = \|\alpha\|^2 + \|\beta\|^2 - 2\|\alpha\|\,\|\beta\| \cos\phi$$

   where ϕ is the angle between α and β.

8. Use the results of this section to verify each of the following inequalities:

   (i) $\left[\int_0^1 f(t)g(t)\,dt\right]^2 \leq \left[\int_0^1 f^2(t)\,dt\right]\left[\int_0^1 g^2(t)\,dt\right]$

   for all functions f, g in $C[0, 1]$

   (ii) $\left[\sum_{k=1}^n (a_k + b_k)^2\right]^{1/2} \leq \left(\sum_{k=1}^n a_k^2\right)^{1/2} + \left(\sum_{k=1}^n b_k^2\right)^{1/2}$

   for all real numbers $a_1, \ldots, a_n$ and $b_1, \ldots, b_n$

   (iii) $\left(\int_0^1 [f(t) + g(t)]^2\,dt\right)^{1/2} \leq \left[\int_0^1 f^2(t)\,dt\right]^{1/2} + \left[\int_0^1 g^2(t)\,dt\right]^{1/2}$

   for all functions f, g in $C[0, 1]$

9. (i) If $A = (a_{ij})$ and $B = (b_{ij})$ are $m \times n$ matrices, show that

   $$\mathrm{tr}\,(B^t A) = \sum_{i=1}^n \sum_{k=1}^m b_{ki} a_{ki}$$

   (The trace of a square matrix was defined in exercise 5 of section 1.4.)

   (ii) Use the formula obtained in part (i) to verify that the function $(A \mid B) = \mathrm{tr}\,(B^t A)$ is an inner product on $_m R_n$.

10. (i) If α satisfies $(\alpha \mid v) = 0$ for every vector v in the real inner product space V, show that $\alpha = \theta$.

    (ii) If $(\alpha \mid \beta) = 0$ and s, t are real numbers, show that $(s\alpha \mid t\beta) = 0$ also.

    (iii) Show that $\|\alpha\|^2 + \|\beta\|^2 = \|\alpha + \beta\|^2$ if and only if $(\alpha \mid \beta) = 0$. What familiar theorem does this generalize?

11. Complete the proof of theorem 2.6 by showing that $(\alpha \mid \alpha)(\beta \mid \beta) = (\alpha \mid \beta)^2$ if and only if $\beta = t\alpha$, where

    $$t = \frac{(\alpha \mid \beta)}{(\alpha \mid \alpha)}$$

    Hint: What does it mean to have equality in inequality (3.5)?

12. Let $A = (a_{ij})$ be an $n \times n$ real matrix, and let $X = (x_j)$ and $Y = (y_j)$ be column n-tuples.

    (i) Verify that $(X \mid Y) = X^t Y$, where $(\ \mid\ )$ is the standard inner product on R_n.

    (ii) Show that $(AX \mid Y) = X^t A^t Y = (X \mid A^t Y)$

(iii) If A is symmetric, show that $(AX|\,Y) = (X|\,AY)$. What can be said if A is skew-symmetric?

13. Let V be an inner product space with inner product (|).
   (i) Show that $\|\theta\| = 0$.
      Hint: Use part (ii) of theorem 3.7 to evaluate $\|2\theta\|$.
   (ii) Show that $d(\alpha, \alpha) = 0$ for each vector α in V.

3.3 ORTHOGONAL BASES AND THE GRAM-SCHMIDT PROCESS

One of the main reasons for using a Cartesian coordinate system to represent 3-space lies in the fact that length, distance, and angle can be easily measured in such a system. By analogy, we would expect to derive similar advantages from representing a given real inner product space V by a basis whose members have unit length and are mutually perpendicular with respect to the inner product of V. The purpose of this section is to explore the implications of such a representation and to show how a basis with the desired characteristics may actually be constructed. We begin by generalizing the terminology introduced in definition 3.4.

Definition 3.5. The vectors $\gamma_1, \gamma_2, \ldots, \gamma_k$ are said to be *mutually orthogonal* with respect to the real inner product (|) if $(\gamma_i|\gamma_j) = 0$ whenever $i \neq j$. If, in addition, each vector γ_j has unit length (that is, $\|\gamma_j\| = 1$), then $\gamma_1, \gamma_2, \ldots, \gamma_k$ are said to be *mutually orthonormal*.

EXAMPLE 1.   The vectors $\gamma_1 = (1, 0, 2, 2)$, $\gamma_2 = (-2, 5, -1, 2)$, $\gamma_3 = (104, 63, 1, -53)$ are mutually orthogonal with respect to the standard inner product in R_4 since

$$(\gamma_1|\gamma_2) = (1)(-2) + (0)(5) + (2)(-1) + (2)(2) = 0$$

$$(\gamma_1|\gamma_3) = (1)(104) + (0)(63) + (2)(1) + (2)(-53) = 0$$

$$(\gamma_2|\gamma_3) = (-2)(104) + (5)(63) + (-1)(1) + (2)(-53) = 0$$

EXAMPLE 2.   The polynomials $p_1(t) = \sqrt{3}\,(t - 1)$ and $p_2(t) = 3t - 1$ are mutually orthonormal with respect to the integral inner product in P_1, for we have

$$(p_1|p_2) = \int_0^1 \sqrt{3}\,(t - 1)(3t - 1)\,dt = 0$$

$$(p_1|p_1) = \|p_1\|^2 = \int_0^1 3(t - 1)^2\,dt = 1$$

$$(p_2|p_2) = \|p_2\|^2 = \int_0^1 (3t - 1)^2\,dt = 1$$

The zero vector in a real inner product space V must be orthogonal to every other vector in V, for the equation

$$(\theta \,|\, \alpha) = (\theta + \theta \,|\, \alpha) = (\theta \,|\, \alpha) + (\theta \,|\, \alpha)$$

implies that $(\theta \,|\, \alpha) = 0$. Turning things around, if γ has the property that $(\gamma \,|\, \alpha) = 0$ for every vector α in V, then, in particular, we have $(\gamma \,|\, \gamma) = 0$, and it follows that $\gamma = \theta$. These simple observations prove to be extremely useful, as does the result contained in the following lemma:

Lemma 3.1.   If $\gamma_1, \gamma_2, \ldots, \gamma_k$ are mutually orthogonal vectors, then

$$\sum_{i=1}^{k} c_i(\gamma_i \,|\, \gamma_j) = c_j(\gamma_j \,|\, \gamma_j)$$

for each index j.

Proof: Since $(\gamma_i \,|\, \gamma_j) = 0$ whenever $i \neq j$, the only possible non-zero term in the sum is $c_j(\gamma_j \,|\, \gamma_j)$. ∎

In 3-space, if three non-zero vectors are mutually perpendicular, they obviously cannot lie in the same plane and must therefore be linearly independent. Our next theorem shows that this property is shared by any finite collection of mutually orthogonal vectors.

Theorem 3.9.   If the non-zero vectors $\gamma_1, \gamma_2, \ldots, \gamma_k$ are mutually orthogonal, they must be linearly independent.

Proof: If $c_1, \ldots, c_k$ are real numbers which satisfy $c_1\gamma_1 + c_2\gamma_2 + \ldots + c_k\gamma_k = \theta$, we have

$$0 = (\gamma_j \,|\, \theta) = \left(\gamma_j \,\Big|\, \sum_{i=1}^{k} c_i\gamma_i\right) = \sum_{i=1}^{k} c_i(\gamma_j \,|\, \gamma_i) = c_j(\gamma_j \,|\, \gamma_j)$$

for $j = 1, 2, \ldots, k$. Since each γ_j is a non-zero vector, we have $(\gamma_j \,|\, \gamma_j) \neq 0$, and it follows that $c_j = 0$ for each index j. Thus, S is linearly independent, as claimed. ∎

If the real inner product space V has dimension n, then any collection of n mutually orthogonal vectors must be a basis. We shall refer to such a collection as an *orthogonal basis* for V or as an *orthonormal basis* in the case where each γ_j has unit length. In the following theorem, we show that whenever V is represented by an orthonormal basis, the inner product, length, and distance functions all have the same appearance as the corresponding functions derived from the standard inner product on R_n.

Theorem 3.10. If $S = \{\gamma_1, \gamma_2, \ldots, \gamma_n\}$ is an orthonormal basis for the real inner product space V and $\alpha = a_1\gamma_1 + \ldots + a_n\gamma_n$ and $\beta = b_1\gamma_1 + \ldots + b_n\gamma_n$ are typical elements of V, then

(i) $(\alpha \,|\, \beta) = \sum\limits_{i=1}^{n} a_i b_i$

(ii) $\|\alpha\| = \left(\sum\limits_{i=1}^{n} a_i^2\right)^{1/2}$

(iii) $d(\alpha, \beta) = \left[\sum\limits_{i=1}^{n} (a_i - b_i)^2\right]^{1/2}$

Proof: Applying corollary 3.2, we find that

$$(\alpha \,|\, \beta) = \left(\sum_{i=1}^{n} a_i\gamma_i \,\Big|\, \sum_{j=1}^{n} b_j\gamma_j\right) = \sum_{i=1}^{n} a_i\left[\sum_{j=1}^{n} b_j(\gamma_i \,|\, \gamma_j)\right]$$

Since the γ's are mutually orthonormal, we have

$$\sum_{j=1}^{n} b_j(\gamma_i \,|\, \gamma_j) = b_i(\gamma_i \,|\, \gamma_i) = b_i$$

for each index i, and it follows that $(\alpha \,|\, \beta) = \sum_{i=1}^{n} a_i b_i$. The other two formulas may be obtained by substituting in definition 3.3. ∎

If $\{\gamma_1, \gamma_2, \ldots, \gamma_n\}$ is a basis for the vector space V, we know that each vector α in V can be expressed in the form $\alpha = a_1\gamma_1 + \ldots + a_n\gamma_n$ in exactly one way, but it may not be easy to find the coefficients a_j. However, in the case where the γ's are mutually orthogonal, we have

$$(\alpha \,|\, \gamma_j) = \left(\sum_{i=1}^{k} a_i\gamma_i \,|\, \gamma_j\right) = \sum_{i=1}^{k} a_i(\gamma_i \,|\, \gamma_j) = a_j(\gamma_j \,|\, \gamma_j)$$

for each index j, and it follows that

$$a_j = \frac{(\alpha \,|\, \gamma_j)}{(\gamma_j \,|\, \gamma_j)}$$

In other words, we can write

$$\alpha = \left[\frac{(\alpha \,|\, \gamma_1)}{(\gamma_1 \,|\, \gamma_1)}\right]\gamma_1 + \ldots + \left[\frac{(\alpha \,|\, \gamma_n)}{(\gamma_n \,|\, \gamma_n)}\right]\gamma_n \tag{3.6}$$

and in the important special case where the γ's are mutually orthonormal, we have

$$\alpha = (\alpha \,|\, \gamma_1)\gamma_1 + \ldots + (\alpha \,|\, \gamma_n)\gamma_n$$

We shall find many important applications for the representation given in (3.6).

Our next goal is to show how to systematically construct an orthogonal basis $\{\gamma_1, \ldots, \gamma_n\}$ for a given real inner product space V. As the first step in this construction, we can choose any non-zero vector of V to be γ_1. For the second step, we first select a vector α_2 which is not a scalar multiple of γ_1. Then of all vectors of the form $\alpha_2 - t\gamma$, we choose γ_2 to be the one which is orthogonal to γ_1 (see Figure 3.6). To determine t (and hence γ_2), we solve the equation

$$0 = (\alpha_2 - t\gamma_1 \,|\, \gamma_1) = (\alpha_2 \,|\, \gamma_1) - t(\gamma_1 \,|\, \gamma_1)$$

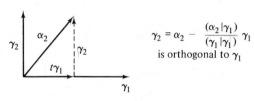

$$\gamma_2 = \alpha_2 - \frac{(\alpha_2 | \gamma_1)}{(\gamma_1 | \gamma_1)}\, \gamma_1$$

is orthogonal to γ_1

FIGURE 3.6

and it follows that

$$\gamma_2 = \alpha_2 - \left[\frac{(\alpha_2 \,|\, \gamma_1)}{(\gamma_1 \,|\, \gamma_1)}\right]\gamma_1$$

Next, we select a vector α_3 from those not contained in the linear span of γ_1 and γ_2 and then choose γ_3 to be that vector of the form $\alpha_3 - t_1\gamma_1 - t_2\gamma_2$ which is orthogonal to both γ_1 and γ_2. Since $(\gamma_1 \,|\, \gamma_2) = 0$, we have

$$0 = (\alpha_3 - t_1\gamma_1 - t_2\gamma_2 \,|\, \gamma_1) = (\alpha_3 \,|\, \gamma_1) - t_1(\gamma_1 \,|\, \gamma_1)$$

$$0 = (\alpha_3 - t_1\gamma_1 - t_2\gamma_2 \,|\, \gamma_2) = (\alpha_3 \,|\, \gamma_2) - t_2(\gamma_2 \,|\, \gamma_2)$$

and it follows that

$$\gamma_3 = \alpha_3 - \left[\frac{(\alpha_3 \,|\, \gamma_1)}{(\gamma_1 \,|\, \gamma_1)}\right]\gamma_1 - \left[\frac{(\alpha_3 \,|\, \gamma_2)}{(\gamma_2 \,|\, \gamma_2)}\right]\gamma_2$$

has the desired characteristics.

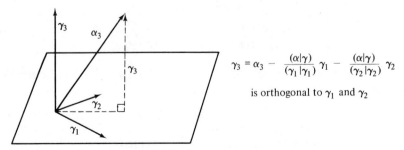

$$\gamma_3 = \alpha_3 - \frac{(\alpha|\gamma)}{(\gamma_1 | \gamma_1)}\, \gamma_1 - \frac{(\alpha|\gamma)}{(\gamma_2 | \gamma_2)}\, \gamma_2$$

is orthogonal to γ_1 and γ_2

FIGURE 3.7

By now the general pattern of construction should be clear. In particular, we begin the kth stage by choosing a vector α_k from those which lie outside the linear span of the mutually orthogonal vectors $\gamma_1, \ldots, \gamma_{k-1}$ obtained during the first $k - 1$ stages of the construction. Then we set

$$\gamma_k = \alpha_k - \sum_{i=1}^{k-1} \left[\frac{(\alpha_k \mid \gamma_i)}{(\gamma_i \mid \gamma_i)}\right]\gamma_i$$

and observe that γ_k is indeed orthogonal to each γ_j, since

$$(\gamma_k \mid \gamma_j) = \left(\alpha_k - \sum_{i=1}^{k-1} \left[\frac{(\alpha_k \mid \gamma_i)}{(\gamma_i \mid \gamma_i)}\right]\gamma_i \,\Big|\, \gamma_j\right)$$

$$= (\alpha_k \mid \gamma_j) - \left(\sum_{i=1}^{k-1} \left[\frac{(\alpha_k \mid \gamma_i)}{(\gamma_i \mid \gamma_i)}\right]\gamma_i \,\Big|\, \gamma_j\right)$$

$$= (\alpha_k \mid \gamma_j) - \sum_{i=1}^{k-1} \left[\frac{(\alpha_k \mid \gamma_i)}{(\gamma_i \mid \gamma_i)}\right](\gamma_i \mid \gamma_j) \qquad (3.7)$$

$$= (\alpha_k \mid \gamma_j) - \left[\frac{(\alpha_k \mid \gamma_j)}{(\gamma_j \mid \gamma_j)}\right](\gamma_j \mid \gamma_j) = 0$$

for $j = 1, 2, \ldots, k - 1$. This procedure, which is known as the *Gram-Schmidt orthogonalization process*, may be continued until a stage is reached where the γ's which have already been generated span all of V. Note that at each stage, the ordered set $\{\gamma_1, \ldots, \gamma_k\}$ is an orthogonal basis for the subspace spanned by $\alpha_1, \ldots, \alpha_k$ (see exercise 7). If V has dimension n, then exactly n stages are necessary to construct an orthogonal basis for V, and an orthonormal basis may be obtained by dividing each γ_j by its length. The Gram-Schmidt process can be used in many different ways, several of which are illustrated in the following solved problems.

PROBLEM 1.   Find a basis for R_3 which is orthogonal with respect to the standard inner product and which includes the vector $(-1, 1, 3)$.

Solution: To begin the Gram-Schmidt process, we set $\gamma_1 = (-1, 1, 3)$. For the second step, we pick $\alpha_2 = (1, 0, 0)$ and compute γ_2 as follows:

$$\gamma_2 = \alpha_2 - \left[\frac{(\alpha_2 \mid \gamma_1)}{(\gamma_1 \mid \gamma_1)}\right]\gamma_1 = (1, 0, 0) - \frac{(-1)}{11}(-1, 1, 3) = \left(\frac{10}{11}, \frac{1}{11}, \frac{3}{11}\right)$$

Finally, we choose $\alpha_3 = (0, 1, 0)$ and substitute in the formula

$$\gamma_3 = \alpha_3 - \left[\frac{(\alpha_3 \mid \gamma_1)}{(\gamma_1 \mid \gamma_1)}\right]\gamma_1 - \left[\frac{(\alpha_3 \mid \gamma_2)}{(\gamma_2 \mid \gamma_2)}\right]\gamma_2$$

to obtain

$$\gamma_3 = (0, 1, 0) - \frac{1}{11}(-1, 1, 3) - \frac{\frac{11}{11}}{\frac{110}{121}}\left(\frac{10}{11}, \frac{1}{11}, \frac{3}{11}\right) = \left(0, \frac{99}{110}, \frac{-33}{110}\right)$$

Therefore, $\{(-1, 1, 3), (\frac{10}{11}, \frac{1}{11}, \frac{3}{11}), (0, \frac{99}{110}, -33/110)\}$ is an orthogonal basis for R_3 which includes the given vector.

Incidentally, since $\gamma_1, \gamma_2, \gamma_3$ are mutually orthogonal, so are the vectors $\gamma_1, 11\gamma_2,$ and $\frac{110}{33}\gamma_3$ (see exercise 10, section 3.2), and it follows that $\{(-1, 1, 3),$ $(10, 1, 3), (0, 3, -1)\}$ is also an orthogonal basis with the desired characteristics.

We shall say that the ordered set $S = \{\alpha_1, \alpha_2, \ldots, \alpha_k\}$ is *orthogonalized* by the Gram-Schmidt process if for each index j, the vector α_j is used to initiate the jth stage of the process. It is important to note that the orthogonal collection produced by orthogonalizing one ordering of a given collection of vectors may be completely different from that obtained by orthogonalizing another ordering of the same set. This observation is illustrated in our next solved problem.

PROBLEM 2.   Find a basis for P_2 which is orthogonal with respect to the integral inner product.

Solution: Applying the Gram-Schmidt process to the basis $\{1, t, t^2\}$ for P_2, we first set $\gamma_1 = 1$ and then obtain

$$\gamma_2 = t - \left[\frac{(t \mid 1)}{(1 \mid 1)}\right]1 = t - \left(\frac{\frac{1}{2}}{1}\right)1 = t - \frac{1}{2}$$

and

$$\gamma_3 = t^2 - \left[\frac{(t^2 \mid 1)}{(1 \mid 1)}\right]1 - \left[\frac{(t^2 \mid t - \frac{1}{2})}{(t - \frac{1}{2} \mid t - \frac{1}{2})}\right]\left(t - \frac{1}{2}\right)$$

$$= t^2 - \left(\frac{\frac{1}{3}}{1}\right)1 - \left(\frac{\frac{1}{12}}{\frac{1}{12}}\right)\left(t - \frac{1}{2}\right) = t^2 - t + \frac{1}{6}$$

Incidentally, if we had orthogonalized the ordered set $\{t^2, t, 1\}$ instead of $\{1, t, t^2\}$, we would have obtained the orthogonal basis $\{t^2, t - \frac{3}{4}t^2, 1 - 4t + \frac{10}{3}t^2\}$ which has little in common with the orthogonal basis $\{1, t - \frac{1}{2}, t^2 - t + \frac{1}{6}\}$ found above.

PROBLEM 3.   Find a basis for the plane $x_1 - 2x_2 + 5x_3 = 0$ which is orthonormal with respect to the standard inner product on R_3.

Solution: The fundamental basis of the plane in question is $\{\eta_1, \eta_2\}$, where $\eta_1 = (2, 1, 0)$ and $\eta_2 = (-5, 0, 1)$. Applying the Gram-Schmidt process to this basis, we find that

$$\gamma_1 = \eta_1 = (2, 1, 0)$$

and

$$\gamma_2 = \eta_2 - \left[\frac{(\eta_2 \mid \gamma_1)}{(\gamma_1 \mid \gamma_1)}\right]\gamma_1 = (-5, 0, 1) - \left(\frac{-10}{5}\right)(2, 1, 0)$$

$$= (-1, 2, 1)$$

Finally, since $\|\gamma_1\| = \sqrt{5}$ and $\|\gamma_2\| = \sqrt{6}$, we conclude that $\left\{\left(\frac{2\sqrt{5}}{5}, \frac{\sqrt{5}}{5}, 0\right), \left(\frac{-\sqrt{6}}{6}, \frac{2\sqrt{6}}{6}, \frac{\sqrt{6}}{6}\right)\right\}$ is an orthonormal basis for the given plane.

The theoretical and computational ideas developed in this section have many important applications. For instance, in the next section, we shall use our knowledge of mutually orthogonal vectors to develop a way of measuring the distance from a vector to a subspace.

Exercise Set 3.3

1. In each of the following cases, use the Gram-Schmidt process to perform the required computations:
   (i) Let $\alpha_1 = (5, 2, -1)$, $\alpha_2 = (0, 1, -1)$, $\alpha_3 = (3, -7, 1)$, and orthogonalize the ordered set $\{\alpha_1, \alpha_2, \alpha_3\}$ with respect to the standard inner product on R_3.
   (ii) Obtain a basis for R_3 which is orthogonal with respect to the standard inner product and contains the vector $(1, 1, -1)$.
   (iii) Find an orthonormal basis for the plane

$$2x_1 - x_2 + 4x_3 = 0$$

2. Find real numbers a, b, c so that the vectors $\alpha_1 = (-3, a, 1)$, $\alpha_2 = (b, 1, 1)$, and $\alpha_3 = (4, 5, c)$ will be mutually orthogonal with respect to the standard inner product.

3. (i) Find a basis for P_1 which is orthogonal with respect to the integral inner product and includes the polynomial $p_1(t) = 3t - 1$.
   (ii) Find a basis for P_1 which is orthogonal with respect to the inner product defined by

$$(p \mid q) = \int_{-1}^{1} p(t)q(t)\, dt$$

Note that the polynomials found in (i) are not mutually orthogonal with respect to the inner product in (ii).

4. Find real numbers a, b so that the polynomial $p(t) = at + b$ will have unit

length and will be orthogonal to $q(t) = 3t + 5$ with respect to the integral inner product on P_1.

5. It can be shown that the function defined by $(\alpha \mid \beta)_* = a_1b_1 + 3a_2b_2 + 7a_3b_3 - a_1b_2 - a_2b_1 + 2a_1b_3 + 2a_3b_1$ is an inner product on R_3.

   (i) Apply the Gram-Schmidt process to the standard basis $\{\epsilon_1, \epsilon_2, \epsilon_3\}$ to obtain a basis $\{\gamma_1, \gamma_2, \gamma_3\}$ for R_3 which is orthonormal with respect to the inner product $(\ \mid\ )_*$.

   (ii) Let $\sigma = 3\gamma_1 - 4\gamma_2 + 12\gamma_3$ and $\tau = 3\gamma_1 + 8\gamma_2 + 7\gamma_3$. Find $(\sigma \mid \tau)_*$, $\|\sigma\|_*$, and $d(\sigma, \tau)_*$.

   Hint: Use theorem 3.10.

6. If $\{\gamma_1, \gamma_2, \ldots, \gamma_n\}$ is an orthonormal basis for the real inner product space V and α, β are vectors in V, show that the following formulas are valid:

   (i) $(\alpha \mid \beta) = \sum_{j=1}^{n} (\alpha \mid \gamma_j)(\beta \mid \gamma_j)$

   Hint: See equation (3.6).

   (ii) $\|\alpha\|^2 = \sum_{j=1}^{n} (\alpha \mid \gamma_j)^2$

*7. Let $\alpha_1, \ldots, \alpha_k$ be linearly independent vectors in the inner product space V, and let $\gamma_1, \ldots, \gamma_k$ be the corresponding vectors produced by the Gram-Schmidt process. Show that for each index j, $S_j = \{\gamma_1, \ldots, \gamma_j\}$ is an orthogonal basis for $\text{Sp}\{\alpha_1, \ldots, \alpha_j\}$.

8. Apply the Gram-Schmidt process to the ordered set $\{\alpha_1, \alpha_2, \alpha_3\}$, where $\alpha_1 = (1, 1, 0)$, $\alpha_2 = (-2, 1, 1)$, and $\alpha_3 = (0, 3, 1)$.

*9. Let $S = \{\alpha_1, \alpha_2, \ldots, \alpha_k\}$ be an ordered set of vectors (not necessarily independent) in the real inner product space V, and let $\gamma_1, \gamma_2, \ldots, \gamma_k$ be the vectors produced by applying the Gram-Schmidt process to S.

   (i) Show that $\gamma_j = \theta$ if and only if α_j is a dependent member of S (see exercise 8.)

   (ii) Design an algorithm for finding an orthogonal basis for the linear span of a given sequence of vectors.

   (iii) Use the algorithm obtained in part (ii) to find an orthogonal basis for the linear span of the vectors $\alpha_1 = (-3, 1, 1, 2)$, $\alpha_2 = (1, 1, 0, 1)$, $\alpha_3 = (-1, 3, 1, 4)$, $\alpha_4 = (0, 4, 1, 5)$, and $\alpha_5 = (1, 0, 0, 0)$.

10. Let S be a finite-dimensional subspace of the real inner product space V, and let $\{\alpha_1, \ldots, \alpha_k\}$ be a basis for S. Show that the vector v is orthogonal to every vector in S if and only if $(\alpha_j \mid v) = 0$ for $j = 1, 2, \ldots, k$.

   Hint: For the "if" part, let $\beta = \sum_{i=1}^{k} c_i\alpha_i$ be a member of S, and verify that $(\beta \mid v) = 0$.

3.4   ORTHOGONAL PROJECTIONS

In this section, we shall use the results of section 3.3 to determine the distance from a vector α to a fixed subspace S and to locate the vector in S which is closest to α. In addition, we shall derive a result known as the *norm approximation theorem*, which has important applications in the theory of approximations and data evaluation.

To illustrate the ideas we plan to explore, let P be a point in 3-space and let S be a plane which contains the origin O. Then, to find the point Q in S which is closest to P, we "project" the vector $\alpha = \overrightarrow{OP}$ onto S by dropping the perpendicular from P down to S. In other words, if $\beta = \overrightarrow{OQ}$, the vector $\alpha - \beta$ is orthogonal to every vector in S (see figure 3.8). According to formula

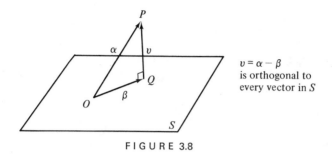

$v = \alpha - \beta$
is orthogonal to
every vector in S

FIGURE 3.8

(3.6) of the last section, if $\{\gamma_1, \gamma_2\}$ is an orthogonal basis for S, we have

$$\beta = \left[\frac{(\beta \mid \gamma_1)}{(\gamma_1 \mid \gamma_1)}\right]\gamma_1 + \left[\frac{(\beta \mid \gamma_2)}{(\gamma_2 \mid \gamma_2)}\right]\gamma_2 \tag{3.8}$$

Furthermore, since $v = \alpha - \beta$ is orthogonal to every vector in S, it certainly must be orthogonal to γ_1 and γ_2. Therefore, we have $(v \mid \gamma_1) = (v \mid \gamma_2) = 0$ and

$$0 = (v \mid \gamma_j) = (\alpha - \beta \mid \gamma_j) = (\alpha \mid \gamma_j) - (\beta \mid \gamma_j)$$

for $j = 1, 2$. It follows that $(\alpha \mid \gamma_1) = (\beta \mid \gamma_1)$ and $(\alpha \mid \gamma_2) = (\beta \mid \gamma_2)$, and we can rewrite (3.8) as

$$\beta = \left[\frac{(\alpha \mid \gamma_1)}{(\gamma_1 \mid \gamma_1)}\right]\gamma_1 + \left[\frac{(\alpha \mid \gamma_2)}{(\gamma_2 \mid \gamma_2)}\right]\gamma_2 \tag{3.9}$$

The vector β is called the *projection of α on S*, while $v = \alpha - \beta$ is called the *normal* from α to S, and the number $\|v\|$ is said to be the *distance from α to S*. The following solved problem shows how these ideas can be used.

PROBLEM 1.   Find the distance from the point $P(-1, 3, -5)$ to the plane S whose equation is $x - 2y + 5z = 0$, and locate the point in S which is closest to P.

Solution: In problem 3 of section 3.3, we found that $\{\gamma_1, \gamma_2\}$ is an orthogonal basis for S, where $\gamma_1 = (2, 1, 0)$ and $\gamma_2 = (-1, 2, 1)$. Therefore, the projection of the vector $\alpha = \overrightarrow{OP}$ on S is

$$\beta = \left[\frac{(\alpha \mid \gamma_1)}{(\gamma_1 \mid \gamma_1)}\right]\gamma_1 + \left[\frac{(\alpha \mid \gamma_2)}{(\gamma_2 \mid \gamma_2)}\right]\gamma_2$$
$$= (\tfrac{1}{5})(2, 1, 0) + (\tfrac{2}{6})(-1, 2, 1) = \tfrac{1}{15}(1, 13, 5)$$

and the normal from α to S is

$$v = \alpha - \beta = (-1, 3, -5) - \tfrac{1}{15}(1, 13, 5) = \tfrac{16}{15}(-1, 2, -5)$$

Hence, if Q is the point in S which is closest to P, we have $\beta = \overrightarrow{OQ}$, and it follows that Q has coordinates $(\tfrac{1}{15}, \tfrac{13}{15}, \tfrac{5}{15})$. Finally, the distance from P to S is

$$\|v\| = \tfrac{16}{15}\sqrt{(-1)^2 + (2)^2 + (-5)^2} = \tfrac{16}{15}\sqrt{30}$$

Note that the desired distance could have been obtained by simply substituting into the formula derived in problem 3 of section 3.1. However, the solution of that problem does not tell us how to identify the point in S which is closest to P.

To generalize the notion of vector projection to inner product spaces, we make the following definition.

Definition 3.6.   Let α be a non-zero vector in the real inner product space V, and let S be a subspace of V. Then the vector α_S of S is said to be an *orthogonal projection* of α on S if $\alpha - \alpha_S$ is orthogonal to every vector in S.

In figure 3.8, it is easy to see that the vector α projects onto exactly one vector in the plane S. However, such "visual" observations are possible only in 3-space, and in the general case, it is necessary to show that each non-zero vector projects onto one and only one vector in a given subspace. To this end, we prove the following result.

Theorem 3.11.   Let S be a finite-dimensional subspace of the real inner product space V, and let α be a non-zero vector in V. Then there exists one and only one orthogonal projection of α on S, and if $\{\gamma_1, \ldots, \gamma_k\}$ is an orthogonal basis for S, we have

$$\alpha_S = \left[\frac{(\alpha \mid \gamma_1)}{(\gamma_1 \mid \gamma_1)}\right]\gamma_1 + \left[\frac{(\alpha \mid \gamma_2)}{(\gamma_2 \mid \gamma_2)}\right]\gamma_2 + \ldots + \left[\frac{(\alpha \mid \gamma_k)}{(\gamma_k \mid \gamma_k)}\right]\gamma_k$$

Proof: Let $\beta = \sum_{i=1}^{k} c_i\gamma_i$ be a member of S, and note that the vector $v = \alpha - \beta$ is orthogonal to every vector in S if and only if it is orthogonal to each member of the basis $\{\gamma_1, \ldots, \gamma_k\}$ (see exercise 10, section 3.3). Since the γ's are mutually orthogonal, it follows that β is an orthogonal projection of α on S whenever

$$0 = (\alpha - \beta \mid \gamma_j) = \left(\alpha - \sum_{i=1}^{k} c_i \gamma_i \mid \gamma_j\right)$$

$$= (\alpha \mid \gamma_j) - \sum_{i=1}^{k} c_i(\gamma_i \mid \gamma_j) = (\alpha \mid \gamma_j) - c_j(\gamma_j \mid \gamma_j)$$

for each index j. Solving, we find that

$$c_j = \left[\frac{(\alpha \mid \gamma_j)}{(\gamma_j \mid \gamma_j)}\right]$$

which means that the vector

$$\alpha_S = \left[\frac{(\alpha \mid \gamma_1)}{(\gamma_1 \mid \gamma_1)}\right]\gamma_1 + \ldots + \left[\frac{(\alpha \mid \gamma_k)}{(\gamma_k \mid \gamma_k)}\right]\gamma_k$$

is the one and only orthogonal projection of α on S. ∎

Henceforth, we shall always denote the orthogonal projection of α on the subspace S by α_S, and we shall denote the vector $\alpha - \alpha_S$ by the symbol $\alpha_{S\perp}$ (pronounced "alpha S perp"). The vector equation $\alpha = \alpha_S + \alpha_{S\perp}$ is sometimes referred to as the *orthogonal decomposition of α along S*, and in our next theorem, we show that this is the only way α can be represented as the sum of a vector in S and a vector which is orthogonal to every vector in S.

Theorem 3.12. Let S be a finite-dimensional subspace of the real inner product space V, and let α be a non-zero vector in V. If $\alpha = \sigma + v$, where σ is an element of S and v is orthogonal to every vector in S, then $\sigma = \alpha_S$ and $v = \alpha_{S\perp}$.

Proof: Since $\alpha = \alpha_S + \alpha_{S\perp}$ and $\alpha = \sigma + v$, we have $\alpha_S - \sigma = v - \alpha_{S\perp}$. The vector $\alpha_S - \sigma$ must be contained in S, for S is a subspace of V which contains both α_S and σ. Moreover, since $\alpha_{S\perp}$ and v are both orthogonal to every vector in S, it can be shown that $v - \alpha_{S\perp}$ has the same property [see exercise 8, part (i)]. Thus, $\alpha_S - \sigma$ is a vector in S which is orthogonal to every vector in S, including itself. In other words, we have

$$0 = (\alpha_S - \sigma \mid \alpha_S - \sigma) = \|\alpha_S - \sigma\|^2$$

and it follows that $\sigma = \alpha_S$, which in turn implies that $v = \alpha_{S\perp}$. ∎

By analogy with the situation in 3-space, it is convenient to think of α, α_S, and $\alpha_{S\perp}$ as the sides of a right triangle (see figure 3.9). In this context, the

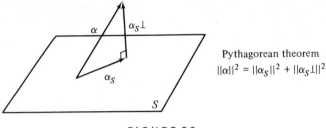

Pythagorean theorem
$$||\alpha||^2 = ||\alpha_S||^2 + ||\alpha_S\bot||^2$$

FIGURE 3.9

following result may be regarded as a generalized version of the pythagorean theorem.

Theorem 3.13. If S is a subspace of the real inner product space V and α is a non-zero vector in V, then

$$||\alpha||^2 = ||\alpha_S||^2 + ||\alpha_{S\perp}||^2$$

Proof: Since $\alpha = \alpha_S + \alpha_{S\perp}$ and $(\alpha_S|\alpha_{S\perp}) = 0$, we have

$$||\alpha||^2 = (\alpha|\alpha) = (\alpha_S + \alpha_{S\perp}|\alpha_S + \alpha_{S\perp})$$
$$= (\alpha_S|\alpha_S) + 2(\alpha_S|\alpha_{S\perp}) + (\alpha_{S\perp}|\alpha_{S\perp}) = ||\alpha_S||^2 + ||\alpha_{S\perp}||^2$$

as desired. ∎

Earlier, we observed that in 3-space, the distance from a point P to the plane S is just the distance from $\alpha = \overrightarrow{OP}$ to the projection of α on S. To show that an analogous result holds in other inner product spaces, we need the following important theorem.

Theorem 3.14. (The norm approximation theorem.) Let S be a finite-dimensional subspace of the real inner product space V, and let α be a non-zero vector in V. Then if σ is any vector in S other than α_S, we have

$$||\alpha - \sigma|| > ||\alpha - \alpha_S||$$

Proof: Let σ be a vector in S. We find that

$$\alpha - \sigma = (\alpha_S + \alpha_{S\perp}) - \sigma = (\alpha_S - \sigma) + \alpha_{S\perp}$$

and since $\alpha_S - \sigma$ is contained in S (why?) the uniqueness of the orthogonal decomposition of $\alpha - \sigma$ guarantees that $\alpha_S - \sigma$ is the orthogonal projection of $\alpha - \sigma$ on S. According to theorem 3.13 we have

$$||\alpha - \sigma||^2 = ||\alpha_S - \sigma||^2 + ||\alpha_{S\perp}||^2 = ||\alpha_S - \sigma||^2 + ||\alpha - \alpha_S||^2$$

where the last equation follows from the fact that $\alpha_{S_\perp} = \alpha - \alpha_S$. Finally, if σ is not α_S, we have $\|\alpha_S - \sigma\| > 0$, and it follows that $\|\alpha - \sigma\| > \|\alpha - \alpha_S\|$, as claimed. ∎

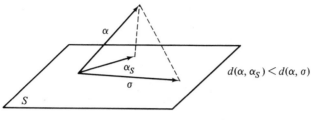

FIGURE 3.10

According to theorem 3.14, the distance from α to α_S is less than that from α to any other vector in S. In this sense, α_S is the element of S which is closest to α, and for this reason, it is natural to use $d(\alpha, \alpha_S)$ as a measure of the distance from α to S. These ideas are illustrated in the following solved problems.

PROBLEM 2.   Let S denote the subspace of R_4 which is spanned by $\beta_1 = (1, 0, 1, 0)$ and $\beta_2 = (1, 1, 0, 1)$. Find the vector in S which is closest to $\alpha = (1, 0, 0, 1)$, and determine the distance from α to S with respect to the standard inner product.

Solution: We wish to find α_S and $d(\alpha, \alpha_S)$. First of all, applying the Gram-Schmidt process to $\{\beta_1, \beta_2\}$, we obtain the orthogonal basis $\{\gamma_1, \gamma_2\}$, where $\gamma_1 = (1, 0, 1, 0)$ and $\gamma_2 = (1, 2, -1, 2)$. Thus, we have

$$\alpha_S = \left[\frac{(\alpha\,|\,\gamma_1)}{(\gamma_1\,|\,\gamma_1)}\right]\gamma_1 + \left[\frac{(\alpha\,|\,\gamma_2)}{(\gamma_2\,|\,\gamma_2)}\right]\gamma_2$$

$$= (\tfrac{1}{2})(1, 0, 1, 0) + (\tfrac{3}{10})(1, 2, -1, 2) = \tfrac{1}{5}(4, 3, 1, 3)$$

and since

$$\alpha - \alpha_S = (1, 0, 0, 1) - \tfrac{1}{5}(4, 3, 1, 3) = \tfrac{1}{5}(1, -3, -1, 2)$$

it follows that

$$d(\alpha, \alpha_S) = \|\alpha - \alpha_S\| = \tfrac{1}{5}\sqrt{1^2 + (-3)^2 + (-1)^2 + (2)^2} = \tfrac{1}{5}\sqrt{15}$$

In numerical analysis, the result established in theorem 3.14 is often used to determine how much a given function f differs from those with certain specified characteristics. The following is an example of the kind of problem that appears in this context.

PROBLEM 3.   Find the linear polynomial $p(t) = at + b$ which best approximates the exponential function $f(t) = e^t$ with respect to the norm derived from the integral inner product on $C[0, 1]$.

Solution: We wish to find the orthogonal projection of f on the subspace P_1. Accordingly, we begin by using the Gram-Schmidt process to convert the basis $\{1, t\}$ into an orthogonal basis $\{\gamma_1, \gamma_2\}$ for P_1, where $\gamma_1 = 1$ and $\gamma_2 = 2t - 1$ (see problem 2, section 3.3). Since we have

$$(\gamma_1 | \gamma_1) = \int_0^1 1^2 \, dt = 1, \qquad (\gamma_2 | \gamma_2) = \int_0^1 (2t - 1)^2 \, dt = \tfrac{1}{3}$$

$$(f | \gamma_1) = \int_0^1 e^t \, dt = e - 1, \qquad (f | \gamma_2) = \int_0^1 e^t(2t - 1) \, dt = 3 - e$$

it follows that the orthogonal projection of f on P_1 is

$$f_{P_1} = \left[\frac{(f | \gamma_1)}{(\gamma_1 | \gamma_1)} \right] \gamma_1 + \left[\frac{(f | \gamma_2)}{(\gamma_2 | \gamma_2)} \right] \gamma_2$$

$$= \left(\frac{e - 1}{1} \right) 1 + \left(\frac{3 - e}{\tfrac{1}{3}} \right)(2t - 1) = (18 - 6e)t + (4e - 10)$$

and this is the linear polynomial which best approximates e^t in the given norm.

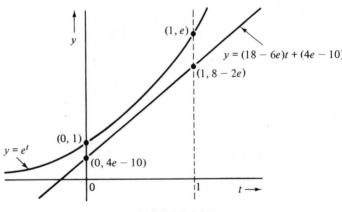

FIGURE 3.11

This completes our survey of the fundamental properties of real inner product spaces. The concepts developed in this chapter have many important applications, two of which are analyzed in detail in the next section.

Exercise Set 3.4

1. In each of the following cases, find the orthogonal projection of the vector $\alpha = (-1, 3, 5)$ on the given subspace of R_3.
   (i) The line S_1 whose equation is

   $$\frac{x}{3} = \frac{y}{-2} = z.$$

   (ii) The plane S_2 whose equation is $2x - y + 3z = 0$.
   (iii) The linear span S_3 of the vectors $\beta_1 = (1, -4, 6)$ and $\beta_2 = (2, -7, 1)$.

2. Find the distance from the vector $\alpha = (1, -1, 1, 1)$ to the linear span S of the vectors $\beta_1 = (1, 0, 1, 1)$, $\beta_2 = (-2, 1, 2, 0)$, and $\beta_3 = (0, 0, 1, 1)$, as measured by the norm derived from the standard inner product on R_4.

3. Find the equation of the sphere in 3-space which is tangent to the plane $3x - 4y + 12z = 0$ and has its center at $P(1, -3, 2)$.

4. Let S denote the subspace of P_1 which is spanned by $p(t) = 3t - 1$. Find the orthogonal projection of $q(t) = 7t + 5$ on S and compute the distance from q to S with respect to the integral inner product.

5. In the inner product space V, let $\{\gamma_1, \gamma_2\}$ be an orthonormal basis for the subspace S and let α be a vector for which $(\alpha \mid \gamma_1) = 3$, $(\alpha \mid \gamma_2) = -5$, and $(\alpha \mid \alpha) = 43$.
   (i) Find real numbers a, b which satisfy $\alpha_S = a\gamma_1 + b\gamma_2$.
   (ii) Find the distance from α to S.
   Hint: Compute $(\alpha - \alpha_S \mid \alpha - \alpha_S)$.

6. Find the quadratic polynomial $p(t) = at^2 + bt + c$ which best approximates the function $f(t) = t^3$ with respect to the integral inner product.

7. Let S denote the linear span of the mutually orthogonal vectors $\gamma_1 = (1, 0, 1)$ and $\gamma_2 = (0, 1, 0)$, and let $\alpha = (-3, 1, 7)$.
   (i) Find the orthogonal projection of α on S.
   (ii) Verify that $\{\eta_1, \eta_2\}$ is an orthogonal basis for S, where $\eta_1 = (1, 1, 1)$ and $\eta_2 = (1, -2, 1)$. Find the orthogonal projection of α on S in terms of this new basis and compare this vector with the projection found in (i).

8. Let S be a subspace of the finite-dimensional real inner product space V, and let $S^\perp$ denote the collection of all vectors v such that $(v \mid \sigma) = 0$ for every vector σ in S. Prove each of the following statements:
   (i) $S^\perp$ is a subspace of V.
   (ii) $(S^\perp)^\perp = S$.
   (iii) $S \cap S^\perp$ is the zero vector.
   (iv) $V = S + S^\perp$.
   Hint: It suffices to verify that each α in V can be expressed as $\alpha = \sigma + \tau$ where σ and τ are contained in S and $S^\perp$, respectively.
   Incidentally, the subspace $S^\perp$ is usually called the *orthogonal complement* of S.

9. Let S be a subspace of the finite-dimensional real inner product space V, and let α be a vector in V. Show that $\alpha_{S^\perp}$ is the orthogonal projection of α on $S^\perp$.

10. (i) Find an orthonormal basis for the plane $x - y + z = 0$ and extend it to an orthonormal basis for all of R_3.

(ii) Find an orthonormal basis for the orthogonal complement of the plane in part (i).

Hint: The result obtained in exercise 9 may help.

11. Let S be a finite-dimensional subspace of the real inner product space V. Verify the following statements regarding orthogonal projections on S:

(i) $(t\alpha)_S = t\alpha_S$ for each real number t and each α in V

(ii) $(\alpha + \beta)_S = \alpha_S + \beta_S$ for α, β in S

Hint: Use the fact that the orthogonal decomposition of a vector is unique (theorem 3.12).

12. Let S be a subspace of the finite-dimensional real inner product space V, and let $\{\gamma_1, \ldots, \gamma_k\}$ and $\{\gamma_{k+1}, \ldots, \gamma_n\}$ be orthogonal bases for S and $S^\perp$, respectively. Show that $\{\gamma_1, \ldots, \gamma_n\}$ is an orthogonal basis for V.

Hint: Recall supplementary exercise 7 of chapter 2 and exercise 8, above.

3.5 (OPTIONAL) TWO APPLICATIONS OF ORTHOGONALITY: FOURIER SERIES AND THE METHOD OF LEAST SQUARES.

The concept of orthogonality is used in many different areas of mathematical inquiry as both a theoretical tool and a method of practical analysis. In this section we examine two of these applications, the method of least squares and the generation of Fourier coefficients.

The Method of Least Squares. In carrying out experimental research, a scientist naturally tries to ask significant questions and to make accurate observations, but in practice, there are always errors of one type or another. Therefore, in order to obtain a proper interpretation of his experimental data, the researcher often employs the methods of the theory of approximations, and one of the simplest and most powerful techniques of this theory is the curve fitting technique known as the method of least squares.

Before developing the general theory behind the method of least squares, let us first consider a problem in which the basic ideas of curve fitting can be made more explicit. Specifically, it is known that if an ideal spring is displaced a distance y from its natural length by a force (weight) x, then x and y are related by the linear equation $y = kx$, where k is a constant which depends only upon the spring. In order to compute this constant for a given spring, a scientist might conduct an experiment in which he attaches various weights to the spring and then measures the resulting displacements. Suppose that as a result of his experiment, the scientist obtains the following data:

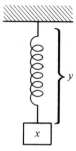

x (lb)	y(in.)
5	11.1
7	15.4
8	17.5
10	22.0
12	26.3

Plugging these numbers into the spring equation, we obtain the following system of five equations in one unknown:

$$11.1 = 5k$$

$$15.4 = 7k$$

$$17.5 = 8k$$

$$22 = 10k$$

$$26.3 = 12k$$

Due to experimental errors this system is inconsistent, and we certainly cannot hope to determine an exact value of k from our data. Thus, we are forced to seek an approximation to k which is "best" in some mathematically significant sense. One of several ways to obtain such an approximation is to find the number m which makes the expression

$$\sum_{i=1}^{5} (y_i - mx_i)^2$$

as small as possible. Geometrically, this means that we seek a line of the form $y = mx$ which is located so as to minimize the sum of the squares of the differences $\Delta_i = y_i - mx_i$ between the ordinate of each data point (x_i, y_i) and the corresponding point (x_i, mx_i) on the line. The data points and a typical line of the desired form are plotted in figure 3.12.

This interpretation of the term "best approximation" may or may not seem natural, but it has the advantage of being extremely compatible with the theory developed in this chapter. To illustrate this fact, consider the 5-tuples $X = (5, 7, 8, 10, 12)$ and $Y = (11.1, 15.4, 17.5, 22, 26.3)$. Since $\sum_{i=1}^{5} (y_i - mx_i)^2 = \| Y - mX \|^2$, our approximation problem can be re-

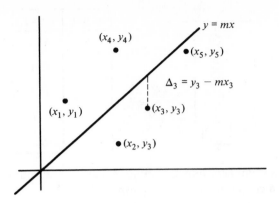

FIGURE 3.12

stated as follows: Find the scalar multiple of X which is closest to Y. The norm approximation theorem tells us that the scalar multiple we seek is the orthogonal projection of Y on the subspace spanned by X. Thus, the desired multiple mX must satisfy the equation

$$0 = (Y - mX \,|\, X) = (Y \,|\, X) - m(X \,|\, X)$$

and it follows that

$$m = \frac{(Y \,|\, X)}{(X \,|\, X)} = \frac{838.9}{382} = 2.19$$

is the best approximation to the spring constant which can be obtained from the given data.

The method we have illustrated in this simple example can be generalized to accommodate a more comprehensive class of problems. In particular, suppose that a certain quantity y is known to have the general form

$$y = k_1 x_1 + k_2 x_2 + \ldots + k_m x_m$$

where the x's are real variables and the k's are scalar constants which we plan to determine by means of a series of experiments. To be specific, assume that in the ith experiment, we obtain the value y_i for y as a result of choosing the values $x_{i1}, x_{i2}, \ldots, x_{im}$ for the variables $x_1, x_2, \ldots, x_m$, respectively. After n such experiments have been carried out, we have the following system of n equations in the m unknowns $k_1, k_2, \ldots, k_m$:

$$y_1 = x_{11}k_1 + x_{12}k_2 + \ldots + x_{1m}k_m$$

$$y_2 = x_{21}k_1 + x_{22}k_2 + \ldots + x_{2m}k_m$$

$$\begin{matrix} \cdot & \cdot & \cdot & & \cdot \\ \cdot & \cdot & \cdot & & \cdot \\ \cdot & \cdot & \cdot & & \cdot \end{matrix} \qquad (3.10)$$

$$y_n = x_{n1}k_1 + x_{n2}k_2 + \ldots + x_{nm}k_m$$

Of course, no definitive information concerning the k's can be obtained from system (3.10) unless we perform at least m experiments (why is this true?). On the other hand, if $n > m$, there is a good chance that experimental errors will prevent the resulting system from being consistent, and in this case, we must seek to approximate the k's in some reasonable way. For this purpose, we first define the n-tuples

$$\alpha_1 = (x_{11}, x_{21}, \ldots, x_{n1})$$

$$\alpha_2 = (x_{12}, x_{22}, \ldots, x_{n2})$$

$$\cdot$$
$$\cdot$$
$$\cdot$$

$$\alpha_m = (x_{1m}, x_{2m}, \ldots, x_{nm})$$

$$\beta = (y_1, y_2, \ldots, y_n)$$

and then seek that vector in the subspace spanned by the set $\{\alpha_1, \alpha_2, \ldots, \alpha_m\}$ which is "closest" to β; in other words, we want to find the orthogonal projection of β on the subspace $S = \text{Sp}\,\{\alpha_1, \alpha_2, \ldots, \alpha_m\}$. Since in our experiment we have control over the components of the α's, it is reasonable to assume that the set $\{\alpha_1, \alpha_2, \ldots, \alpha_m\}$ is linearly independent, and as such, it is a basis for S. Therefore, the orthogonal projection β_S can be expressed as a unique linear combination of the α's—say, $\beta_S = \bar{k}_1\alpha_1 + \bar{k}_2\alpha_2 + \ldots + \bar{k}_m\alpha_m$ —and it is then natural to regard the number $\bar{k}_j$ as the best approximation of k_j in terms of the given data. Since the numbers $\bar{k}_j$ also minimize the quantity

$$\| \beta_S - \beta \|^2 = \|(\bar{k}_1\alpha_1 + \ldots + \bar{k}_m\alpha_m) - \beta\|^2$$

$$= \sum_{i=1}^{n} [\bar{k}_1 x_{i1} + \ldots + \bar{k}_m x_{im}) - y_i]^2$$

it is easy to see why the approximation technique we have just described is called the method of least squares.

In order to actually compute the $\bar{k}_j$, we could orthogonalize the basis

$\{\alpha_1, \alpha_2, \ldots, \alpha_m\}$ by means of the Gram-Schmidt process and then use the formula established in theorem 3.11 to find β_S, but this method is not always computationally feasible. However, since $\beta_S - \beta$ must be orthogonal to every vector in S, we know that

$$0 = (\beta_S - \beta \,|\, \alpha_i) = (\bar{k}_1\alpha_1 + \ldots + \bar{k}_m\alpha_m - \beta \,|\, \alpha_i)$$

$$= \left[\sum_{j=1}^{m} \bar{k}_j(\alpha_j \,|\, \alpha_i) \right] - (\beta \,|\, \alpha_i)$$

for each α_i, and it follows that the $\bar{k}_j$ must satisfy the $m \times m$ linear system

$$(\alpha_1|\alpha_1)\bar{k}_1 + (\alpha_1|\alpha_2)\bar{k}_2 + \ldots + (\alpha_1|\alpha_m)\bar{k}_m = (\alpha_1|\beta)$$

$$(\alpha_2|\alpha_1)\bar{k}_1 + (\alpha_2|\alpha_2)\bar{k}_2 + \ldots + (\alpha_2|\alpha_m)\bar{k}_m = (\alpha_2|\beta)$$

$$\vdots \qquad \qquad \vdots \qquad \qquad \qquad \vdots \qquad \qquad \vdots \tag{3.11}$$

$$(\alpha_m|\alpha_1)\bar{k}_1 + (\alpha_m|\alpha_2)\bar{k}_2 + \cdots + (\alpha_m|\alpha_m)\bar{k}_m = (\alpha_m|\beta)$$

Since the vectors $\alpha_1, \alpha_2, \ldots, \alpha_m$ are linearly independent, it can be shown that system (3.11) must have a unique solution (see supplementary exercise 7 at the end of this chapter), which in turn provides the desired approximations.

Incidentally, in the mathematical literature, the type of approximation problem we have just analyzed is often called a *multiple regression*, while the linear expression $k_1x_1 + k_2x_2 + \ldots + k_mx_m$ is referred to as a *regression surface*, and the derived equations displayed in (3.11) are called the *normal equations* of the approximation (or regression). The quantity $1/n \| \beta_S - \beta \|^2$ is called the *variance* of the approximation and can be regarded as a measure of the consistency of the data.

PROBLEM 1.   A researcher has found that whenever his pet monkey, Banana-breath, eats x_1 oz of fruit and x_2 oz of garlic, his body assimilates y units of vitamin B_{52} in the amounts listed in the following table:

x_1	x_2	y
7	3	1.6
9	2	2.1
5	5	2.0
4	6	2.2
3	1	.8
3	2	1.1

If it is known that $y = k_1 x_1 + k_2 x_2$, where k_1 and k_2 are constants, approximately how many units of the vitamin will be assimilated by Bananabreath if he eats 6 oz of fruit and 4 oz of garlic?

Solution: First of all, we use the method of least squares to approximate k_1 and k_2. Accordingly, we define the vectors $\alpha_1 = (7, 9, 5, 4, 3, 3)$, $\alpha_2 = (3, 2, 5, 6, 1, 2)$, and $\beta = (1.6, 2.1, 2, 2.2, .8, 1.1)$, and then substitute in the general form displayed in (3.11) to obtain the normal equations

$$189\bar{k}_1 + 97\bar{k}_2 = 54.6$$

$$97\bar{k}_1 + 79\bar{k}_2 = 35.2$$

Solving, we obtain the values $\bar{k}_1 = .16$ and $\bar{k}_2 = .25$ as the best approximations to k_1 and k_2 in terms of the given data. Thus, we would expect approximately $.16(6) + .25(4) = 1.96$ units of the vitamin to be assimilated when $x_1 = 6$ and $x_2 = 4$.

II The Fourier Coefficients of a Function. A function $\tau(x)$ is called a trigonometric polynomial of degree $2m + 1$ if

$$\tau(x) = a_0 + a_1 \cos x + a_2 \cos 2x + \ldots + a_m \cos mx$$
$$+ b_1 \sin x + b_2 \sin 2x + \ldots + b_m \sin mx$$

where the a's and b's are real numbers and the leading coefficients a_m and b_m are not both zero. For example,

$$p(x) = 3 - 2 \cos x + 5 \cos 2x - \sin x + 2 \sin 3x$$

is a trigonometric polynomial of degree $2(3) + 1 = 7$. We shall investigate the problem of approximating a given continuous function f by a trigonometric polynomial.

It is a fairly routine matter to verify that the collection T_n of all trigonometric polynomials of degree at most $2n + 1$ together with the zero polynomial forms a $(2n + 1)$-dimensional subspace of $C[-\pi, \pi]$, the vector space of all functions continuous on $[-\pi, \pi]$. Let $C[-\pi, \pi]$ be regarded as a real inner product space with the integral inner product

$$(f \mid g) = \int_{-\pi}^{\pi} f(x)g(x)\,dx$$

Then the trigonometric polynomial which best approximates f in the norm derived from the integral inner product is just the orthogonal projection of f on T_n, which we denote by f_n. To find this projection, we need an orthonormal basis for T_n, and it can be shown (see exercise 1) that the ordered set

$\{\gamma_0, \gamma_1, \ldots, \gamma_n, \gamma_{n+1}, \ldots, \gamma_{2n}\}$ satisfies this requirement, where

$$\gamma_0 = \frac{1}{\sqrt{2\pi}}, \qquad \gamma_1 = \frac{\cos x}{\sqrt{\pi}}, \qquad \ldots, \qquad \gamma_n = \frac{\cos nx}{\sqrt{\pi}}$$

$$\gamma_{n+1} = \frac{\sin x}{\sqrt{\pi}}, \qquad \ldots, \qquad \gamma_{2n} = \frac{\sin nx}{\sqrt{\pi}}$$

According to theorem 3.11, we have

$$f_n = \sum_{k=0}^{2n} (f \,|\, \gamma_k)\gamma_k$$

and reshaping this sum to fit the general form of a trigonometric polynomial, we find that

$$f_n(x) = c_0 + \sum_{k=1}^{n} c_k \cos(kx) + \sum_{k=1}^{n} d_k \sin(kx)$$

where coefficients c_k and d_k of the projection polynomial are given by

$$c_0 = (f \,|\, \gamma_0)\gamma_0 = \frac{1}{\sqrt{2\pi}} \int_{-\pi}^{\pi} f(x) \cdot \frac{1}{\sqrt{2\pi}}\, dx = \frac{1}{2\pi} \int_{-\pi}^{\pi} f(x)\, dx$$

$$c_k = \frac{1}{\sqrt{\pi}}(f \,|\, \gamma_k) = \frac{1}{\sqrt{\pi}} \int_{-\pi}^{\pi} f(x)\frac{\cos kx}{\sqrt{\pi}}\, dx = \frac{1}{\pi} \int_{-\pi}^{\pi} f(x) \cos kx\, dx$$

$$d_k = \frac{1}{\sqrt{\pi}}(f \,|\, \gamma_{n+k}) = \frac{1}{\sqrt{\pi}} \int_{-\pi}^{\pi} f(x)\frac{\sin kx}{\sqrt{\pi}}\, dx = \frac{1}{\pi} \int_{-\pi}^{\pi} f(x) \sin kx\, dx$$

for $k = 1, 2, \ldots, n$.

The numbers we have just computed are known as the *Fourier coefficients* of f, and our theory tells us that f_n is that trigonometric polynomial in T_n which best approximates f in the sense that if τ is any other element of T_n, then $\| f - \tau \| > \| f - f_n \|$. Thus, we have

$$\| f - \tau \|^2 = \int_{-\pi}^{\pi} [f(x) - \tau(x)]^2\, dx > \int_{-\pi}^{\pi} [f(x) - f_n(x)]^2\, dx = \| f - f_n \|^2$$

The value of the integral $\int_{\pi}^{\pi} [f(x) - f_n(x)]^2\, dx$ is often called the *mean square deviation* of f_n from f, and can be regarded as a measure of the accuracy of the approximation. In more advanced texts it is shown that this deviation tends to zero as n becomes arbitrarily large. In symbols,

$$\lim_{n \to \infty} \int_{-\pi}^{\pi} [f(x) - f_n(x)]^2\, dx = 0$$

In other words, a given function f in $C[-\pi, \pi]$ can be approximated by a suitable trigonometric polynomial to any degree of accuracy as measured by the mean square deviation, and for this reason, we say that the sequence of trigonometric polynomials $\{f_1, f_2, \ldots, f_n, \ldots\}$ *converges in the mean* to f on the interval $[-\pi, \pi]$. Although we may write

$$f(x) = c_0 + \sum_{k=1}^{\infty} c_k \cos kx + \sum_{k=1}^{\infty} d_k \sin kx \qquad (3.12)$$

it is important to note that convergence in the mean is not the same as the pointwise convergence or uniform convergence encountered in calculus. In fact, if x_0 is an arbitrary point in $[-\pi, \pi]$, there is no guarantee that the point sequence $f_1(x_0), f_2(x_0), \ldots, f_n(x_0), \ldots$ will even converge.

The form of representation given in formula (3.12) is generally known as the Fourier series expansion of f, and the results we have mentioned form a small but crucial part of an important area of mathematics known as Fourier analysis. Although we have focused our attention on functions which are continuous on the interval $[-\pi, \pi]$ the whole theory of the Fourier series expansion can be easily modified to apply to other intervals and to a class of functions which are not everywhere continuous. In this more general form, the Fourier series representation turns out to be very useful in describing certain phenomena in mathematical physics. The reader who is interested in knowing more about Fourier analysis can find an excellent treatment of the subject in *Introduction to Linear Analysis*, by D. L. Kreider et al., Addison-Wesley, Reading, Massachusetts, 1966.

PROBLEM 2. Find the Fourier series representation of the function $f(x) = |x|$ on the interval $[-\pi, \pi]$.

Solution: Computing the Fourier coefficients of f, we find that

$$c_0 = \frac{1}{2\pi} \int_{-\pi}^{\pi} |x| \, dx = \frac{1}{2\pi} (\pi^2) = \frac{\pi}{2}$$

and for $k = 1, 2, \ldots, n$

$$c_k = \frac{1}{\pi} \int_{-\pi}^{\pi} |x| \cos kx \, dx = \begin{cases} \dfrac{-4}{\pi k^2} & \text{if } k \text{ is odd} \\ 0 & \text{if } k \text{ is even} \end{cases}$$

$$d_k = \frac{1}{\pi} \int_{-\pi}^{\pi} |x| \sin kx \, dx = 0$$

Therefore, for each fixed positive integer n, we find that the orthogonal projection, f_n, of f on T_n is

$$f_n = \frac{\pi}{2} - \frac{4}{\pi}\left(\cos x + \frac{\cos 3x}{3^2} + \frac{\cos 5x}{5^2} + \ldots + \frac{\cos mx}{m^2}\right)$$

where $m = \begin{cases} n \text{ if } n \text{ is odd} \\ n - 1 \text{ if } n \text{ is even} \end{cases}$. Thus, the Fourier series expansion of f is

$$|x| = \frac{\pi}{2} - \frac{4}{\pi}\sum_{k=1}^{\infty}\frac{\cos(2k-1)}{(2k-1)^2}$$

In figure 3.13, we display the graph of

$$f_3(x) = \frac{\pi}{2} - \frac{4}{\pi}\left(\cos x + \frac{\cos 3x}{9}\right)$$

superimposed on that of $f(x) = |x|$.

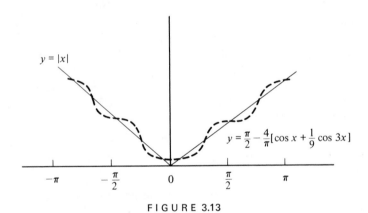

FIGURE 3.13

Exercise Set 3.5

1. Show that the functions

$$\frac{1}{\sqrt{2\pi}}, \frac{1}{\sqrt{\pi}}\cos x, \frac{1}{\sqrt{\pi}}\cos 2x, \ldots, \frac{1}{\sqrt{\pi}}\cos nx,$$

$$\frac{1}{\sqrt{\pi}}\sin x, \frac{1}{\sqrt{\pi}}\sin 2x, \ldots, \frac{1}{\sqrt{\pi}}\sin nx$$

are mutually orthonormal with respect to the integral inner product

$$(f|g) = \int_{-\pi}^{\pi} f(x)g(x)\,dx$$

Hint: The following trigonometric identities may be useful:

$$\sin (px) \cos (qx) = \tfrac{1}{2}[\sin (p + q)x + \sin (p - q)x]$$

$$\cos (px) \cos (qx) = \tfrac{1}{2}[\cos (p + q)x + \cos (p - q)x]$$

$$\sin (px) \sin (qx) = \tfrac{1}{2}[\cos (p - q)x - \cos (p + q)x]$$

2. A researcher knows that the quantity y is related to x_1 and x_2 by a formula whose general form is $y = k_1 x_1 + k_2 x_2$, where k_1 and k_2 are physical constants. To determine these constants, he conducts a series of experiments, the results of which are tabulated below:

x_1	x_2	y
1	0	3
0	1	5
1	1	2
1	-1	0
-1	1	-1

Find the best approximation to k_1 and k_2 in the least squares sense.

*3. Find the cubic equation of the form $y = ax^3 + bx + c$ which best fits the following points in the least squares sense:

$$(-3, -5), \qquad (-1, -1), \qquad (0, 1), \qquad (1, 3)$$

*4. If $f(x) = a_0 + \sum_{k=1}^{n} (c_k \cos kx + d_k \sin kx)$ for $-\pi \le x \le \pi$, show that

$$\frac{1}{\pi} \int_{-\pi}^{\pi} f^2(x)\, dx = 2a_0^2 + \sum_{k=1}^{n} (c_k^2 + d_k^2)$$

Hint: Use the pythagorean theorem (theorem 3.13) or theorem 3.10.

SUPPLEMENTARY EXERCISES

1. (i) Show that $|a \cos x + b \sin x|^2 \le a^2 + b^2$.
   Hint: Apply the Cauchy-Schwarz inequality with $\alpha = (a, b)$ and $\beta = (\cos x, \sin x)$.

   (ii) Verify the inequality

   $$\frac{1}{n} \left| \sum_{k=1}^{n} (a_k \cos kx + b_k \sin kx) \right|^2 \le \sum_{k=1}^{n} (a_k^2 + b_k^2)$$

2. Use vector methods to prove that an angle inscribed in a semicircle must be a right angle.

Hint: It suffices to show that $\overrightarrow{AB} = \alpha + \beta$ and $\overrightarrow{BC} = \alpha - \beta$ are orthogonal:

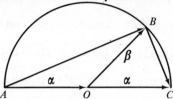

3. An $n \times n$ real matrix P is called *orthogonal* if $P^t = P^{-1}$. For instance, the matrix

$$P = \begin{bmatrix} \dfrac{1}{\sqrt{2}} & \dfrac{1}{\sqrt{2}} \\[2ex] -\dfrac{1}{\sqrt{2}} & \dfrac{1}{\sqrt{2}} \end{bmatrix}$$

has this property. Prove that the n row vectors of such a matrix are mutually orthonormal with respect to the standard inner product in R_n. Can a similar statement be made about the n column vectors of the matrix? Explain.

4. Find the solution of the differential equation $y'' + y = 0$ which best approximates the function $F(x) = x$ with respect to the norm derived from the inner product

$$(f \mid g) = \int_{-\pi}^{\pi} f(x)g(x)\ dx$$

on $C[-\pi, \pi]$.

Hint: The solution space S of the differential equation is spanned by $y_1 = \sin x$ and $y_2 = \cos x$.

5. Let $\gamma_1, \gamma_2, \ldots, \gamma_k, \ldots$ be mutually orthonormal vectors in the real inner product space V, and let α be a vector in V.
   (i) Set $\beta_n = \alpha - \sum_{j=1}^{n} (\alpha \mid \gamma_j)\gamma_j$ and find $\|\beta_n\|^2$.
   (ii) Show that $\sum_{j=1}^{n} (\alpha \mid \gamma_j)^2 \leq \|\alpha\|^2$ for each positive integer n. This result is known as *Bessel's inequality*.
   *(iii) Show that the infinite series $\sum_{j=1}^{\infty} (\alpha \mid \gamma_j)^2$ converges and that

$$\sum_{j=1}^{\infty} (\alpha \mid \gamma_j)^2 \leq \|\alpha\|^2$$

*6. If S and T are subspaces of the finite-dimensional real inner product space V, show that

$$(S + T)^{\perp} = S^{\perp} \cap T^{\perp} \quad \text{and} \quad (S \cap T)^{\perp} = S^{\perp} + T^{\perp}$$

Note: The orthogonal complement of a subspace was defined and discussed in exercise 8 of section 3.4.

7. If $\alpha_1, \alpha_2, \ldots, \alpha_m$ are linearly independent vectors in the real inner product space V, then the $m \times m$ matrix $A = (a_{ij})$ whose (i, j) entry is $a_{ij} = (\alpha_i \mid \alpha_j)$ is called the *Gram matrix* associated with the α's. Prove that such a matrix must be non-singular by completing the following steps:

(i) Let γ_j denote the jth column vector of A, and suppose that $c_1\gamma_1 + c_2\gamma_2 + \ldots + c_m\gamma_m = \theta$. By comparing components in this equation, show that

$$0 = \sum_{j=1}^{m} c_j(\alpha_i \,|\, \alpha_j) \quad \text{for } i = 1, 2, \ldots, m$$

(ii) Let $\beta = c_1\alpha_1 + c_2\alpha_2 + \ldots + c_m\alpha_m$, and show that $(\alpha_i \,|\, \beta) = 0$ for $i = 1, 2, \ldots, m$. Explain why this means that β is the zero vector.

(iii) From (ii) it follows that $c_1 = c_2 = \ldots = c_m = 0$ (why?). Hence, the γ's are linearly independent, and A must be non-singular (why?).

An immediate consequence of this result is the fact that any $m \times m$ linear system whose coefficient matrix is a Gram matrix must have a unique solution.

Exercises 8–13 provide an outline of the basic features of a complex inner product space, the definition of which is as follows:

Definition Let V be a complex vector space. Then a *complex inner product* on V is a function (|) which associates each ordered pair $\{\alpha, \beta\}$ of vectors in V with a unique complex number $(\alpha \,|\, \beta)$ in such a way that the following conditions are satisfied:

(i) $(\alpha \,|\, \alpha) > 0$ for each non-zero vector α in V.

(ii) $(\alpha \,|\, \beta) = \overline{(\beta \,|\, \alpha)}$ for all α, β in V, where as usual, $\overline{a + bi}$ denotes the complex conjugate $a - bi$

(iii) $(s\alpha + t\beta \,|\, \gamma) = s(\alpha \,|\, \gamma) + t(\beta \,|\, \gamma)$ for all α, β, γ in V and complex numbers s, t.

V together with a specified complex inner product is called a *complex inner product space.*

---

8. Show that the following properties are valid in a complex inner product space V:

(i) $(\alpha \,|\, s\beta) = \bar{s}(\alpha \,|\, \beta)$

(ii) $\left(\sum_{i=1}^{n} a_i\alpha_i \,\middle|\, \sum_{j=1}^{m} b_j\beta_j \right) = \sum_{i=1}^{n} \sum_{j=1}^{m} a_i\bar{b}_j(\alpha_i \,|\, \beta_j)$

(iii) The Cauchy-Schwarz inequality:

$$(\alpha \,|\, \beta)\overline{(\alpha \,|\, \beta)} \le (\alpha \,|\, \alpha)(\beta \,|\, \beta)$$

for all α, β in V.

(iv) If $\{\gamma_1, \ldots, \gamma_n\}$ is an orthonormal basis for V, then

$$(\alpha \,|\, \beta) = \sum_{k=1}^{n} (\alpha \,|\, \gamma_k)\overline{(\beta \,|\, \gamma_k)}$$

for all α, β in V.

9. Verify that the function (|) defined by

$$(\alpha \,|\, \beta) = \sum_{k=1}^{n} a_k\bar{b}_k \quad \text{for } \alpha = (a_1, \ldots, a_n) \text{ and } \beta = (b_1, \ldots, b_n)$$

is a complex inner product on $\mathbb{C}_n$, the space of all complex n-tuples. We shall call this the *standard complex inner product on* $\mathbb{C}_n$.

10. Let S denote the subspace of $\mathbb{C}_3$ which is spanned by the vectors $\alpha_1 = (2 + i, 1, i)$ and $\alpha_2 = (-1, 0, 3 - i)$. Use the Gram-Schmidt process to obtain a basis for S which is orthogonal with respect to the standard complex inner product.

11. If $A = (a_{ij})$ is an $n \times n$ complex matrix, let A^* denote the matrix whose (i, j) entry is $\bar{a}_{ji}$, the complex conjugate of the (j, i) entry of A. Show that

    (i) $(A^*)^* = A$

    (ii) $(sA)^* = \bar{s}A^*$ for each complex number s

    (iii) $(A + B)^* = A^* + B^*$

    (iv) $(AB)^* = B^*A^*$

Incidentally, A^* is called the *conjugate transpose* of A. We shall encounter this notation again in chapter 7.

12. Let A be an $n \times n$ complex matrix and let $(\ \ |\ \ )$ be the standard complex inner product on $\mathbb{C}_n$ (see exercise 9.)

    (i) Show that

$$(A\gamma \,|\, \eta) = (\gamma \,|\, A^*\eta)$$

    for all γ, η in $\mathbb{C}_n$. (See exercise 12, section 3.2.)

    (ii) A is said to be *unitary* if $AA^* = I_n = A^*A$. If A has this property, show that

$$(A\gamma \,|\, A\eta) = (\gamma \,|\, \eta)$$

    for all γ, η in $\mathbb{C}_n$.

13. Show that the $n \times n$ complex matrix U is unitary if and only if its column vectors are mutually orthonormal with respect to the standard complex inner product on $\mathbb{C}_n$. Can a similar statement be made about the row vectors of U?

4

Linear Transformations

4.1 BASIC PROPERTIES OF A LINEAR TRANSFORMATION

The study of functional relationships is one of the most important features of modern mathematics, and a substantial number of topics in linear algebra deal with a special kind of function known as a linear transformation. In this section, we shall define linear transformations and examine their basic properties, but before we begin, let us pause to review some of the conventions used in connection with functions of all kinds.

First of all, recall that a function F from the set $\mathfrak{D}$ to the set $\mathfrak{S}$ is a correspondence (that is, a "rule") which associates each member x of $\mathfrak{D}$ with exactly one element y of $\mathfrak{S}$. Functions are also known as "mappings" and "transformations," and to denote such a relationship, we write $F: \mathfrak{D} \rightarrow \mathfrak{S}$ and $F(x) = y$. The set $\mathfrak{D}$ which appears in this context is called the *domain* of F, while $\mathfrak{S}$ is known as the *codomain*. Moreover, y is called the *image* of x under F and x is said to be a *pre-image* of y. The general situation is depicted in figure 4.1. If each member of $\mathfrak{S}$ happens to have a pre-image in $\mathfrak{D}$, we say that F is an *onto function,* and in the case where no two elements of $\mathfrak{D}$ have the same image in $\mathfrak{S}$, the function F is said to be 1-1. These terms are illustrated in figure 4.2.

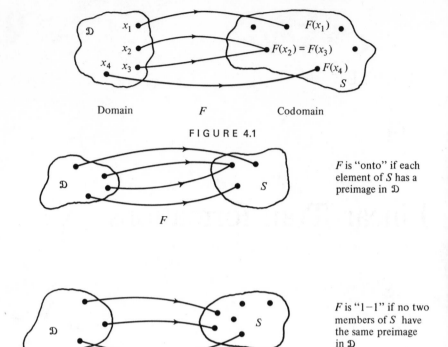

FIGURE 4.1

F is "onto" if each element of S has a preimage in $\mathfrak{D}$

F is "1−1" if no two members of S have the same preimage in $\mathfrak{D}$

FIGURE 4.2

For the particular kind of function we plan to study, both the domain and codomain are vector spaces and the functional correspondence preserves sums and scalar multiples. These statements are given precise mathematical form in the following definition.

Definition 4.1. Let U and V be vector spaces with the same scalar field. Then the function $L: U \longrightarrow V$ is said to be a *linear transformation* from U to V if the following conditions are satisfied:

(i) $L(\alpha + \beta) = L(\alpha) + L(\beta)$ for all α, β in U
(ii) $L(t\alpha) = tL(\alpha)$ for each scalar t and vector α in U

According to condition (i), the image under L of the sum $\alpha + \beta$ is the sum of the separate image vectors $L(\alpha)$ and $L(\beta)$. Similarly, condition (ii) says that the image of the scalar multiple $t\alpha$ is t times the image vector $L(\alpha)$. Before saying anything more about the implications of these linearity conditions, let us examine a few examples of functions which are linear.

EXAMPLE 1.   The rotation R_φ of the plane through φ degrees in a counter-clockwise direction about the origin is a linear transformation.

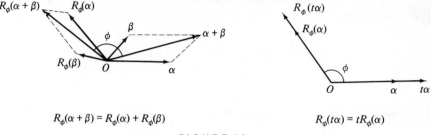

$$R_\phi(\alpha + \beta) = R_\phi(\alpha) + R_\phi(\beta) \qquad\qquad R_\phi(t\alpha) = tR_\phi(\alpha)$$

FIGURE 4.3

To show that the linearity conditions are satisfied, it suffices to give precise form to the geometric relationships suggested in figure 4.3. Note that since R_ϕ merely rotates the plane, the vector α and its image $R_\phi(\alpha)$ always have the same length. We shall have more to say about linear transformations which preserve length in section 4.7.

EXAMPLE 2.   The function $L: R_3 \longrightarrow R_3$ defined by

$$L[(c_1, c_2, c_3)] = (c_1 - c_2, c_2 - c_3, c_3 - c_1)$$

is a linear transformation. To prove this statement, note that if $\alpha = (a_1, a_2, a_3)$ and $\beta = (b_1, b_2, b_3)$, then we have

$$\begin{aligned}
L(\alpha + \beta) &= L[(a_1 + b_1, a_2 + b_2, a_3 + b_3)] \\
&= (a_1 + b_1 - a_2 - b_2, a_2 + b_2 - a_3 - b_3, a_3 + b_3 - a_1 - b_1) \\
&= (a_1 - a_2, a_2 - a_3, a_3 - a_1) + (b_1 - b_2, b_2 - b_3, b_3 - b_1) \\
&= L(\alpha) + L(\beta)
\end{aligned}$$

and

$$\begin{aligned}
L(t\alpha) &= L[(ta_1, ta_2, ta_3)] \\
&= (ta_1 - ta_2, ta_2 - ta_3, ta_3 - ta_1). \\
&= t(a_1 - a_2, a_2 - a_3, a_3 - a_1) = tL(\alpha)
\end{aligned}$$

for each scalar t.

EXAMPLE 3.   If A is an $m \times n$ real matrix, then the function $L: R_n \longrightarrow R_m$ defined as follows is a linear transformation:

$$L(\gamma) = A\gamma \quad \text{for each column } n\text{-tuple } \gamma.$$

To show this, let α, β be column n-tuples and let t be a real number. Then

$$L(\alpha + \beta) = A(\alpha + \beta) = (A\alpha) + (A\beta) = L(\alpha) + L(\beta)$$

and

$$L(t\alpha) = A(t\alpha) = t(A\alpha) = tL(\alpha)$$

Note that the linear transformation analyzed in example 2 has this form if we set

$$A = \begin{bmatrix} 1 & -1 & 0 \\ 0 & 1 & -1 \\ -1 & 0 & 1 \end{bmatrix} \quad \text{and} \quad \gamma = \begin{bmatrix} c_1 \\ c_2 \\ c_3 \end{bmatrix}$$

EXAMPLE 4.   The differentiation operator D defined by

$$D(f) = f'$$

is a linear transformation (see exercise 2). More generally, the operator

$$P = a_n D^n + a_{n-1} D^{n-1} + \ldots + a_1 D + a_0$$

is a linear transformation, where I is the identity mapping (that is, $I(f) = f$), and for each k, $D^k(f)$ denotes the kth derivative of f. For instance, if $P = D^2 + 3D - 2I$, then

$$P(f) = (D^2 + 3D - 2I)(f) = f'' + 3f' - 2f$$

EXAMPLE 5.   Let S be a finite-dimensional subspace of the real inner product space V. Then the function $P_S: V \longrightarrow V$ which maps each vector α onto the orthogonal projection α_S is a linear transformation on V (see exercise 9).

Now that we know what a linear transformation is, our next goal is to develop the basic properties of such a function, and we begin this development with the following theorem.

Theorem 4.1.   If $L: U \longrightarrow V$ is a linear transformation, then
(i) $L(\theta_U) = \theta_V$
(ii) $L(-\alpha) = -L(\alpha)$ for each vector α in U
(iii) $L(\alpha - \beta) = L(\alpha) - L(\beta)$ for all α, β in U
Proof: (i) Since L is linear and $\theta_U = \theta_U + \theta_U$, we have

$$L(\theta_U) = L(\theta_U + \theta_U) = L(\theta_U) + L(\theta_U)$$

from which it follows that $L(\theta_U) = \theta_V$.
(ii) Using part (i) and the fact that $\alpha + (-\alpha) = \theta_U$, we obtain

$$L(\alpha) + L(-\alpha) = L[\alpha + (-\alpha)] = L(\theta_U) = \theta_V$$

Therefore, $L(-\alpha)$ is the additive inverse of $L(\alpha)$, which means that $L(-\alpha) = -L(\alpha)$.
(iii) Exercise. ∎

In order to show that a given function f is linear, it is always necessary to verify both conditions of definition 4.1. However, if we suspect that f is not linear, we can often substantiate our suspicious by simply showing that one of the properties listed in theorem 4.1 does not hold. This technique is illustrated in the following solved problem.

PROBLEM 1.   In each of the following cases, show that the given function is not linear:

   (i) $T(p) = t^3 p'(0) + t$ for each polynomial p
   (ii) $G(\alpha) = \|\alpha\|$ for each vector α in the real inner product space V
   (iii) $f(x) = x^3$ for each real number x

Solution: (i) If T were linear, then according to theorem 4.1, it would have to map the zero polynomial p_0 into itself. However, since $p_0(t) = 0$ for all t, we have $p'_0(0) = 0$ and $T(p_0) = t^3 \cdot 0 + t = t$. Hence, T is non-linear.

   (ii) If G were linear, we would have $G(-\alpha) = -G(\alpha)$ for each vector α in V. However, $G(-\alpha) = \|-\alpha\| = \|\alpha\|$, while $-G(\alpha) = -\|\alpha\|$, and we conclude that G cannot be linear.

   (iii) Even though $f(0) = 0$ and $f(-x) = f(-x)^3 = -x^3 = -f(x)$, the function f is still not linear, for we have

$$f(x + y) = (x + y)^3 \neq x^3 + y^3 = f(x) + f(y)$$
$$f(tx) = (tx)^3 \neq tx^3 = tf(x)$$

If $L: U \to V$ is a linear transformation, then by combining the two conditions of definition 4.1, we find that

$$L(s_1\alpha_1 + s_2\alpha_2) = s_1 L(\alpha_1) + s_2 L(\alpha_2)$$

for vectors α_1, α_2 in U, and more generally, using mathematical induction, it can be shown that

$$L\left(\sum_{i=1}^{m} s_i\alpha_i\right) = \sum_{i=1}^{m} s_i L(\alpha_i) \tag{4.1}$$

for vectors $\alpha_1, \ldots, \alpha_m$ in U. Moreover, if we happen to know the images $L(\alpha_1), \ldots, L(\alpha_n)$ of the vectors in a basis $\{\alpha_1, \ldots, \alpha_n\}$ for U, then the image of each vector $\alpha = a_1\alpha_1 + \ldots + a_n\alpha_n$ in U can be obtained by substituting in (4.1). In other words, *the effect of the linear transformation $L: U \to V$ is completely determined by its effect on a basis for U.* This principle is illustrated in the solved problem which follows:

PROBLEM 2.   If $L: R_2 \to R_3$ is a linear transformation such that

$$L(-1, 2) = (1, 1, -2) \quad \text{and} \quad L(3, -5) = (0, 4, 7)$$

what is $L(9, -17)$?

Solution: Let $\alpha_1 = (-1, 2)$, $\alpha_2 = (3, -5)$, and $\beta = (9, -17)$. Then it is easy to see that $\{\alpha_1, \alpha_2\}$ is a basis for R_2 and that $\beta = -6\alpha_1 + \alpha_2$. Therefore, according to formula (4.1), we have

$$L(\beta) = L(-6\alpha_1 + \alpha_2) = -6L(\alpha_1) + L(\alpha_2)$$
$$= -6(1, 1, -2) + (0, 4, 7) = (-6, -2, 19)$$

It is appropriate to close this introductory section by examining some of the geometric features of linear transformations. First, recall that in 3-space, the point P at the tip of the vector $\gamma = \overrightarrow{OP}$ lies on the line l determined by points P_1, P_2 if and only if there exists a real number t such that $\gamma = (1 - t)\alpha + t\beta$, where $\alpha = \overrightarrow{OP_1}$ and $\beta = \overrightarrow{OP_2}$. Therefore, if L is a linear transformation on 3-space, we have

$$L(\gamma) = L[(1 - t)\alpha + t\beta] = (1 - t)L(\alpha) + tL(\beta) \qquad (4.2)$$

and it follows that the image of the line l under L is the line l' which passes through the points Q_1 and Q_2 located at the tips of the vectors $L(\alpha)$ and $L(\beta)$, respectively. These observations are depicted in figure 4.4. It is also of

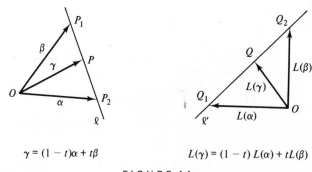

$$\gamma = (1 - t)\alpha + t\beta \qquad\qquad L(\gamma) = (1 - t)\,L(\alpha) + tL(\beta)$$

FIGURE 4.4

interest to note that the parameter t which fixes the position of P in relation to P_1 and P_2, also fixes the position of the image point Q in relation to Q_1 and Q_2. For instance, in figure 4.4, if $t = \frac{1}{3}$ then P is located $\frac{1}{3}$ the distance from P_1 to P_2 and Q is located $\frac{1}{3}$ the distance from Q_1 to Q_2.

Note that in the special case where $L(\alpha) = L(\beta)$, vector equation (4.2) becomes

$$L(\gamma) = (1 - t)L(\alpha) + tL(\alpha) = L(\alpha)$$

and it follows that the entire line l is mapped by L into the point at the tip of $L(\alpha)$.

Summarizing these results, we observe that a linear transformation L must map each line in 3-space into either another line or a single point. It is also of interest to note that L must map each plane in 3-space into either another plane, a line, or a point. The proof of this last statement is outlined in exercise 13.

According to theorem 4.1, the linear transformation $L: U \rightarrow V$ must map θ_U on θ_V, but L may also map many other vectors of U onto the zero vector of V. In the next section, we shall show that the set of all vectors which are mapped by L into θ_V forms a subspace of U, and we shall then use this so-called "null space" of L to obtain additional information about L itself.

Exercise Set 4.1

1. Show that none of the following functions on R_1 is a linear transformation.
    (i) $f(x) = 5$
    (ii) $g(x) = x + 1$
    (iii) $h(x) = x^2$
    (iv) $F(x) = \sin x$
2. Show that the differentiation operator D defined by

$$D(f) = f'$$

and the integral operator

$$T(f) = \int_0^1 f(x)\, dx$$

are both linear.
3. Verify that the function $L: R_2 \rightarrow R_3$ defined by

$$L[(a_1, a_2)] = (-2a_1 + a_2, 3a_1 - 7a_2, 5a_1 - a_2)$$

is a linear transformation, and then find a 3×2 matrix A such that $L(\gamma) = A\gamma$ for each vector

$$\gamma = \begin{bmatrix} a_1 \\ a_2 \end{bmatrix}$$

in R_2.
4. In each of the following cases, determine whether or not the given function is linear.
    (i) $f(x, y) = (2x - y, 3x)$ for each pair (x, y) of real numbers
    (ii) $F[(a_1, a_2, a_3)] = (a_1 - a_2, a_1 a_3)$
    (iii) $T(x) = \int_0^x \dfrac{e^t}{t}\, dt$ for each real number x
    (iv) $L(p) = p'(0)t^2 + p(1)t$ for each polynomial p
    (v) $G(A) = -2A$ for each $m \times n$ matrix A
5. Show that the integral operator K is linear, where

$$K(f) = f(x) - \int_0^1 e^{x+y} f(y)\, dy$$

Incidentally, operators of this general type play an important role in the theory of integral equations.

6. (i) Verify that $\{\alpha_1, \alpha_2, \alpha_3\}$ is a basis for R_3, where $\alpha_1 = (-1, 2, 1)$, $\alpha_2 = (3, 1, -2)$, $\alpha_3 = (1, -1, -1)$.

   (ii) Express $\gamma = (-3, 20, 6)$ as a linear combination of the α's.

   (iii) If $L: R_3 \longrightarrow R_4$ is the linear transformation which satisfies $L(\alpha_1) = (1, 0, 1, 0)$, $L(\alpha_2) = (1, 1, -1, 1)$, $L(\alpha_3) = (1, 1, 0, 1)$, what is $L(\gamma)$?

7. Let $L: R_2 \longrightarrow P_2$ be the linear transformation such that

$$L[(-1, 4)] = t^2 - 3 \quad \text{and} \quad L[(-2, 9)] = t + 1$$

   (i) Find $L[(7, -2)]$.

   (ii) Find a vector $\alpha = (a_1, a_2)$ for which $L(\alpha) = 3t^2 - 2t - 11$.

   (iii) Find a polynomial p in P_2 which has no pre-image in R_2.

8. Let A be a fixed $n \times n$ real matrix, and let T be the function defined by

$$T(X) = AX - XA \quad \text{for each real } n \times n \text{ matrix } X$$

   Show that T is a linear transformation on $_nR_n$.

9. Let S be a finite-dimensional subspace of the real inner product space V, and let $P_S: V \longrightarrow V$ be the function defined by

$$P_S(\alpha) = \alpha_S \quad \text{for each vector } \alpha \text{ in } S$$

   where, as usual, α_S denotes the orthogonal projection of α on S. Show that P_S is a linear transformation.
   Hint: See exercise 11, section 3.4.

10. Let l denote the line in 3-space whose equation is

$$\frac{x - 1}{2} = \frac{y + 2}{-3} = z - 3$$

   and let $L: R_3 \longrightarrow R_3$ denote the linear transformation defined by

$$L[(a_1, a_2, a_3)] = (-a_1 + a_3, a_1, a_2 + 2a_3)$$

   (i) Verify that $\gamma = (x, y, z)$ lies on l if and only if $\gamma = (1 - t)\alpha + t\beta$, where $\alpha = (1, -2, 3)$, $\beta = (3, -5, 4)$, and t is a real parameter.

   (ii) Use the vector equation $L(\gamma) = (1 - t)L(\alpha) + tL(\beta)$ to show that the line l is mapped by L into the line l' whose equation is

$$\frac{x - 2}{-1} = \frac{y - 1}{2} = \frac{z - 4}{-1}$$

   (iii) Describe the image of l under the linear transformation defined by

$$L_1[(a_1, a_2, a_3)] = (-3a_1 - 2a_2, 4a_1 + 3a_2 + a_3, 11a_1 + 8a_2 + 2a_3)$$

*11. Let R_l denote the reflection of the plane in the line l whose equation is $y = mx$. In other words, R_l maps each vector γ in the plane into its mirror image with respect to the line *l*, as indicated in the following figure:

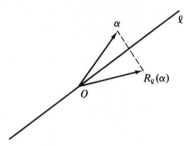

(i) Show that R_l is a linear transformation.

(ii) If the vector α lies in the line *l*, show that $R_l(\alpha) = \alpha$.

(iii) If the vector β lies in the perpendicular line $x = -my$, show that $R_l(\beta)$
$= -\beta$.

*12. Let *F* be a function with domain $\mathfrak{D}$ and codomain $\mathfrak{S}$.

(i) Show that *F* is 1-1 if and only if the equation $F(\alpha) = F(\beta)$ is satisfied only
when $\alpha = \beta$.

(ii) Show that *F* is onto if for each element *y* of $\mathfrak{S}$, there exists an *x* in $\mathfrak{D}$ such
that $F(x) = y$.

*13. Let *L* be a linear transformation of 3-space, and let Π denote the plane deter-
mined by the points $P_1, P_2,$ and P_3. Furthermore, let $\alpha = \overrightarrow{OP_1}$, $\beta = \overrightarrow{OP_2}$, and
$\gamma = \overrightarrow{OP_3}$, where *O* is the origin.

(i) If $L(\alpha), L(\beta),$ and $L(\gamma)$ are linearly independent, show that $L(\Pi)$, the image
of the plane Π under *L*, is another plane.
Hint: Use the parametric form of the plane which is given in figure 2.11
of section 2.2.

(ii) If there exist scalars *r, s* such that $L(\gamma) = rL(\alpha) + sL(\beta)$, show that *L*
maps Π onto a line.

(iii) If $L(\alpha) = L(\beta) = L(\gamma)$, show that *L* maps Π into a single point.

4.2   THE RANGE AND NULL SPACE OF A LINEAR
TRANSFORMATION

In this section, we shall investigate two subspaces which play an important
role in the study of linear transformations. We begin with the following
definition.

Definition 4.2.   Let $L: U \longrightarrow V$ be a linear transformation. Then:

(i) The *kernel of L* (denoted by $\mathfrak{N}_L$) is the set of all vectors γ in *U* which
satisfy $L(\gamma) = \theta_V$.

(ii) The *range* of *L* (denoted by $\mathfrak{R}_L$) is the collection of all vectors in *V*
which have pre-images in *U*.

Thus, the kernel of L is the collection of all vectors in U which are mapped into the zero vector of V, while the vector ρ of V belongs to the range of L if there exists a vector α in U for which $L(\alpha) = \rho$. In our next theorem, we show that $\mathfrak{N}_L$ and $\mathfrak{R}_L$ are subspaces of U and V, respectively.

Theorem 4.2.   If L is a linear transformation from U to V, then the kernel $\mathfrak{N}_L$ is a subspace of U and the range $\mathfrak{R}_L$ is a subspace of V.

Proof: Recall that a subset S of a linear vector space is a subspace if every linear combination of vectors in S is also contained in S. In particular, in order to show that $\mathfrak{N}_L$ is a subspace of U, it suffices to verify that whenever γ_1 and γ_2 are elements of the kernel of L, then so is the linear expression $s_1\gamma_1 + s_2\gamma_2$. For this purpose, note that since L is linear, we have

$$L(s_1\gamma_1 + s_2\gamma_2) = s_1L(\gamma_1) + s_2L(\gamma_2) = s_1\theta_V + s_2\theta_V = \theta_V$$

and thus $s_1\gamma_1 + s_2\gamma_2$ is contained in $\mathfrak{N}_L$, as desired.

Similarly, if ρ_1 and ρ_2 are contained in $\mathfrak{R}_L$, there exist vectors α_1 and α_2 of U such that $L(\alpha_1) = \rho_1$ and $L(\alpha_2) = \rho_2$. Since we have

$$s_1\rho_1 + s_2\rho_2 = s_1L(\alpha_1) + s_2L(\alpha_2) = L(s_1\alpha_1 + s_2\alpha_2)$$

it follows that the linear expression $s_1\rho_1 + s_2\rho_2$ is the image of the vector $s_1\alpha_1 + s_2\alpha_2$ in U, which means that $s_1\rho_1 + s_2\rho_2$ is contained in the range of L. Thus, we conclude that $\mathfrak{R}_L$ is a subspace of V. ∎

Since the kernel $\mathfrak{N}_L$ is composed of all vectors which are "nullified" by L (that is, mapped into θ_V), this set is also known as the *null space* of L, and we shall use these two names interchangeably. If $\mathfrak{N}_L$ is finite-dimensional, its dimension is called the *nullity* of L and is denoted by $\nu(L)$. If $\mathfrak{R}_L$ is finite-dimensional, its dimension is known as the *rank* of L and is denoted by $\rho(L)$. The following solved problems show how to obtain bases for the null space and range of a given linear transformation.

PROBLEM 1.   Find a basis for the null space (kernel) of the linear transformation $L: R_3 \rightarrow R_2$ defined by

$$L[(a_1, a_2, a_3)] = (2a_1 - a_2 + a_3, a_1 - 4a_3)$$

Solution: The vector $\alpha = (a_1, a_2, a_3)$ will be mapped onto the zero vector $(0, 0)$ of R_2 if and only if its components satisfy the linear system

$$2a_1 - a_2 + a_3 = 0$$
$$a_1 \qquad - 4a_3 = 0$$

Solving, we find that this system has one fundamental solution, $\eta = (4, 9, 1)$, and it follows that $\{\eta\}$ is a basis for $\mathfrak{N}_L$. Note that since the nullity of L is the dimension of $\mathfrak{N}_L$, we have $v(L) = 1$.

Geometrically, the null space of L is the line in 3-space whose equation is

$$\frac{x}{4} = \frac{y}{9} = z$$

PROBLEM 2.   Find a basis for the range of the linear transformation $L: R_3 \rightarrow R_3$ defined by

$$L[(a_1, a_2, a_3)] = (a_1 - 3a_2, -2a_1 + 4a_2 + a_3, -2a_2 + a_3)$$

Solution: The vector $\beta = (b_1, b_2, b_3)$ belongs to $\mathfrak{R}_L$ if and only if there exists a vector $\alpha = (a_1, a_2, a_3)$ such that $L(\alpha) = \beta$, and this occurs whenever the following linear system is consistent:

$$\begin{aligned}
a_1 - 3a_2 \quad\quad &= b_1 \\
-2a_1 + 4a_2 + a_3 &= b_2 \\
- 2a_2 + a_3 &= b_3
\end{aligned} \quad\quad (4.2)$$

Next, we perform the reduction

$$\begin{bmatrix} 1 & -3 & 0 & \vdots & b_1 \\ -2 & 4 & 1 & \vdots & b_2 \\ 0 & -2 & 1 & \vdots & b_3 \end{bmatrix} \longrightarrow \begin{bmatrix} 1 & -3 & 0 & \vdots & b_1 \\ 0 & 1 & -\frac{1}{2} & \vdots & -\frac{1}{2}(2b_1 + b_2) \\ 0 & 0 & 0 & \vdots & 2b_1 + b_2 - b_3 \end{bmatrix}$$

and conclude that system (4.2) is consistent if and only if the components of β satisfy $2b_1 + b_2 - b_3 = 0$. Thus, $\mathfrak{R}_L$ is the subspace of R_3 comprised of all 3-tuples (b_1, b_2, b_3) which satisfy $2b_1 + b_2 - b_3 = 0$, and we find that a basis for this subspace is $\{\eta_1, \eta_2\}$, where $\eta_1 = (-1, 2, 0)$ and $\eta_2 = (1, 0, 2)$.

Note that the geometric configuration which corresponds to $\mathfrak{R}_L$ is the plane $2x + y - z = 0$.

It should be fairly clear that when finding a basis for the range of a given linear transformation $L: R_m \rightarrow R_n$, the method of problem 2 is useful only when m and n are both very small. Indeed, even in the case where L maps R_5 into itself, the algebra involved in finding the consistency condition which characterizes $\mathfrak{R}_L$ is almost unmanageable. Fortunately, however, our next result enables us to obtain a basis for $\mathfrak{R}_L$ using only the computational methods developed in chapter 2.

Lemma 4.1. If $L: U \to V$ is a linear transformation and the vectors $\alpha_1, \alpha_2, \ldots, \alpha_n$ span U, then the image vectors $L(\alpha_1), L(\alpha_2), \ldots, L(\alpha_n)$ span $\mathfrak{R}_L$.

Proof: Suppose β is an element of $\mathfrak{R}_L$, and let α be a vector in U such that $L(\alpha) = \beta$. Since the α_k's span U, there exist scalars $a_1, a_2, \ldots, a_n$ for which $\alpha = \sum_{i=1}^{n} a_i \alpha_i$, and the linearity of L guarantees that $\beta = L(\alpha) = \sum_{i=1}^{n} a_i L(\alpha_i)$. Thus, β is contained in the linear span of $L(\alpha_1), \ldots, L(\alpha_n)$, and the proof is complete. ∎

If $S = \{\alpha_1, \alpha_2, \ldots, \alpha_n\}$ is a basis for U, then a basis for the range of $L: U \to V$ may be obtained by eliminating the dependent members from the ordered set $T = \{L(\alpha_1), L(\alpha_2), \ldots, L(\alpha_n)\}$. Usually, this "weeding out" process requires the use of one of the computational methods of chapter 2, but our next theorem provides an easy way to determine the independent members of T in the special case where S contains a basis for $\mathfrak{N}_L$.

Theorem 4.3. Let $L: U \to V$ be a linear transformation. If $\{\alpha_1, \ldots, \alpha_k\}$ is a basis for $\mathfrak{N}_L$ and $S = \{\alpha_1, \ldots, \alpha_k, \alpha_{k+1}, \ldots, \alpha_n\}$ is a basis for all of U, then $T = \{L(\alpha_{k+1}), \ldots, L(\alpha_n)\}$ is a basis for $\mathfrak{R}_L$.

Proof: We must show that the vectors in T span $\mathfrak{R}_L$ and are linearly independent. For the first part, suppose that β is a vector in $\mathfrak{R}_L$, and let $\alpha = \sum_{j=1}^{n} a_j \alpha_j$ be a vector in U which satisfies $L(\alpha) = \beta$. Since $L(\alpha_j) = \theta_V$ for $j = 1, 2, \ldots, k$, we find that

$$\beta = L(\alpha) = L\left(\sum_{j=1}^{n} a_j \alpha_j\right) = \sum_{j=1}^{n} a_j L(\alpha_j) = \sum_{j=k+1}^{n} a_j L(\alpha_j)$$

Since the vector β, which was chosen arbitrarily from $\mathfrak{R}_L$, can be expressed as a linear combination of the vectors $L(\alpha_{k+1}), \ldots, L(\alpha_n)$, we conclude that these vectors span $\mathfrak{R}_L$.

To show that the vectors in T are linearly independent, let $c_{k+1}, \ldots, c_n$ be scalars which satisfy

$$c_{k+1} L(\alpha_{k+1}) + \ldots + c_n L(\alpha_n) = \theta_V \tag{4.3}$$

Then since L is linear and $\alpha_1, \ldots, \alpha_k$ belong to $\mathfrak{N}_L$, we have

$$\theta_V = \sum_{j=k+1}^{n} c_j L(\alpha_j) = L\left(\sum_{j=k+1}^{n} c_j \alpha_j\right)$$

From this equation, we see that $\gamma = \sum_{j=k+1}^{n} c_j \alpha_j$ belongs to the null space of L, and since $\alpha_1, \ldots, \alpha_k$ is a basis for $\mathfrak{N}_L$, there exist scalars $d_1, \ldots, d_k$ such that

$$\gamma = \sum_{j=k+1}^{n} c_j \alpha_j = \sum_{j=k}^{n} d_j \alpha_j$$

Rewriting this equation, we find that

$$d_1\alpha_1 + \ldots + d_k\alpha_k - c_{k+1}\alpha_{k+1} - \ldots - c_n\alpha_n = \theta_V$$

and since the α's are linearly independent (why?), it follows that $d_1 = \ldots = d_k = c_{k+1} = \ldots = c_n = 0$. Returning to equation (4.3), we conclude that the vectors $L(\alpha_{k+1}), \ldots, L(\alpha_n)$ are linearly independent, and this completes the proof. ∎

In the solved problems which follow, we illustrate the use of theorem 4.3.

PROBLEM 3.   Let $L: R_4 \longrightarrow R_3$ denote the linear transformation defined by

$$L[(a_1, a_2, a_3, a_4)] = (a_1 - 2a_3, \, -2a_1 + a_2 + 3a_4, \, -a_1 + 2a_2 - 6a_3 + 6a_4)$$

Find an basis for the null space of L and then use it to obtain a basis for the range.

Solution: We find that (a_1, a_2, a_3, a_4) belongs to the null space of L if and only if the components a_j satisfy

$$
\begin{array}{rcl}
a_1 \quad\quad - 2a_3 \quad\quad &=& 0 \\
-2a_1 + a_2 \quad\quad + 3a_4 &=& 0 \\
-a_1 + 2a_2 - 6a_3 + 6a_4 &=& 0
\end{array}
\qquad (4.4)
$$

The reduction

$$
\begin{bmatrix} 1 & 0 & -2 & 0 \\ -2 & 1 & 0 & 3 \\ -1 & 2 & -6 & 6 \end{bmatrix} \longrightarrow \begin{bmatrix} 1 & 0 & -2 & 0 \\ 0 & 1 & -4 & 3 \\ 0 & 0 & 0 & 0 \end{bmatrix}
$$

tells us that system (4.4) has the fundamental solutions $\eta_1 = (2, 4, 1, 0)$ and $\eta_2 = (0, -3, 0, 1)$, and we conclude that $\{\eta_1, \eta_2\}$ is a basis for $\mathfrak{N}_L$. Moreover, it is easy to verify that $\{\eta_1, \eta_2, \epsilon_1, \epsilon_2\}$ is a basis for all of R_4, where $\epsilon_1 = (1, 0, 0, 0)$ and $\epsilon_2 = (0, 1, 0, 0)$. Transforming the vectors ϵ_1 and ϵ_2 by L, we obtain

$$L(\epsilon_1) = L[(1, 0 \; 0, 0)] = (1, -2, -1)$$
$$L(\epsilon_2) = L[(0, 1, 0, 0)] = (0, 1, 2)$$

and from theorem 4.3, it follows that $\{(1, -2, -1), (0, 1, 2)\}$ is a basis for $\mathfrak{R}_L$.

PROBLEM 4.   Find bases for the null space and range of the linear transformation $L: P_2 \longrightarrow P_2$ defined by

$$L[at^2 + bt + c] = (4a - b + 3c)t^2 + (a - 2c)t + (11a - 2b)$$

Solution: The polynomial $p(t) = at^2 + bt + c$ belongs to $\mathfrak{N}_L$ whenever $L(p) = 0t^2 + 0t + 0$, and this condition is satisfied if and only if a, b, c satisfy the linear system

$$4a - b + 3c = 0$$
$$a \quad\quad - 2c = 0 \quad\quad\quad (4.5)$$
$$11a - 2b \quad\quad = 0$$

Solving, we find that system (4.5) has the fundamental solution $\eta = (2, 11, 1)$, and it follows that $\{2t^2 + 11t + 1\}$ is a basis for $\mathfrak{N}_L$ (why?). Furthermore, it is easy to see that $\{2t^2 + 11t + 1, t, 1\}$ is a basis for P_2, and since

$$L(t) = -t^2 - 2$$
$$L(1) = 3t^2 - 2t$$

theorem 4.3 enables us to conclude that $\{-t^2 - 2, 3t^2 - 2t\}$ is a basis for $\mathfrak{R}_L$.

By counting the elements in each of the three bases which appear in theorem 4.3, we find that the dimension (n) of the domain of $L: U \to V$ is always the sum of the nullity (k) and the rank $(n - k)$ of L. For future reference, we state this useful observation in the form of a corollary.

Corollary 4.1.   If $L: U \to V$ is a linear transformation and dim $U = n$, then

$$\nu(L) + \rho(L) = n$$

So far, we have used the word "rank" in two different contexts: as the number of linearly independent rows (or columns) of a matrix, and as the dimension of the range of a linear transformation. Our next theorem shows that these two concepts are not as unrelated as they may appear.

Theorem 4.4.   Let A be an $m \times n$ matrix, and let $L: R_n \to R_m$ be the linear transformation defined by

$$L(\gamma) = A\gamma$$

for each column n-tuple γ. Then the rank of L is the same as the rank of A.

Proof: The null space of L is the same as the solution space S of the homogeneous $m \times n$ linear system $AX = 0$, and so, dim $S = \nu(L)$. In chapter 2, we showed that dim $S = n - r$, where r is the rank of A, and according to corollary 4.1, we also have $v(L) = n - \rho(L)$. It follows that

$$n - r = \dim S = v(L) = n - \rho(L)$$

and we conclude that $r = \rho(L)$. ∎

Theorem 4.4 provides one example of the close relationship between matrices and linear transformations. In the next few sections, we shall examine this relationship in more detail.

Exercise Set 4.2

1. In each of the following cases, find a basis for the null space of the given linear transformation.
   (i) $L[(a_1, a_2, a_3)] = (-3a_1 + 2a_2, a_1 + 4a_2, 6a_2 + a_3)$
   (ii) $L[(a_1, a_2, a_3)] = (2a_1 - 4a_2 + 12a_3, -a_1 + 2a_2 - 6a_3)$
   (iii) $L[(a_1, a_2, a_3)] = (a_2 + 2a_3, -a_1 + a_3, a_1 + a_2 + a_3, -2a_1 - a_2)$

2. Let $L: R_4 \longrightarrow R_3$ be the linear transformation defined by

$$L[(a_1, a_2, a_3, a_4)] = (2a_1 - 3a_2 + 4a_3, a_1 + 2a_3 - a_4, -3a_2 + 2a_4)$$

   (i) Find a basis for $\mathfrak{N}_L$ and extend it to a basis for all of R_4.
   (ii) Use theorem 4.3 to find a basis for the range of L.
   (iii) Show that $\beta = (15, -2, 19)$ belongs to $\mathfrak{R}_L$, and find a vector α in R_4 which satisfies $L(\alpha) = \beta$.

3. Find bases for the null space and range of each of the following linear transformations.
   (i) $L[at^2 + bt + c] = (-a + 2c)t^2 + (3a + 4b)t + (a + 2b + c)$
   (ii) $L[at^2 + bt + c] = (a - 2b + 5c, -3a + c, -3b + 8c)$

4. Let L denote the linear transformation on 3-space defined by

$$L[(a_1, a_2, a_3)] = (3a_1 + a_2 - 2a_3, -2a_1 + 5a_2 + a_3, a_1 + 6a_2 - a_3)$$

   (i) Show that the null space of L is the line whose equation is

$$\frac{x}{11} = \frac{y}{1} = \frac{z}{17}$$

   (ii) Show that the range of L is a plane and find its equation.

5. Let $L: U \longrightarrow V$ be a linear transformation.
   (i) Show that L is 1-1 if and only if its null space contains only θ_U.
   (ii) Show that L is onto if and only if its range is all of V.
   Note: The definitions of 1-1 and onto are given in section 4.1. See also exercise 12 of that section.

6. The linear transformation $L: V \longrightarrow V$ is said to be *non-singular* if the equation $L(\gamma) = \theta_V$ is satisfied only by $\gamma = \theta_V$. Otherwise, L is said to be *singular*. If V has dimension n, show that the following statements are equivalent to one another:
   (i) L is non-singular.

   (ii) $\nu(L) = 0$.

   (iii) L has rank n.

   (iv) L is 1-1 and onto.

7. Determine which (if any) of the following linear transformations is non-singular (see exercise 6).

   (i) $L[(a_1, a_2, a_3)] = (-5a_1 + a_2 + a_3, 2a_2 - 3a_3, a_1 + 4a_3)$

   (ii) $L[(a_1, a_2, a_3)] = (-2a_1 + 3a_2 + 7a_3, a_1 + a_2, 5a_2 + 7a_3)$

   (iii) $L[at^2 + bt + c] = (2a - 3c)t^2 + (a + b)t + (a - c)$

*8. Let $\{\alpha_1, \alpha_2, \ldots, \alpha_n\}$ be a basis for the vector space V, and let $L: V \longrightarrow V$ be a linear transformation.

   (i) If $\sum_{j=1}^{n} c_j L(\alpha_j) = \theta_V$, show that the vector $\gamma = \sum_{j=1}^{n} c_j \alpha_j$ belongs to $\mathfrak{N}_L$.

   (ii) Show that the vectors $L(\alpha_1)$, $L(\alpha_2)$, $\ldots$, $L(\alpha_n)$ are linearly independent if L is non-singular.

   (iii) Show that $\{L(\alpha_1), \ldots, L(\alpha_n)\}$ is a basis for V if and only if L is non-singular.

9. Let S be a finite-dimensional subspace of the real inner product space V, and let $P_S: V \longrightarrow V$ denote the linear transformation which maps each vector α in V onto its orthogonal projection on S. In other words,

$$P_S(\alpha) = \alpha_S \quad \text{for each } \alpha \text{ in } V$$

   (i) Show that S is the range of P_S.

   (ii) Show that $S^\perp$, the orthogonal complement of S, is the null space of P_S (see exercises 8 and 12, section 3.4.)

10. Let $L: U \longrightarrow V$ be a linear transformation, and let α, β be fixed vectors in $\mathfrak{R}_L$. If X is a vector in U such that $L(X) = \alpha + \beta$, show that there exist vectors X_1 and X_2 such that $L(X_1) = \alpha$, $L(X_2) = \beta$, and $X = X_1 + X_2$.

   Note that this result enables us to solve the functional equation $L(X) = \alpha + \beta$ by obtaining solutions for the related equations $L(X) = \alpha$ and $L(X) = \beta$ (why ?). This feature of a linear mapping is known as the *principle of superposition*.

11. Let $L: V \longrightarrow V$ be a linear transformation on the finite-dimensional vector space V. Show that exactly one of the following alternatives must hold:

   (1) The equation $L(\alpha) = \beta$ is solvable for every vector β in V.

   (2) $\nu(L) > 0$.

This result, which is known as the *Fredholm Alternative*, is an extremely useful tool in functional analysis. A generalized form of the Alternative can be used to analyze functional equations in infinite-dimensional spaces, but such a generalization lies beyond the scope of this text.

4.3   MATRIX REPRESENTATION OF A LINEAR TRANSFORMATION

The objective of this section is to develop a natural correspondence between matrices and linear transformations. Before providing a general description of this correspondence, let us first examine a typical case. Specifically, let $L: U \longrightarrow V$ be a linear transformation, and suppose that $S = \{\alpha_1, \alpha_2, \alpha_3\}$ and

$T = \{\beta_1, \beta_2\}$ are bases for U and V, respectively. Since the range of L is a subspace of V, the image of each vector in U can be expressed as a unique linear combination of the β's. In particular, there exist scalars a_{ij} such that

$$L(\alpha_1) = a_{11}\beta_1 + a_{21}\beta_2$$
$$L(\alpha_2) = a_{12}\beta_1 + a_{22}\beta_2 \qquad\qquad (4.6)$$
$$L(\alpha_3) = a_{13}\beta_1 + a_{23}\beta_2$$

Thus, if $\gamma = c_1\alpha_1 + c_2\alpha_2 + c_3\alpha_3$ is a typical vector in U, we have

$$L(\gamma) = L[c_1\alpha_1 + c_2\alpha_2 + c_3\alpha_3] = c_1 L(\alpha_1) + c_2 L(\alpha_2) + c_3 L(\alpha_3)$$
$$= c_1(a_{11}\beta_1 + a_{21}\beta_2) + c_2(a_{12}\beta_1 + a_{22}\beta_2) + c_3(a_{13}\beta_1 + a_{23}\beta_2)$$
$$= (a_{11}c_1 + a_{12}c_2 + a_{13}c_3)\beta_1 + (a_{21}c_1 + a_{22}c_2 + a_{23}c_3)\beta_2$$

Since $\{\beta_1, \beta_2\}$ is a basis for V, there exist unique scalars d_1, d_2 for which $L(\gamma) = d_1\beta_1 + d_2\beta_2$, and the above computation shows that

$$d_1 = a_{11}c_1 + a_{12}c_2 + a_{13}c_3$$
$$d_2 = a_{21}c_1 + a_{22}c_2 + a_{23}c_3$$

This linear system has the vector form

$$\begin{bmatrix} d_1 \\ d_2 \end{bmatrix} = \begin{bmatrix} a_{11} & a_{12} & a_{13} \\ a_{21} & a_{22} & a_{23} \end{bmatrix} \begin{bmatrix} c_1 \\ c_2 \\ c_3 \end{bmatrix} \qquad\qquad (4.7)$$

which can also be written as $[L(\gamma)]_T = A[\gamma]_S$, where $[L(\gamma)]_T$ and $[\gamma]_S$ denote the coordinatized representations of $L(\gamma)$ and γ with respect to the bases T and S, respectively, and

$$A = \begin{bmatrix} a_{11} & a_{12} & a_{13} \\ a_{21} & a_{22} & a_{23} \end{bmatrix}$$

is the matrix formed by *transposing* the array of scalar coefficients which appear in vector system (4.6). Thus, the T-coordinates of $L(\gamma)$ can be determined from (4.7) once we know the S-coordinates of γ and the entries of the matrix A, and for this reason, the properties of L are closely related to those of A. This example encourages us to make the following definition.

Definition 4.3. Let $L: U \rightarrow V$ be a linear transformation, and let $S = \{\alpha_1, \ldots, \alpha_n\}$ and $T = \{\beta_1, \ldots, \beta_m\}$ be bases for U and V, respectively. If the scalars a_{ij} satisfy the $n \times m$ vector system

$$
\begin{aligned}
L(\alpha_1) &= a_{11}\beta_1 + a_{21}\beta_2 + \ldots + a_{m1}\beta_m \\
L(\alpha_2) &= a_{12}\beta_1 + a_{22}\beta_2 + \ldots + a_{m2}\beta_m \\
&\;\;\vdots \\
L(\alpha_n) &= a_{1n}\beta_1 + a_{2n}\beta_2 + \ldots + a_{mn}\beta_m
\end{aligned}
\tag{4.8}
$$

then the $m \times n$ matrix

$$
A = \begin{bmatrix}
a_{11} & a_{12} & \ldots & a_{1n} \\
a_{21} & a_{22} & \ldots & a_{2n} \\
\vdots & \vdots & & \vdots \\
a_{m1} & a_{m2} & \ldots & a_{mn}
\end{bmatrix}
$$

is said to *represent L with respect to the chosen bases S and T.*

The following mapping diagram is a useful device for remembering the various features of definition 4.3:

$$
\{S\}U \xrightarrow[\;A\;]{\;L\;} V\{T\}
$$

When discussing the correspondence between L and its matrix representative A, we shall often find it convenient to write $A = \mu(L)$. This notational convention will be used extensively in sections 4.4 and 4.5.

Using the summation notation, we find that $A = (a_{ij})$ represents L with respect to the given choice of bases whenever

$$
L(\alpha_i) = \sum_{j=1}^{m} a_{ji}\beta_j \quad \text{for} \quad i = 1, 2, \ldots, n
$$

It is extremely important to note that A is the *transpose* of the coefficient array of system (4.8) and not the array itself. In the examples which follow, we show how the matrix representative of a given linear transformation may be computed in certain selected cases.

EXAMPLE 1.   In example 1 of section 4.1, we observed that R_φ, the rotation of the plane through φ degrees, is a linear transformation. By examining the trigonometric relationships in figure 4.5, we can determine the effect of R_φ

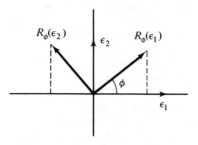

FIGURE 4.5

on the standard basis vectors ϵ_1 and ϵ_2. Specifically, we have

$$R_\varphi(\epsilon_1) = (\cos \varphi)\epsilon_1 + (\sin \varphi)\epsilon_2$$
$$R_\varphi(\epsilon_2) = (-\sin \varphi)\epsilon_1 + (\cos \varphi)\epsilon_2$$

and it follows that R_φ is represented by the matrix

$$\begin{bmatrix} \cos \varphi & -\sin \varphi \\ \sin \varphi & \cos \varphi \end{bmatrix}$$

with respect to the standard basis $\{\epsilon_1, \epsilon_2\}$.

EXAMPLE 2.   Suppose we wish to find the matrix which represents the linear transformation

$$L(a_1, a_2, a_3) = (3a_1 - a_2 + a_3, 7a_1 + 2a_2 - 3a_3)$$

with respect to the bases $S = \{(1, 0, -1), (0, 1, 3), (0, 1, -1)\}$ for R_3 and $T = \{(2, 1), (5, 3)\}$ for R_2.
   Using the formula which defines L, we find that

$$L(1, 0, -1) = (2, 10)$$
$$L(0, 1, 3) = (2, -7)$$
$$L(0, 1, -1) = (-2, 5)$$

In order to apply definition 4.3, we must express each of the vectors $\gamma_1 = (2, 10), \gamma_2 = (2, -7), \gamma_3 = (-2, 5)$ as a linear combination of $\beta_1 = (2, 1)$ and $\beta_2 = (5, 3)$. To this end, we perform the combined reduction

$$\begin{matrix} \beta_1 & \beta_2 & \gamma_1 & \gamma_2 & \gamma_3 \\ \begin{bmatrix} 2 & 5 & 2 & 2 & -2 \\ 1 & 3 & 10 & -7 & 5 \end{bmatrix} & & \longrightarrow & \begin{bmatrix} 1 & 0 & -44 & 41 & -31 \\ 0 & 1 & 18 & -16 & 12 \end{bmatrix} \end{matrix} \qquad (4.9)$$

and conclude that

$$L(1, 0, -1) = \gamma_1 = -44\beta_1 + 18\beta_2$$
$$L(0, 1, 3) = \gamma_2 = 41\beta_1 - 16\beta_2$$
$$L(0, 1, -1) = \gamma_3 = -31\beta_1 + 12\beta_2$$

Therefore, the desired matrix representative of L is

$$\mu(L) = \begin{bmatrix} -44 & 41 & -31 \\ 18 & -16 & 12 \end{bmatrix}$$

Note that $\mu(L)$ appears as a block in the reduced row echelon matrix on the right of reduction (4.9).

In general, the matrix $\mu(L)$ which represents the linear transformation $L: R_n \rightarrow R_m$ with respect to the basis $S = \{\alpha_1, \ldots, \alpha_n\}$ for R_n and $T = \{\beta_1, \ldots, \beta_m\}$ for R_m can be found as follows:

1. Use the defining condition for L to obtain the image vectors $\gamma_1 = L(\alpha_1), \gamma_2 = L(\alpha_2), \ldots, \gamma_n = L(\alpha_n)$.
2. Form the $m \times m$ matrix B whose jth column vector is β_j and the $m \times n$ matrix C whose jth column vector is γ_j. Then perform the reduction

$$\begin{array}{ccccc} \beta_1 & \cdots & \beta_m & \gamma_1 & \cdots & \gamma_n \end{array}$$
$$\left[\begin{array}{c|c} B & C \end{array} \right] \longrightarrow \left[\begin{array}{c|c} I_m & A \end{array} \right]$$

and the desired representative of L is the matrix A.

The solved problem which follows shows how a variation of this method can be used to find the matrix representative of a linear transformation whose domain and codomain are abstract vector spaces.

PROBLEM 1. Let $L: P_2 \rightarrow P_1$ denote the linear transformation defined by

$$L[at^2 + bt + c] = (3a - b + c)t + (7a + 2b - 3c)$$

Find the matrix which represents L with respect to the bases $\{t^2 - 1, t + 3, t - 1\}$ for P_2 and $\{2t + 1, 5t + 3\}$ for P_1.

Solution: First of all, note that by identifying each quadratic polynomial $p(t) = c_2 t^2 + c_1 t + c_0$ with the ordered 3-tuple (c_2, c_1, c_0) and each linear polynomial $q(t) = d_1 t + d_0$ with (d_1, d_0), the problem in question can be restated as follows:

Find the matrix which represents the linear transformation

$$L[(a, b, c)] = (3a - b + c, 7a + 2b - 3c)$$

with respect to the bases $S = \{(1, 0, -1), (0, 1, 3), (0, 1, -1)\}$ for R_3 and $T = \{(2, 1), (5, 3)\}$ for R_2.

We recognize this as the problem we solved in example 2, and it follows that

$$\begin{bmatrix} -44 & 41 & -31 \\ 18 & -16 & 12 \end{bmatrix}$$

is the desired matrix representative of L.

The matrix which represents a given linear transformation with respect to a fixed choice of bases is uniquely determined by (4.8). Turning things around, if we are given the $m \times n$ matrix A and bases S for U and T for V, there exists one and only one linear transformation $L: U \longrightarrow V$ for which $A = \mu(L)$. The following solved problem shows how L may be determined from such information.

PROBLEM 2. Describe the linear transformation $L: R_3 \longrightarrow R_2$ which is represented by the matrix

$$A = \begin{bmatrix} -9 & 1 & 4 \\ 3 & 1 & 7 \end{bmatrix}$$

with respect to the standard basis for R_3 and the basis $\{(4, -1), (-3, 1)\}$ for R_2.

Solution: According to definition 4.3, we have

$$L[(1, 0, 0)] = -9(4, -1) + 3(-3, 1) = (-45, 12)$$
$$L[(0, 1, 0)] = \quad 1(4, -1) + 1(-3, 1) = (1, 0)$$
$$L[(0, 0, 1)] = \quad 4(4, -1) + 7(-3, 1) = (-5, 3)$$

and it follows that

$$L[(a, b, c)] = aL[(1, 0, 0)] + bL[(0, 1, 0)] + cL[(0, 0, 1)]$$
$$= a(-45, 12) + b(1, 0) + c(-5, 3)$$
$$= (-45a + b - 5c, 12a + 3c)$$

Before stating definition 4.3, we derived the formula $[L(\gamma)]_T = A[\gamma]_S$ for the special case where dim $U = 3$ and dim $V = 2$, and in our next theorem, we show that this useful formula is always valid.

Theorem 4.5.   If $L: U \to V$ is a linear transformation and A is the $m \times n$ matrix which represents L with respect to the bases $S = \{\alpha_1, \ldots, \alpha_n\}$ for U and $T = \{\beta_1, \ldots, \beta_m\}$ for V, then we have

$$[L(\gamma)]_T = A[\gamma]_S \tag{4.10}$$

for each vector γ in U.

Proof: Let $\gamma = c_1\alpha_1 + \ldots + c_n\alpha_n$ be a vector in U and suppose that $L(\gamma) = d_1\beta_1 + \ldots + d_m\beta_m$. Since A represents L with respect to the given choice of bases, we know that $L(\alpha_j) = \sum_{i=1}^{m} a_{ij}\beta_i$ for $j = 1, 2, \ldots, n$. [See (4.8).] Thus, we have

$$L(\gamma) = L\left(\sum_{j=1}^{n} c_j\alpha_j\right) = \sum_{j=1}^{n} c_j L(\alpha_j)$$

$$= \sum_{j=1}^{n} c_j\left(\sum_{i=1}^{m} a_{ij}\beta_i\right) = \sum_{i=1}^{m}\left(\sum_{j=1}^{n} a_{ij}c_j\right)\beta_i$$

and since the coordinatized representation of the vector $L(\gamma)$ with respect to the basis T is unique, we find that $d_i = \sum_{j=1}^{n} a_{ij}c_j$ for $i = 1, 2, \ldots, m$. Hence, we conclude that

$$[L(\gamma)]_T = \begin{bmatrix} d_1 \\ d_2 \\ \cdot \\ \cdot \\ \cdot \\ d_m \end{bmatrix} = \begin{bmatrix} a_{11} & a_{12} & \cdots & a_{1n} \\ a_{21} & a_{22} & \cdots & a_{2n} \\ \cdot & \cdot & & \cdot \\ \cdot & \cdot & & \cdot \\ \cdot & \cdot & & \cdot \\ a_{m1} & a_{m2} & \cdots & a_{mn} \end{bmatrix} \begin{bmatrix} c_1 \\ c_2 \\ \cdot \\ \cdot \\ \cdot \\ c_n \end{bmatrix} = A[\gamma]_S$$

which is the formula we set out to establish. ∎

Roughly speaking, formula (4.10) is important because it enables us to deal with the linear transformation $L: U \to V$ as if it were the transformation $L_A: R_n \to R_m$ defined by $L_A(X) = AX$ for each n-tuple X. Thus, in many cases, questions concerning L can be answered by analyzing an appropriate $m \times n$ linear system. In particular, the vector γ belongs to the null space of L if and only if the n-tuple $X = [\gamma]_S$ is a solution of the homogeneous linear system $AX = 0$, and it follows that the nullity of L is the same as the dimension of the solution space of $AX = 0$. These observations enable us to prove the following generalization of theorem 4.4.

Corollary 4.2. If A is a matrix representative of the linear transformation $L: U \longrightarrow V$, then the rank of A equals the rank of L.

Proof: Exercise.

To see how this result may be used, suppose that $L: R_3 \longrightarrow R_3$ is the linear transformation defined by

$$L[(a_1, a_2, a_3)] = (a_1 + a_2 - a_3, a_1 - a_2 - 7a_3, 2a_1 + a_2 - 5a_3)$$

When the standard basis is used in R_3, the matrix representative of L is

$$A = \begin{bmatrix} 1 & 1 & -1 \\ 1 & -1 & -7 \\ 2 & 1 & -5 \end{bmatrix}$$

Using the reduction algorithm, we find that A has rank 2, and corollary 4.2 enables us to conclude that L also has rank 2.

In the next section, we shall continue our study of linear transformations by examining the algebraic properties of such functions and their matrix representatives.

Exercise Set 4.3

1. In each of the following cases, find the matrix which represents the given linear transformation with respect to the bases $\{(1, 1, 0), (-1, 3, 5), (2, -7, 1)\}$ for R_3 and $\{(9, 2), (4, 1)\}$ for R_2.
    (i) $L[(a_1, a_2, a_3)] = (-3a_1 + a_2 - a_3, a_1 + 5a_3)$
    (ii) $L[(a_1, a_2, a_3)] = (-a_1 + 4a_2, a_2 - a_3)$
    (iii) $L[(a_1, a_2, a_3)] = (7a_1 - 2a_2 - a_3, a_1 + 4a_2 + 3a_3)$
2. Let $L: R_2 \longrightarrow R_3$ denote the linear transformation

$$L[(a_1, a_2)] = (a_1 - 2a_2, 3a_1 + a_2, -7a_1 + a_2)$$

In each of the following cases, find the matrix which represents L with respect to the given bases S for R_2 and T for R_3.
    (i) $S = \{(1, 0), (0, 1)\}$ and $T = \{(1, 0, 0), (0, 1, 0), (0, 0, 1)\}$
    (ii) $S = \{(-1, 5), (3, -4)\}$ and $T = \{(1, 0, 0), (0, 1, 0), (0, 0, 1)\}$
    (iii) $S = \{(2, 1), (1, 7)\}$ and $T = \{(-3, 1, -6), (1, 0, 1), (5, 2, -2)\}$
3. Describe the linear transformation $L: R_3 \longrightarrow R_2$ which is represented by the matrix

$$A = \begin{bmatrix} -2 & 3 & 1 \\ 5 & -6 & 0 \end{bmatrix}$$

with respect to the standard basis for R_3 and the basis $\{(8, 3), (3, 1)\}$ for R_2.

4. Let $L: R_2 \longrightarrow R_3$ be the linear transformation which is represented by the matrix

$$A = \begin{bmatrix} 7 & -3 \\ 2 & 1 \\ 8 & 0 \end{bmatrix}$$

with respect to the bases $\{(2, 1), (5, 3)\}$ for R_2 and $\{(1, 0, 1), (1, 1, 0), (1, 1, 1)\}$ for R_3.
   (i) Show that $L[(2, 1)] = (17, 10, 15)$. What is $L[(5, 3)]$?
   (ii) Express $(1, 0)$ and $(0, 1)$ as linear combinations of $(2, 1)$ and $(5, 3)$.
   (iii) Find the matrix which represents L with respect to the standard bases in R_2 and R_3.

5. Find the matrix which represents the linear transformation

$$L[at^2 + bt + c] = (-3a + b - c)t + (a + b + 7c)$$

with respect to the bases $\{2t^2 + t - 1, t + 2, t^2 - 3t - 7\}$ for P_2 and $\{2t + 7, t + 4\}$ for P_1.

6. A certain linear transformation $L: U \longrightarrow V$ is known to be represented by the matrix

$$\begin{bmatrix} -7 & 4 & -5 & 1 & 2 \\ 5 & 7 & -1 & 1 & 3 \\ 1 & -4 & 7 & 5 & -2 \\ -1 & 7 & 1 & 7 & 3 \end{bmatrix}$$

with respect to fixed bases S for U and T for V.
   (i) Determine dim U and dim V.
   (ii) Find the rank and nullity of L.
   (iii) If γ is that vector in U whose S-coordinates are $[\gamma]_S = (1, 1, 2, -7, -1)$, what are the T-coordinates of $L(\gamma)$?

7. Let $L: R_3 \longrightarrow R_3$ be the linear transformation

$$L[(a_1, a_2, a_3)] = (a_1 - a_2 + 3a_3, 2a_1 + a_3, 2a_1 + 3a_2 - 6a_3)$$

   (i) Find the matrix A which represents L when the standard basis is used in both the domain and codomain.
   (ii) Show that A is non-singular, and find A^{-1}.
   (iii) Prove that L is non-singular (see exercise 6, section 4.2).
       Hint: What is the rank of L?

8. Let $L: R_3 \longrightarrow R_3$ be the linear transformation defined by

$$L[(a_1, a_2, a_3)] = \begin{bmatrix} 5 & -12 & 3 \\ 3 & -10 & 3 \\ 6 & -24 & 8 \end{bmatrix} \begin{bmatrix} a_1 \\ a_2 \\ a_3 \end{bmatrix}$$

(i) Show that the collection S of all vectors γ which satisfy $L(\gamma) = 2\gamma$ is a subspace of R_3.

(ii) Find a basis $\{\alpha_1, \alpha_2\}$ for S.

(iii) Let $\alpha_3 = (1, 1, 2)$. Verify that $\{\alpha_1, \alpha_2, \alpha_3\}$ is a basis for R_3.

(iv) Find the matrix which represents L when the basis $\{\alpha_1, \alpha_2, \alpha_3\}$ is used in both the domain and codomain.

*9. The *identity transformation* $I_V \colon V \longrightarrow V$ is the function which satisfies $I_V(\alpha) = \alpha$ for each vector α in V.

(i) Verify that I_V is a linear transformation.

(ii) If V is finite-dimensional, prove that I_V is non-singular. (See exercise 6, section 4.2.)

(iii) Show that I_V is represented by the identity matrix I_n if dim $V = n$ and the same basis $S = \{\alpha_1, \ldots, \alpha_n\}$ is used in both the domain and codomain.

(iv) Show that I_V is represented by the S to T change of basis matrix (see section 2.7) when the basis $S = \{\alpha_1, \ldots, \alpha_n\}$ is used in the domain and $T = \{\beta_1, \ldots, \beta_n\}$ is used in the codomain.

Note: Compare formula (4.10) with formula (2.24) of section 2.7.

10. Let V denote the vector space whose basis is $S = \{\sin x, \cos x, e^{-x}, e^{2x}, xe^{2x}\}$, and let $L \colon V \longrightarrow V$ denote the linear transformation defined by

$$L(f) = f''' - 3f'' + 4f \quad \text{for each } f \text{ in } V$$

(i) Find the matrix A which represents L when the basis S is used in both the domain and codomain.

(ii) Find bases for the null space and range of L.

(iii) Find a function f in V for which

$$L(f) = -17 \cos x + 19 \sin x$$

Is there only one such function? Explain.

11. Let R_l denote the reflection of the plane in the line l whose equation is $y = mx$ (see exercise 11, section 4.1), and let φ denote the angle of inclination of l, as indicated in the figure below:

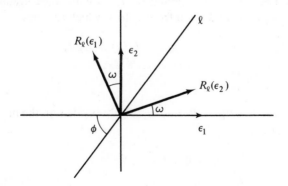

Show that R_l is represented by the following matrix with respect to the standard basis $S = \{\epsilon_1, \epsilon_2\}$

$$\begin{bmatrix} \cos 2\varphi & \sin 2\varphi \\ \sin 2\varphi & -\cos 2\varphi \end{bmatrix}$$

Hint: Note that $2\varphi - \omega = 90°$, where ω is the angle labeled in the figure.

4.4 SUMS, MULTIPLES, AND PRODUCTS OF LINEAR TRANSFORMATIONS

Our next major objective is the development of the algebra of linear transformations. In this section, we begin by examining three different ways of combining linear transformations to form new functions which are also linear.

First, recall that if F and G are functions with the same domain $\mathfrak{D}$ and codomain $\mathfrak{S}$, then the sum $F + G$ and scalar multiple sF are those functions from $\mathfrak{D}$ to $\mathfrak{S}$ defined by

$$(F + G)(x) = F(x) + G(x)$$
$$(sF)(x) = sF(x)$$

for each element x of $\mathfrak{D}$. Therefore, since a linear transformation is just a function with a few special features, it is natural to make the following definitions.

Definition 4.4. If $L: U \longrightarrow V$ is a linear transformation and s is a scalar, then the *scalar multiple* sL is that function from U to V such that

$$(sL)(\alpha) = sL(\alpha)$$

for each vector α in U.

Definition 4.5. If $L_1: U \longrightarrow V$ and $L_2: U \longrightarrow V$ are linear transformations, the *sum* $L_1 + L_2$ is that function from U to V which satisfies

$$(L_1 + L_2)(\alpha) = L_1(\alpha) + L_2(\alpha)$$

for each vector α in U.

EXAMPLE 1.   Let $L_1: R_2 \longrightarrow R_2$ and $L_2: R_2 \longrightarrow R_2$ denote the linear transformations defined by

$$L_1[(a_1, a_2)] = (a_2, a_1 + 3a_2)$$
$$L_2[(a_1, a_2)] = (2a_1 - a_2, -a_1 + a_2)$$

Then we have

$$(-5L_1)[(a_1, a_2)] = -5L_1[(a_1, a_2)]$$
$$= -5(a_2, a_1 + 3a_2) = (-5a_2, -5a_1 - 15a_2)$$

and

$$(L_1 + L_2)[(a_1, a_2)] = L_1[(a_1, a_2)] + L_2[(a_1, a_2)]$$
$$= (a_2, a_1 + 3a_2) + (2a_1 - a_2, -a_1 + a_2) = (2a_1, 4a_2)$$

Note that the functions $-5L_1$ and $L_1 + L_2$ described in example 1 are both linear. Our next theorem shows that sums and scalar multiples of linear transformations are always linear.

Theorem 4.6. If L_1 and L_2 are linear transformations from U to V, then so are $L_1 + L_2$ and sL_1.

Proof: We shall prove that $L_1 + L_2$ is linear and leave the parallel proof for sL_1 as an exercise. Accordingly, let α and β be vectors in U and let t be a scalar. Then, since L_1 and L_2 are themselves linear, we have

$$(L_1 + L_2)(\alpha + \beta) = L_1(\alpha + \beta) + L_2(\alpha + \beta)$$
$$= [L_1(\alpha) + L_1(\beta)] + [L_2(\alpha) + L_2(\beta)]$$
$$= [L_1(\alpha) + L_2(\alpha)] + [L_1(\beta) + L_2(\beta)]$$
$$= (L_1 + L_2)(\alpha) + (L_1 + L_2)(\beta)$$

and

$$(L_1 + L_2)(t\alpha) = L_1(t\alpha) + L_2(t\alpha) = tL_1(\alpha) + tL_2(\alpha)$$
$$= t[L_1(\alpha) + L_2(\alpha)] = t[(L_1 + L_2)(\alpha)]$$

and it follows that $L_1 + L_2$ is linear. ∎

Henceforth, we shall write $\mathcal{L}(U, V)$ to denote the system comprised of all linear transformations with domain U and codomain V together with the operations of functional addition and multiplication of a function by a scalar. Thanks to theorem 4.6, it is a fairly simple matter to prove the following result.

Theorem 4.7. The system $\mathcal{L}(U, V)$ is a vector space.

Proof: Exercise.

In $\mathcal{L}(U, V)$ the zero vector is the linear transformation L_0, which satisfies $L_0(\alpha) = \theta_V$ for each vector α in U. In keeping with the notational conventions of chapter 2, we shall denote the additive inverse of the linear transformation L by $-L$ and write $L_1 - L_2$ to denote the sum of the transformations L_1 and $-L_2$.

If the vector spaces U and V are both finite-dimensional, we know that each linear transformation $L: U \to V$ can be represented by a unique matrix with respect to a fixed choice of bases for U and V. Our next theorem shows how the matrices which represent the scalar multiple sL and the sum $L_1 + L_2$ are related to the matrix representatives of L, L_1, and L_2.

Theorem 4.8. Let $L_1: U \to V$ and $L_2: U \to V$ be linear transformations and let A and B be the $m \times n$ matrices which represent L_1 and L_2, respectively, with respect to the fixed bases $S = \{\alpha_1, \ldots, \alpha_n\}$ for U and $T = \{\beta_1, \ldots, \beta_m\}$ for V. Then:
  (i) The scalar multiple sL_1 is represented by sA.
  (ii) The sum $L_1 + L_2$ is represented by $A + B$.

Proof: We shall prove (ii) and leave (i) as an exercise. Recall from definition 4.3 that $C = (c_{ij})$ is the matrix which represents $L: U \to V$ with respect to the S-T choice of bases if $L(\alpha_i) = \sum_{j=1}^{m} c_{ji}\beta_j$ for $i = 1, 2, \ldots, n$. In particular, since $\mu(L_1) = A$ and $\mu(L_2) = B$, we have

$$(L_1 + L_2)(\alpha_i) = L_1(\alpha_i) + L_2(\alpha_i)$$

$$= \sum_{j=1}^{m} a_{ji}\beta_j + \sum_{j=1}^{m} b_{ji}\beta_i = \sum_{j=1}^{m} (a_{ji} + b_{ji})\beta_j$$

for each index j, and it follows that $L_1 + L_2$ is represented by the $m \times n$ matrix

$$\begin{bmatrix} a_{11} + b_{11} & \cdots & a_{1n} + b_{1n} \\ \cdot & & \cdot \\ \cdot & & \cdot \\ \cdot & & \cdot \\ a_{m1} + b_{m1} & \cdots & a_{mn} + b_{mn} \end{bmatrix}$$

which, of course is $A + B$. Thus, we have shown that $\mu(L_1 + L_2) = A + B$, as desired. ∎

We have seen that functions may be added together and multiplied by scalars, and under certain conditions, they may even be multiplied by one another. In particular, if F and G are functions and the domain of G contains the range of F, then the *composition* (that is, the product) of G with F is the function (denoted by $G \circ F$) which satisfies

$$(G \circ F)(x) = G[F(x)] \tag{4.11}$$

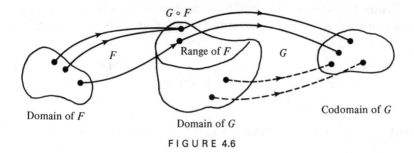

FIGURE 4.6

for each element x of the domain of F. In the special case where F and G are linear transformations, our next theorem shows that the composite function $G \circ F$ is also linear.

Theorem 4.9. If $L_1: U \longrightarrow V$ and $L_2: V \longrightarrow W$ are linear transformations, then the composite function $L_2 \circ L_1$ is a linear transformation from U to W.

Proof: Let α and β be vectors in U and let t be a scalar. Then since L_1 and L_2 are linear, we have

$$(L_2 \circ L_1)(\alpha + \beta) = L_2[L_1(\alpha + \beta)] = L_2[L_1(\alpha) + L_1(\beta)]$$
$$= L_2[L_1(\alpha)] + L_2[L_1(\beta)] = (L_2 \circ L_1)(\alpha) + (L_2 \circ L_1)(\beta)$$

and

$$(L_2 \circ L_1)(t\alpha) = L_2[L_1(t\alpha)] = L_2[tL_1(\alpha)]$$
$$= tL_2[L_1(\alpha)] = t[(L_2 \circ L_1)(\alpha)]$$

and it follows that $L_2 \circ L_1$ is linear, as claimed. ∎

EXAMPLE 2.   If $L_1: R_3 \longrightarrow R_2$ and $L_2: R_2 \longrightarrow R_3$ are the linear transformations defined by

$$L_1[(a_1, a_2, a_3)] = (a_1 - 2a_2, 3a_2 - a_3)$$
$$L_2[(b_1, b_2)] = (3b_1 + b_2, 5b_1 - b_2, 2b_1)$$

then the composite function $L_2 \circ L_1$ is the linear transformation on R_3 which satisfies

$$(L_2 \circ L_1)[(a_1, a_2, a_3)] = L_2[L_1(a_1, a_2, a_3)] = L_2[(\underbrace{a_1 - 2a_2}_{b_1}, \underbrace{3a_2 - a_3}_{b_2})]$$

$$= (3(a_1 - 2a_2) + (3a_2 - a_3), 5(a_1 - 2a_2) - (3a_2 - a_3), 2(a_1 - 2a_2))$$
$$= (3a_1 - 3a_2 - a_3, 5a_1 - 13a_2 + a_3, 2a_1 - 4a_2)$$

Note that when the standard bases are used in both R_2 and R_3, the linear transformations L_1 and L_2 described in example 2 are represented by the matrices

$$\mu(L_1) = \begin{bmatrix} 1 & -2 & 0 \\ 0 & 3 & -1 \end{bmatrix} \quad \text{and} \quad \mu(L_2) = \begin{bmatrix} 3 & 1 \\ 5 & -1 \\ 2 & 0 \end{bmatrix}$$

and the composite transformation $L_2 \circ L_1$ is represented by

$$\mu(L_2 \circ L_1) = \begin{bmatrix} 3 & -3 & -1 \\ 5 & -13 & 1 \\ 2 & -4 & 0 \end{bmatrix}$$

It is easy to verify that $\mu(L_2 \circ L_1) = \mu(L_2)\mu(L_1)$, and in our next theorem, we show that this formula is always valid.

Theorem 4.10. Let $L_1: U \to V$ and $L_2: V \to W$ be linear transformations, and let A and B be the matrices which represent L_1 and L_2, respectively, with respect to a fixed choice of bases for U, V, and W. Then the composite transformation $(L_2 \circ L_1): U \to W$ is represented by the product matrix BA.

Proof: Let $R = \{\alpha_1, \ldots, \alpha_n\}$, $S = \{\beta_1, \ldots, \beta_p\}$, and $T = \{\gamma_1, \ldots, \gamma_m\}$ be bases for U, V, and W, respectively, and assume that $A = (a_{ij})$ and $B = (b_{ij})$ represent L_1 and L_2, respectively, with respect to these bases. Then for $i = 1, 2, \ldots, n$, we have

$$(L_2 \circ L_1)(\alpha_i) = L_2[L_1(\alpha_i)]$$
$$= L_2\left(\sum_{j=1}^{p} a_{ji}\beta_j\right) = \sum_{j=1}^{p} a_{ji}L_2(\beta_j)$$
$$= \sum_{j=1}^{p} a_{ji}\left[\sum_{k=1}^{m} b_{kj}\gamma_k\right] = \sum_{k=1}^{m}\left[\sum_{j=1}^{p} b_{kj}a_{ji}\right]\gamma_k$$

According to definition 4.3, the transformation $L_2 \circ L_1$ is represented by the transpose of the $n \times m$ matrix whose (i, k) entry is $\sum_{j=1}^{p} b_{kj}a_{ji}$, and it is easy to verify that this matrix is BA. ∎

Figure 4.7 provides a convenient device for remembering the result obtained in theorem 4.10.

In example 1 of section 4.3, we showed that R_φ, the rotation of the plane through φ degrees, is represented by the matrix

$$\mu(R_\varphi) = \begin{bmatrix} \cos\varphi & -\sin\varphi \\ \sin\varphi & \cos\varphi \end{bmatrix}$$

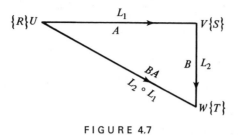

FIGURE 4.7

with respect to the standard basis. In the example which follows, we shall use theorem 4.10 in conjunction with this representation to derive two important trigonometric identities.

EXAMPLE 3.   It is easy to see that a rotation through φ_2 degrees followed by one through φ_1 degrees is the same as a single rotation through $\varphi_1 + \varphi_2$ degrees. In other words, $R_{\varphi_1} \circ R_{\varphi_2} = R_{\varphi_1 + \varphi_2}$, and we find that

$$\mu(R_{\varphi_1 + \varphi_2}) = \mu(R_{\varphi_1} \circ R_{\varphi_2}) = \mu(R_{\varphi_1})\mu(R_{\varphi_2})$$

where the second equation is a consequence of theorem 4.10. Therefore, we have

$$\begin{bmatrix} \cos(\varphi_1 + \varphi_2) & -\sin(\varphi_1 + \varphi_2) \\ \sin(\varphi_1 + \varphi_2) & \cos(\varphi_1 + \varphi_2) \end{bmatrix} = \mu(R_{\varphi_1 + \varphi_2}) = \mu(R_{\varphi_1})\mu(R_{\varphi_2})$$

$$= \begin{bmatrix} \cos\varphi_1 & -\sin\varphi_1 \\ \sin\varphi_1 & \cos\varphi_1 \end{bmatrix}\begin{bmatrix} \cos\varphi_2 & -\sin\varphi_2 \\ \sin\varphi_2 & \cos\varphi_2 \end{bmatrix}$$

$$= \begin{bmatrix} \cos\varphi_1\cos\varphi_2 - \sin\varphi_1\sin\varphi_2 & -(\cos\varphi_1\sin\varphi_2 + \sin\varphi_1\cos\varphi_2) \\ (\sin\varphi_1\cos\varphi_2 + \cos\varphi_1\sin\varphi_2) & -\sin\varphi_1\sin\varphi_2 + \cos\varphi_1\cos\varphi_2 \end{bmatrix}$$

and by equating the corresponding entries in this matrix equation, we obtain the trigonometric identities

$$\cos(\varphi_1 + \varphi_2) = \cos\varphi_1\cos\varphi_2 - \sin\varphi_1\sin\varphi_2$$

$$\sin(\varphi_1 + \varphi_2) = \cos\varphi_1\sin\varphi_2 + \sin\varphi_1\cos\varphi_2$$

The technique used in example 3 demonstrates once again the value of being able to represent a given linear transformation by a matrix. In the next section, we shall examine the relationship between a linear transformation and its matrix representatives in more detail.

Exercise Set 4.4

1. Let $L_1: R_2 \longrightarrow R_3$ and $L_2: R_2 \longrightarrow R_3$ denote the linear transformations defined by

$$L_1[(a_1, a_2)] = (-3a_1 + 5a_2, a_2, a_1 - 7a_2)$$

$$L_2[(a_1, a_2)] = (a_2, 2a_1 + 3a_2, -3a_1 + 5a_2)$$

Describe each of the following transformations and find its matrix representative with respect to the standard bases for R_1 and R_2:
(i) $L_1 + L_2$ (ii) $3L_1$ (iii) $3L_1 - 2L_2$

2. Let $L_1: R_2 \longrightarrow R_2$ and $L_2: R_2 \longrightarrow R_3$ denote the linear transformations

$$L_1[(a_1, a_2)] = (2a_1 - 3a_2, a_1 + 5a_2)$$

$$L_2[(b_1, b_2)] = (b_1 - 2b_2, b_1 + 3b_2, 7b_1 - 5b_2)$$

(i) Describe $L_1 \circ L_1$ and $L_2 \circ L_1$.
(ii) Find the matrices which represent $L_1, L_2, L_1 \circ L_1$ and $L_2 \circ L_1$ with respect to the standard bases for R_2 and R_3.
(iii) Explain why $L_2 \circ L_2$ and $L_1 \circ L_2$ are not defined.

3. Let $L_1: R_3 \longrightarrow R_3$ and $L_2: R_3 \longrightarrow R_3$ be the linear transformations defined by

$$L_1[(a_1, a_2, a_3)] = (3a_1 - 5a_2 + 3a_3, -a_1 - 2a_2 + 3a_3, a_2 - a_3)$$

$$L_2[(b_1, b_2, b_3)] = (b_1 + 2b_2 + 9b_3, b_1 + 3b_2 + 12b_3, b_1 + 3b_2 + 11b_3)$$

(i) Find the matrices A and B which represent L_1 and L_2, respectively, when the standard basis for R_3 is used in both the domain and codomain.
(ii) Show that $AB = I_3 = BA$.
(iii) Describe the transformations $L_1 \circ L_2$ and $L_2 \circ L_1$.

4. Let $L_1: R_3 \longrightarrow R_2$ and $L_2: R_3 \longrightarrow R_2$ be linear transformations, and assume that when the standard bases are used in R_2 and R_3, the linear transformations $2L_1 + L_2$ and $3L_1 + 2L_2$ are represented by the matrices C and D, respectively, where

$$C = \begin{bmatrix} -1 & 3 & 7 \\ 2 & 7 & 5 \end{bmatrix} \quad \text{and} \quad D = \begin{bmatrix} 0 & -2 & 1 \\ 4 & 1 & 3 \end{bmatrix}$$

(i) Show that $\mu(L_1)$ and $\mu(L_2)$ must satisfy

$$2\mu(L_1) + \mu(L_2) = C \quad \text{and} \quad 3\mu(L_1) + 2\mu(L_2) = D$$

(ii) By solving the matrix equations in part (i), find $\mu(L_1)$ and $\mu(L_2)$.
(iii) Find the matrices which represent L_1 and L_2 with respect to the bases $S = \{(-1, 1), (2, -1)\}$ for R_2 and $T = \{(1, -1, 0), (0, 1, 1), (1, -1, -1)\}$ for R_3.

5. Complete the proofs of theorems 4.6 and 4.8 by showing that the scalar multiple sL of the linear transformation $L: U \longrightarrow V$ is a linear transformation which is represented by $s\mu(L)$.

6. If $L: U \longrightarrow V$ is a linear transformation, show that $I_V \circ L = L = L \circ I_U$, where I_U and I_V are the identity transformations on U and V, respectively (see exercise 9, section 4.3).

7. Find linear transformations $L_1: R_2 \longrightarrow R_2$ and $L_2: R_2 \longrightarrow R_2$ for which $L_1 \circ L_2 \neq L_2 \circ L_1$.

*8. Let $L_1: U \longrightarrow V$ and $L_2: V \longrightarrow W$ be linear transformations, and assume that dim $U = n$ and dim $V = m$.

   (i) Show that the null space of L_1 is contained in the null space of $L_2 \circ L_1$.
   (ii) Show that $\nu(L_2 \circ L_1) \geq \nu(L_1)$ and use this inequality to show that $\rho(L_2 \circ L_1) \leq \rho(L_1)$.
   (iii) Show that the range of $L_2 \circ L_1$ is contained in that of L_2 and then show that $\rho(L_2 \circ L_1) \leq \rho(L_2)$.
   (iv) Is the inequality $\nu(L_2 \circ L_1) \leq \nu(L_2)$ always valid? Either prove that it is, or exhibit transformations L_1 and L_2 for which it is false.

*9. If $L: V \longrightarrow V$ is a linear transformation whose rank is 1, show that $L \circ L = sL$ for some scalar s.

4.5 THE CORRESPONDENCE PRINCIPLE AND ITS APPLICATIONS

In corollary 4.2 of section 4.3, we showed that the rank of the linear transformation L is the same as that of its matrix representative $\mu(L)$, and in theorems 4.8 and 4.10 of the last section, we derived the formulas

$$\mu(L_1 + L_2) = \mu(L_1) + \mu(L_2)$$

$$\mu(sL) = s\mu(L) \tag{4.12}$$

$$\mu(L_2 \circ L_1) = \mu(L_2)\mu(L_1)$$

These results illustrate an important feature of linear algebra known as the *correspondence principle*, which states that the general algebraic properties of transformations are the same as those of the matrices by which they are represented, and vice versa. The purpose of this section is to examine a few of the many applications of this principle.

The correspondence principle is valuable primarily because it enables us to derive functional properties with matrices and matrix properties with transformations. We have already seen that the rank and nullity of a given linear transformation L may be found most easily by analyzing the matrix representative of L. On the other hand, certain general algebraic rules may be obtained more readily using transformations than matrices. For instance, in section 1.5 we were able to show that matrix multiplication is associative, but only after wading through the intricacies of an identity involving double summations. However, it is easy to see that functional composition is an associative operation; indeed, the transformations $L_3 \circ (L_2 \circ L_1)$ and $(L_3 \circ L_2) \circ L_1$ are equal whenever they are defined since

$$[L_3 \circ (L_2 \circ L_1)](\alpha) = L_3[(L_2 \circ L_1)(\alpha)] = L_3[L_2(L_1(\alpha))]$$

and

$$[(L_3 \circ L_2) \circ L_1](\alpha) = [L_3 \circ L_2](L_1(\alpha)) = L_3[L_2(L_1(\alpha))]$$

for each vector α in the domain of L_1. To apply this fact to the algebra of matrices, we pick three matrices A, B, C and then find the transformations L_1, L_2, L_3 for which $\mu(L_3) = A$, $\mu(L_2) = B$, and $\mu(L_1) = C$ with respect to a fixed choice of bases. According to (4.12), we have

$$A(BC) = \mu(L_3)[\mu(L_2)\mu(L_1)] = \mu[L_3 \circ (L_2 \circ L_1)]$$

$$= \mu[(L_3 \circ L_2) \circ L_1] = [\mu(L_3)\mu(L_2)]\mu(L_1) = (AB)C$$

and it follows that matrix multiplication is associative. The main ingredients of this argument are displayed in figure 4.8.

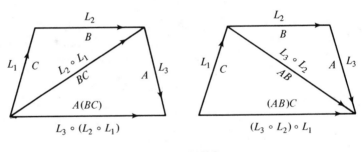

$$A(BC) = (AB)C$$

FIGURE 4.8

When the correspondence principle is stated in formal mathematical terms, it assumes the following form. *Note:* Recall that the concept of vector space isomorphism was defined in section 2.7.

Theorem 4.11. If dim $U = n$ and dim $V = m$, then the vector space $\mathcal{L}(U, V)$ is isomorphic to $_mR_n$.

Proof: Let $S = \{\alpha_1, \ldots, \alpha_n\}$ and $T = \{\beta_1, \ldots, \beta_m\}$ be bases for U and V, respectively, and define the correspondence $\mu(L) = A$, where A is the $m \times n$ matrix which represents $L: U \longrightarrow V$ with respect to the S-T choice of bases. Then μ is an isomorphism between $\mathcal{L}(U, V)$ and $_mR_n$, for each matrix is associated with exactly one transformation and the required linearity conditions

$$\mu(L_1 + L_2) = \mu(L_1) + \mu(L_2)$$

$$\mu(sL) = s\mu(L)$$

are also satisfied, thanks to the result of theorem 4.8. ∎

Corollary 4.3. If dim $U = n$ and dim $V = m$, then $\mathfrak{L}(U, V)$ has dimension mn.

Proof: Isomorphic vector spaces have the same dimension, and $_mR_n$ has dimension mn. ∎

The linear transformations $L_1, L_2, \ldots, L_k$ are linearly independent in $\mathfrak{L}(U, V)$ if the zero transformation L_0 can be expressed in the form $L_0 = s_1L_1 + \ldots + s_kL_k$ only when $s_1 = s_2 = \ldots = s_k = 0$. The correspondence principle tells us that this occurs if and only if the matrix representatives $\mu(L_1), \mu(L_2), \ldots, \mu(L_k)$ are linearly independent in $_mR_n$. Thus, we have the following result.

Corollary 4.4. If dim $U = n$ and dim $V = m$, then $\{L_1, \ldots, L_k\}$ is a basis for $\mathfrak{L}(U, V)$ if and only if $\{\mu(L_1), \ldots, \mu(L_k)\}$ is a basis for $_mR_n$.

Using corollary 4.4, it is easy to find a basis for $\mathfrak{L}(U, V)$. For instance, suppose U is P_2 and V is P_1. Then, according to the corollary. $\{L_{11}, L_{12}, L_{13}, L_{21}, L_{22}, L_{23}\}$ is a basis for $\mathfrak{L}(P_2, P_1)$, where L_{ij} is the linear transformation which is represented by the standard basis matrix E_{ij} with respect to the natural bases $\{t^2, t, 1\}$ for P_2 and $\{t, 1\}$ for P_1. For example, L_{13} is the transformation which is represented by

$$E_{13} = \begin{bmatrix} 0 & 0 & 1 \\ 0 & 0 & 0 \end{bmatrix}$$

Thus, we have

$$L_{13}(t^2) = 0t + 0$$

$$L_{13}(t) = 0t + 0$$

$$L_{13}(1) = 1t + 0$$

and it follows that

$$L_{13}[at^2 + bt + c] = aL_{13}(t^2) + bL_{13}(t) + cL_{13}(1)$$

$$= a(0t + 0) + b(0t + 0) + c(t + 0) = ct$$

Descriptions of the other five "basic" transformations may be obtained by similar means, and we find that

$$L_{11}(p) = at, \qquad L_{12}(p) = bt, \qquad L_{13}(p) = ct$$
$$L_{21}(p) = a, \qquad L_{22}(p) = b, \qquad L_{23}(p) = c \tag{4.13}$$

where $p(t) = at^2 + bt + c$.

We have already observed that it is often useful to know whether or not a given matrix has an inverse, and our next objective is to define and analyze the notion of an inverse mapping. If L is a mapping of the vector space V onto itself, it is natural to expect the inverse mapping L^{-1} to also map V onto V and to "undo" the work of L, that is, if $L(\alpha) = \beta$, then $L^{-1}(\beta)$ should be α. Note that L^{-1} can have these properties only if for each vector β in V, there exists exactly one vector α such that $L(\alpha) = \beta$, and this occurs whenever L is 1-1 and onto. These observations lead to the following definition.

Definition 4.6. Let $L: V \to V$ be a 1-1 mapping of the vector space V onto itself. Then L^{-1} is the mapping such that for each vector β in V, the image $L^{-1}(\beta)$ is the (unique) vector α in V for which $L(\alpha) = \beta$.

EXAMPLE 1.   The inverse of a rotation through φ degrees is a rotation through $-\varphi$ degrees (why?). In symbols, $R_{\varphi}^{-1} = R_{-\varphi}$. Similarly, if l is a line through the origin, then the reflection R_l of the plane in the line l is its own inverse. That is, $R_l^{-1} = R_l$. (See exercises 10 and 11.)

EXAMPLE 2.   Let V denote the collection of all functions f which are differentiable and for which $f(0) = 0$, and let D denote the differentiation operator, that is, $D(f) = f'$. Then D is 1-1 and onto (why?), and D^{-1} is that mapping such that for each function g in V, we have $D^{-1}(g) = f$ where $g = D(f) = f'$. In other words, $D^{-1}(g) = \int g$, and we recognize D^{-1} as the integration operator.

We know that the $n \times n$ matrix A is non-singular (and thus has an inverse) if and only if the homogeneous linear system $AX = O$ has only the trivial solution $X = O$. Analogously, we shall say that the linear transformation $L: V \to V$ is non-singular if the vector equation $L(\gamma) = \theta_V$ is satisfied only when $\gamma = \theta_V$, and it can be shown that this occurs if and only if L is 1-1 and onto (see exercise 6, section 4.2.) Thus, a linear mapping has an inverse whenever it is non-singular, and our next theorem shows that if L^{-1} exists, then it, too, is linear and non-singular.

Theorem 4.12. If $L: V \to V$ is a non-singular linear transformation, then its inverse L^{-1} is also non-singular and linear.

Proof: First, to show that L^{-1} is linear, let α, β be vectors in V, and let α_1, β_1 be the (unique) vectors in V for which $L(\alpha_1) = \alpha$ and $L(\beta_1) = \beta$. According to definition 4.6, $\alpha_1 = L^{-1}(\alpha)$ and $\beta_1 = L^{-1}(\beta)$. Therefore, we have

$$L^{-1}(s\alpha) = L^{-1}[sL(\alpha_1)] = L^{-1}[L(s\alpha_1)]$$

$$= [L^{-1} \circ L](s\alpha_1) = I_V(s\alpha_1) = s\alpha_1 = sL^{-1}(\alpha)$$

for each scalar s, and

$$L^{-1}(\alpha + \beta) = L^{-1}[L(\alpha_1) + L(\beta_1)] = L^{-1}[L(\alpha_1 + \beta_1)]$$

$$= [L^{-1} \circ L](\alpha_1 + \beta_1) = I_V(\alpha_1 + \beta_1)$$

$$= \alpha_1 + \beta_1 = L^{-1}(\alpha) + L^{-1}(\beta)$$

and it follows that L^{-1} is linear. To show that L^{-1} is non-singular, note that γ satisfies $L^{-1}(\gamma) = \theta_V$ if and only if $L(\theta_V) = \gamma$, which means that γ must be θ_V. ∎

Using definition 4.6, it is easy to show that $L^{-1} \circ L = I_V = L \circ L^{-1}$ (see exercise 4.) In the theorem which follows, we use this formula in conjunction with the correspondence principle to establish an important link between non-singular matrices and transformations.

Theorem 4.13. Let S be a fixed basis for the n-dimensional vector space V. Then the linear transformation $L: V \to V$ is non-singular if and only if its matrix representative $\mu(L)$ is non-singular. Furthermore, in the case where L is nonsingular, $\mu(L^{-1}) = [\mu(L)]^{-1}$.

Proof: If L is non-singular, then L^{-1} exists and we have $L \circ L^{-1} = I_V = L^{-1} \circ L$. Applying the third formula in (4.12) to this equation and using the fact that $\mu(I_V) = I_n$ (see exercise 9, section 4.3), we obtain the matrix equation

$$\mu(L)\mu(L^{-1}) = \mu(I_V) = I_n = \mu(L^{-1})\mu(L)$$

which tells us that $\mu(L)$ is non-singular and that $\mu(L^{-1}) = [\mu(L)]^{-1}$.

Conversely, suppose the matrix $\mu(L)$ is non-singular. Then $\mu(L)$ has rank n, and according to corollary 4.2, the transformation L also has rank n. This is the same as saying that the null space of L contains only θ_V, and it follows that L is non-singular. ∎

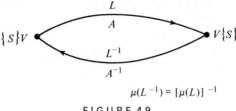

$$\mu(L^{-1}) = [\mu(L)]^{-1}$$

FIGURE 4.9

PROBLEM 1. Determine whether or not the linear transformation
$L: R_3 \longrightarrow R_3$ defined by

$$L[(a_1, a_2, a_3)] = (2a_1 + a_2 + a_3, a_1 - a_3, 3a_1 + a_2 - a_3)$$

is non-singular, and if it is, find its inverse.

Solution: When the standard basis is used in R_3, we find that the matrix
representative of L is

$$\mu(L) = A = \begin{bmatrix} 2 & 1 & 1 \\ 1 & 0 & -1 \\ 3 & 1 & -1 \end{bmatrix}$$

By applying the matrix inversion algorithm, we find that A is non-singular
and that

$$A^{-1} = \begin{bmatrix} 1 & 2 & -1 \\ -2 & -5 & 3 \\ 1 & 1 & -1 \end{bmatrix}$$

According to theorem 4.13, L is also non-singular, and L^{-1} is the linear transformation which is represented by A^{-1} with respect to the standard basis.
Thus, we have

$$L^{-1}(\epsilon_1) = 1\epsilon_1 - 2\epsilon_2 + 1\epsilon_3$$

$$L^{-1}(\epsilon_2) = 2\epsilon_1 - 5\epsilon_2 + 1\epsilon_3$$

$$L^{-1}(\epsilon_3) = -1\epsilon_1 + 3\epsilon_2 - 1\epsilon_3$$

where as usual, $\epsilon_1, \epsilon_2, \epsilon_3$ denote the standard basis vectors. Finally, since
$(a_1, a_2, a_3) = a_1\epsilon_1 + a_2\epsilon_2 + a_3\epsilon_3$, we find that L^{-1} may be described in the
following manner:

$$L^{-1}[(a_1, a_2, a_3)] = a_1 L^{-1}(\epsilon_1) + a_2 L^{-1}(\epsilon_2) + a_3 L^{-1}(\epsilon_3)$$

$$= (a_1 + 2a_2 - a_3, -2a_1 - 5a_2 + 3a_3, a_1 + a_2 - a_3)$$

Non-singular transformations and matrices have important special properties which make them both interesting and useful. For instance, in our next theorem, we show that if L, L_1, and L_2 are linear transformations and L is non-singular, then $L \circ L_1$ has the same null space as L_1 and $L_2 \circ L$ has the same range as L_2.

Theorem 4.14. If $L: V \longrightarrow V$, $L_1: U \longrightarrow V$, $L_2: V \longrightarrow W$ are linear transformations and L is non-singular, then

$$\mathfrak{N}_{L \circ L_1} = \mathfrak{N}_{L_1} \quad \text{and} \quad \mathfrak{R}_{L_2 \circ L} = \mathfrak{R}_{L_2}$$

Proof: It is fairly easy to see that $\mathfrak{N}_{L_1}$ is contained in $\mathfrak{N}_{L \circ L_1}$ and that $\mathfrak{R}_{L_2 \circ L}$ is contained in $\mathfrak{R}_{L_2}$, and we leave this part of the proof to the reader (see exercise 8, section 4.4).

If α is a vector in $\mathfrak{N}_{L \circ L_1}$, we have

$$\theta_V = [L \circ L_1](\alpha) = L[L_1(\alpha)]$$

and it follows that $L_1(\alpha)$ belongs to the null space of L. Since L is non-singular, we conclude that $L_1(\alpha) = \theta_V$, which means that α belongs to the null space of L_1. Since each vector in $\mathfrak{N}_{L_1}$ belongs to $\mathfrak{N}_{L \circ L_1}$ and vice versa, we have $\mathfrak{N}_{L_1} = \mathfrak{N}_{L \circ L_1}$.

Finally, if β is a vector in $\mathfrak{R}_{L_2}$, there exists a vector α in V for which $L_2(\alpha) = \beta$. Since L is nonsingular, there exists a γ in V such that $L(\gamma) = \alpha$, and we find that

$$\beta = L_2(\alpha) = L_2[L(\gamma)] = [L_2 \circ L](\gamma)$$

This means that β is also contained in the range of $L_2 \circ L$, and we conclude that $\mathfrak{R}_{L_2} = \mathfrak{R}_{L_2 \circ L}$, as desired. $\blacksquare$

As an immediate consequence of theorem 4.14, we have $\nu(L \circ L_1) = \nu(L_1)$ and $\rho(L_2 \circ L) = \rho(L_2)$, and in the special case where U, V, and W are the same finite-dimensional space, we can say even more.

Corollary 4.5. Let $L: V \longrightarrow V$ and $L_1: V \longrightarrow V$ be linear transformations on the finite-dimensional vector space V. If L is non-singular, then

$$\rho(L_1 \circ L) = \rho(L_1) = \rho(L \circ L_1)$$

and

$$\nu(L_1 \circ L) = \nu(L_1) = \nu(L \circ L_1)$$

Proof: We already know that $v(L \circ L_1) = v(L_1)$ and $\rho(L_1 \circ L) = \rho(L_1)$. To establish the equality $\rho(L \circ L_1) = \rho(L_1)$, we note that

$$\rho(L \circ L_1) = \dim V - v(L \circ L_1) = \dim V - v(L_1) = \rho(L_1)$$

The equality $v(L_1 \circ L) = v(L_1)$ may be obtained by similar means. ∎

We invite the reader to test his understanding of the correspondence principle by providing a proof for the following matrix version of corollary 4.5.

Corollary 4.6.   If A and B are $n \times n$ matrices and A is non-singular, then $\rho(AB) = \rho(B) = \rho(BA)$.

We now have a fairly complete picture of the relationship between linear transformations and matrices, and in the next section, we conclude our study of this relationship by examining the effect of a change in basis on the matrix representation of a given transformation.

Exercise Set 4.5

1. Let $L_1: R_3 \longrightarrow R_3$ and $L_2: R_3 \longrightarrow R_3$ denote the linear transformations defined by

$$L_1[(a_1, a_2, a_3)] = (-a_1 + a_2 + a_3, a_1 - 2a_3, -2a_1 - a_2)$$

$$L_2[(a_1, a_2, a_3)] = (-2a_2 - 3a_3, 3a_1 + 3a_2 + a_3, a_1 - 2a_2)$$

   (i) Find the matrices A and B which represent L_1 and L_2, respectively, when the standard basis is used in R_3.
   (ii) Show that $AB = BA$.
   (iii) Is it true that $L_1 \circ L_2 = L_2 \circ L_1$? Explain your reasoning.

2. In each of the following cases, determine whether or not the given linear transformation is non-singular and if it is, find its inverse.
   (i) $L[(a_1, a_2)] = (7a_1 + 4a_2, 2a_1 + a_2)$
   (ii) $L[(a_1, a_2, a_3)] = (-5a_1 + 2a_2 + 3a_3, 4a_1 - 3a_2 + a_3, 3a_1 - 4a_2 + 5a_3)$
   (iii) $L[(a_1, a_2, a_3)] = (3a_1 - 2a_2, a_1 + a_2 + 4a_3, -a_1 + a_2 + a_3)$

3. Let $L: P_2 \longrightarrow P_2$ be the linear transformation defined by

$$L[at^2 + bt + c] = (a - b)t^2 + (2b + c)t + (2a - b + c)$$

   (i) Find the matrix A which represents L with respect to the basis $\{t^2, t, 1\}$.
   (ii) Show that A is non-singular and find A^{-1}.
   (iii) Find $L^{-1}(t^2)$, $L^{-1}(t)$, and $L^{-1}(1)$.
   (iv) Describe $L^{-1}[at^2 + bt + c]$.

4. (i) If $L: V \longrightarrow V$ is a non-singular linear transformation, show that

$$(L^{-1} \circ L)(\gamma) = \gamma = (L \circ L^{-1})(\gamma)$$

for each vector γ in V.

(ii) Use part (i) to show that $L^{-1} \circ L = I_V = L \circ L^{-1}$ and then verify the formula $(L^{-1})^{-1} = L$.

5. Let A, B, C be $n \times n$ matrices, and let L_1, L_2, L_3 be those linear transformations on R_n for which $\mu(L_1) = A$, $\mu(L_2) = B$, and $\mu(L_3) = C$, when the standard basis is used in R_n.

   (i) Show that $L_1 \circ (L_2 + L_3) = (L_1 \circ L_2) + (L_1 \circ L_3)$.

   (ii) Use part (i) and the formulas in (4.12) to prove the distributive law

$$A(B + C) = (AB) + (AC)$$

6. A certain non-singular linear transformation $L: R_3 \longrightarrow R_3$ is known to have these two properties:

   (1) $L[(1, 0, 0)] = (-3, 1, 7)$ and $L[(0, 0, 1)] = (0, 1, 8)$
   (2) $L^{-1}[(1, 0, 0)] = (-1, -1, 1)$

Use this information to answer the following questions:

   (i) What is $L[(0, 1, 0)]$?

   (ii) Find the matrix A which represents L with respect to the standard basis in R_3. Compute A^{-1}.

   (iii) Find the 3-tuple $\alpha = (a_1, a_2, a_3)$ which satisfies $L(\alpha) = (9, 1, 11)$.

7. If A and B are $m \times n$ and $n \times p$ matrices, respectively, show that $\rho(AB) \leq \rho(A)$ and $\rho(AB) \leq \rho(B)$.

   Hint: Use the correspondence principle in conjunction with the result of exercise 8, section 4.4.

8. Let A be an $m \times n$ matrix, and let P and Q be $m \times m$ and $n \times n$ non-singular matrices, respectively.

   (i) Use theorem 4.14 to show that $\rho(AQ) = \rho(A)$.

   (ii) Show that $\rho(PA) = \rho(A)$.

   Hint: Use the fact that row equivalent matrices have the same rank.

   (iii) Combine parts (i) and (ii) to prove that $\rho(PAQ) = \rho(A)$.

9. Show that the identity transformation $I: V \longrightarrow V$ is non-singular. (See exercise 9, section 4.3.)

10. Show that R_φ, the rotation of the plane through φ degrees, is non-singular and verify that $R_\varphi^{-1} = R_{-\varphi}$. (See example 1, section 4.3.)

11. Show that R_l, the reflection of the plane in the line l, is non-singular and that $R_l^{-1} = R_l$. (See exercise 11, section 4.3.)

*12. Let $S = \{\alpha_1, \ldots, \alpha_n\}$ and $T = \{\beta_1, \ldots, \beta_n\}$ be bases for the vector space V, and let A denote the S to T change of basis matrix.

   (i) Show that A is non-singular.

   Hint: Use the result of exercise 9, section 4.3.

   (ii) Show that A^{-1} is the T to S change of basis matrix.

   Hint: A proof can be constructed by putting into precise mathematical form the relationships suggested in the following mapping diagram:

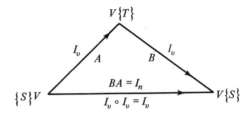

*13. Let A, B be $n \times n$ matrices, and assume that A is non-singular.
   (i) Show that the homogeneous linear systems $BX = O$ and $(AB)X = O$ have the same solution space.
   (ii) Show that the solution space of the system $(BA)X = O$ has the same dimension as that of $BX = O$. Are these two spaces always the same?

4.6 SIMILARITY

In this section, we shall study the relationship between the various matrices which represent the same linear transformation $L: V \longrightarrow V$ with respect to different bases for V. We begin by deriving a formula for the relationship in question.

Theorem 4.15.   Let $L: V \longrightarrow V$ be a linear transformation, and let S and T be bases for V. If

A represents L with respect to the basis S

B represents L with respect to the basis T

P is the T to S change of basis matrix

then B is related to A by the equation

$$B = P^{-1}AP$$

Proof: Note that L may be expressed as the composite function $L = I_V \circ L \circ I_V$, where I_V is the identity transformation on V. According to theorem 4.10, the matrix B which represents L with respect to the basis T can be expressed as the product of the matrices which represent I_V, L, and I_V when the basis pairs T-S, S-S, and S-T are used in the domain and codomain, respectively. We know that L is represented by A with respect to the basis S, and it can be shown that I_V is represented by P with respect to the T-S choice of bases and by P^{-1} with respect to the S-T choice (see exercise 9, section 4.3

and exercise 12, section 4.5). Therefore, we have

$$B = \mu(I_V)\ \mu(L)\ \mu(I_V) = P^{-1}AP$$
$$\{S-T\}\ \{S-S\}\ \{T-S\}$$

as claimed. ∎

The ideas behind this theorem and its proof are depicted schematically in the mapping diagram of Figure 4.10.

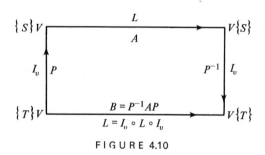

FIGURE 4.10

EXAMPLE 1. A certain linear transformation $L: R_3 \longrightarrow R_3$ is represented by the matrix

$$A = \begin{bmatrix} 17 & 9 & -27 \\ 14 & 6 & -21 \\ 16 & 8 & -25 \end{bmatrix}$$

with respect to the basis $S = \{(1, 1, 0), (0, 1, 1), (1, 0, 0)\}$. To compute the matrix B which represents L with respect to the basis $T = \{(2, 2, 1), (27, 25, 11), (-17, -16, -7)\}$, we need to have the T to S change of basis matrix P. Applying the computational method developed in section 2.7, we perform the reduction

$$\begin{bmatrix} 1 & 0 & 1 & | & 2 & 27 & -17 \\ 1 & 1 & 0 & | & 2 & 25 & -16 \\ 0 & 1 & 0 & | & 1 & 11 & -7 \end{bmatrix} \longrightarrow \begin{bmatrix} 1 & 0 & 0 & | & 1 & 14 & -9 \\ 0 & 1 & 0 & | & 1 & 11 & --7 \\ 0 & 0 & 1 & | & 1 & 13 & -8 \end{bmatrix}$$

to find P and then compute P^{-1} by the matrix inversion algorithm. As a result of these computations, we have

$$P = \begin{bmatrix} 1 & 14 & -9 \\ 1 & 11 & -7 \\ 1 & 13 & -8 \end{bmatrix} \quad \text{and} \quad P^{-1} = \begin{bmatrix} -3 & 5 & -1 \\ -1 & -1 & 2 \\ -2 & -1 & 3 \end{bmatrix}$$

and it follows that

$$B = P^{-1}AP = \begin{bmatrix} -1 & 0 & 0 \\ 0 & -1 & 0 \\ 0 & 0 & 0 \end{bmatrix}$$

In chapter 6, we shall examine in detail the special kind of linear transformation which has a diagonal matrix representative. There are many reasons for seeking such a representation, most of which are related to the fact that diagonal matrices are so easy to manipulate.

EXAMPLE 2. Let R_l denote the reflection of the plane in the line l whose equation is $y = mx$. As a second application of theorem 4.15, we shall find the matrix B which represents R_l with respect to the standard basis $T = \{\epsilon_1, \epsilon_2\}$. First of all, let α_1 and α_2 be vectors with unit length which point along the lines $y = mx$ and $x = -my$, respectively, as shown in Figure 4.11.

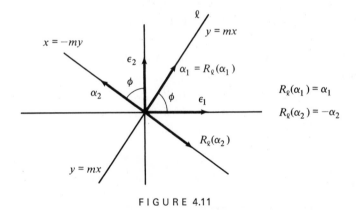

FIGURE 4.11

Since $R_l(\alpha_1) = \alpha_1$ and $R_l(\alpha_2) = -\alpha_2$, we find that R_l is represented by the matrix

$$A = \begin{bmatrix} 1 & 0 \\ 0 & -1 \end{bmatrix}$$

with respect to the basis $S = \{\alpha_1, \alpha_2\}$. Note also that

$$\alpha_1 = (\cos \varphi)\epsilon_1 + (\sin \varphi)\epsilon_2$$

$$\alpha_2 = (-\sin \varphi)\epsilon_1 + (\cos \varphi)\epsilon_2$$

where φ is the angle of inclination of l—that is, $\varphi = \tan^{-1} m$. Therefore, if

P and P^{-1} are the T to S and S to T change of basis matrices, respectively, we have

$$P^{-1} = \begin{bmatrix} \cos\varphi & -\sin\varphi \\ \sin\varphi & \cos\varphi \end{bmatrix} \quad \text{and} \quad P = \begin{bmatrix} \cos\varphi & \sin\varphi \\ -\sin\varphi & \cos\varphi \end{bmatrix}$$

and theorem 4.15 enables us to conclude that

$$B = P^{-1}AP = \begin{bmatrix} \cos\varphi & -\sin\varphi \\ \sin\varphi & \cos\varphi \end{bmatrix} \begin{bmatrix} 1 & 0 \\ 0 & -1 \end{bmatrix} \begin{bmatrix} \cos\varphi & \sin\varphi \\ -\sin\varphi & \cos\varphi \end{bmatrix}$$

$$= \begin{bmatrix} \cos^2\varphi - \sin^2\varphi & \cos\varphi\sin\varphi + \sin\varphi\cos\varphi \\ \sin\varphi\cos\varphi + \cos\varphi\sin\varphi & \sin^2\varphi - \cos^2\varphi \end{bmatrix}$$

$$= \begin{bmatrix} \cos 2\varphi & \sin 2\varphi \\ \sin 2\varphi & -\cos 2\varphi \end{bmatrix}$$

Note that this is the same result we obtained by direct computation in exercise 11 of section 4.3.

Thanks to theorem 4.15, the matrix relationship $B = P^{-1}AP$ plays an extremely important role in linear algebra, and it is convenient to have the terminology introduced in the following definition.

Definition 4.7.   The $n \times n$ matrix B is said to be *similar* to the matrix A if there exists a non-singular matrix P such that $B = P^{-1}AP$.

It is easy to show that A is similar to B if and only if B is similar to A (see exercise 7), and henceforth, we shall say simply that A and B are similar, without indicating a "direction" to the relationship.

In theorem 4.15, we showed that two matrices which represent the same linear transformation with respect to different bases are similar, and in our next theorem, we show that the converse of this result is also true.

Theorem 4.16.   Let V be an n-dimensional vector space. Then the $n \times n$ matrices A and B both represent the same linear transformation $L: V \rightarrow V$ (with respect to different bases) if and only if they are similar.

Proof: The "only if" part is a restatement of theorem 4.15. For the "if" part, assume that $B = P^{-1}AP$ and that A represents the linear transformation $L: V \rightarrow V$ with respect to the basis $S = \{\alpha_1, \alpha_2, \ldots, \alpha_n\}$. For each index j, let $\beta_j = \sum_{i=1}^{n} p_{ij}\,\alpha_i$. Since $P = (p_{ij})$ is non-singular, the vectors $\beta_1, \ldots, \beta_n$

are linearly independent (see Supplementary Exercise 10, Chapter 2), and it follows that $T = \{\beta_1, \beta_2, \ldots, \beta_n\}$ is a basis for V. We leave it to the reader to complete the proof by verifying that L is represented by B with respect to the basis T. ∎

It is reasonable to expect similar matrices to share many properties since they represent the same linear transformation. Some of these properties are listed in the following theorem.

Theorem 4.17. If the $n \times n$ matrices A and B are similar, then
   (i) A and B have the same rank.
   (ii) A^t and B^t are similar.
   (iii) A^k and B^k are similar, for each positive integer k.
   (iv) A is non-singular if and only if B is non-singular, and in this case, A^{-1} and B^{-1} are similar.

Proof: (i) If $L: R_n \rightarrow R_n$ is the linear transformation which is represented by A with respect to the standard basis for R_n, then theorem 4.16 tells us that B is also a matrix representative of L. According to corollary 4.2, A and B both have the same rank as L, which means that their ranks must be equal.

(ii) Let P be the non-singular matrix which satisfies $P^{-1}AP = B$. Then A^t and B^t are similar since

$$B^t = P^t A^t (P^{-1})^t = Q^{-1} A^t Q$$

where $Q = (P^t)^{-1}$.

Similarly, it can be shown that $B^k = P^{-1}A^kP$ and $B^{-1} = P^{-1}A^{-1}P$ (see exercise 4), and these relationships may be used to prove parts (iii) and (iv). ∎

Incidentally, although similar matrices have the same rank, the converse statement is not necessarily true. For instance, the 2×2 matrices

$$A = \begin{bmatrix} 1 & 0 \\ 0 & 0 \end{bmatrix} \quad \text{and} \quad B = \begin{bmatrix} 0 & 1 \\ 0 & 0 \end{bmatrix}$$

both have rank 1, but they are not similar (see exercise 9). Actually, there is no simple way of determining whether or not two given matrices are similar. A great deal of mathematical research has been devoted to the theory and applications of similarity, and we shall examine some of the results of these efforts in chapter 6. In the next section, we conclude the work of this chapter by investigating the special kind of linear transformation that preserves metric relationships in a real inner product space.

Exercise Set 4.6

1. Let $L: R_3 \longrightarrow R_3$ denote the linear transformation defined by

$$L(a_1, a_2, a_3) = (2a_1 - a_2 + a_3, a_1 + 5a_3, a_1 + 3a_2 - 2a_3)$$

(i) Find the matrix A which represents L with respect to the standard basis S and then compute the matrix B which represents L with respect to the basis $T = \{(1, 1, 0), (3, -2, 1), (3, -1, 1)\}$.

(ii) Compute the T to S change of basis matrix P and verify that $B = P^{-1}AP$.

2. A certain linear transformation $L: R_3 \longrightarrow R_3$ is known to be represented by the matrix

$$A = \begin{bmatrix} 3 & -1 & 5 \\ -4 & 2 & 9 \\ 10 & -4 & 1 \end{bmatrix}$$

with respect to the basis $S = \{(1, 1, 2), (0, 1, 0), (1, -1, 1)\}$.

(i) What matrix represents L with respect to the standard basis for R_3?

(ii) Find the matrix B which represents L with respect to the basis $T = \{(2, 1, 3), (17, 68, 36), (2, 11, 5)\}$.

3. Let $L: P_2 \longrightarrow P_2$ be the linear transformation which is represented by the matrix

$$A = \begin{bmatrix} 1 & 1 & 3 \\ -1 & 2 & -5 \\ 0 & 1 & 2 \end{bmatrix}$$

with respect to the basis $S = \{t^2 + t + 2, t, t^2 - t + 1\}$. Find the matrix B which represents L with respect to the natural basis $T = \{t^2, t, 1\}$.

4. If $B = P^{-1}AP$, show that

(i) $sB = P^{-1}(sA)P$ for each scalar s

(ii) $B^k = P^{-1}A^kP$ for each positive integer k

(iii) $B^{-1} = P^{-1}A^{-1}P$ if A is non-singular

5. Let A be an $n \times n$ matrix, and let $q(A)$ denote the polynomial matrix

$$q(A) = a_nA^n + a_{n-1}A^{n-1} + \ldots + a_1A + a_0I_n$$

If $B = P^{-1}AP$, show that $q(B) = P^{-1}[q(A)]P$.

6. Let A be an $n \times n$ matrix.

(i) If A is similar to the zero matrix O, show that $A = O$.

(ii) If A is similar to sI_n, a scalar multiple of the identity matrix, show that $A = sI_n$.

7. Let A, B, and C be $n \times n$ matrices.

(i) Show that A is similar to itself.

(ii) If A is similar to B, show that B is similar to A.

(iii) If A is similar to B and B is similar to C, show that A is also similar to C.

8. Construct an alternative proof of the fact that similar matrices have the same rank by using an argument based on corollary 4.6.

9. Let

$$A = \begin{bmatrix} 1 & 0 \\ 0 & 0 \end{bmatrix} \quad \text{and} \quad B = \begin{bmatrix} 0 & 1 \\ 0 & 0 \end{bmatrix}$$

and let

$$P = \begin{bmatrix} p_{11} & p_{12} \\ p_{21} & p_{22} \end{bmatrix}$$

satisfy $PB = AP$.
(i) Show that $p_{11} = 0 = p_{21}$.
(ii) Use part (i) to prove that A cannot be similar to B.

10. If A and B are similar $n \times n$ matrices, show that tr $A = $ tr B.
Hint: Recall that tr $(CD) = $ tr (DC) for $n \times n$ matrices C and D.

4.7   ORTHOGONAL TRANSFORMATIONS AND MATRICES

For certain applications, it is important to know how metric characteristics such as distance, length, and angle are affected within the inner product space V by a given linear transformation $L: V \rightarrow V$. In this section, we shall investigate the special kind of transformation that leaves all such metric characteristics unchanged. Also, as a natural part of this study, we shall examine the properties of the matrix representatives of such transformations.

Since all metric functions are defined in terms of the inner product, it is appropriate to begin our work by defining the notion of a linear transformation which preserves inner products.

Definition 4.8.   The linear transformation $L: V \rightarrow V$ is said to be *orthogonal* on the real inner product space V if $(L(\alpha)\,|\,L(\beta)) = (\alpha\,|\,\beta)$ for all vectors α, β in V.

EXAMPLE 1.   The linear transformations $L_1: R_2 \rightarrow R_2$ and $L_2: R_2 \rightarrow R_2$ defined by

$$L_1[(c_1, c_2)] = \left(\frac{1}{2}c_1 - \frac{\sqrt{3}}{2}c_2, \; \frac{\sqrt{3}}{2}c_1 + \frac{1}{2}c_2 \right)$$

$$L_2[(c_1, c_2)] = (c_2, c_1)$$

are both orthogonal with respect to the standard inner product on R_2. To prove this statement, let $\alpha = (a_1, a_2)$ and $\beta = (b_1, b_2)$ and observe that

$$(L_1(\alpha) \,|\, L_1(\beta)) = \left[\frac{1}{2}a_1 - \frac{\sqrt{3}}{2}a_2 \right]\left[\frac{1}{2}b_1 - \frac{\sqrt{3}}{2}b_2 \right]$$

$$+ \left[\frac{\sqrt{3}}{2}a_1 + \frac{1}{2}a_2 \right]\left[\frac{\sqrt{3}}{2}b_1 + \frac{1}{2}b_2 \right]$$

$$= a_1 b_1 + a_2 b_2 = (\alpha \,|\, \beta)$$

and

$$(L_2(\alpha) \,|\, L_2(\beta)) = a_2 b_2 + a_1 b_1 = a_1 b_1 + a_2 b_2 = (\alpha \,|\, \beta)$$

Geometrically, L_1 is a rotation of the plane through $60°$, while L_2 is a reflection in the line $y = x$ (see example 1, section 4.1, and exercise 11, section 4.3). Later in this section, we shall show that any orthogonal transformation of the plane must be either a rotation or a reflection.

Since an orthogonal transformation $L: V \to V$ preserves the inner product of V, it also preserves lengths, distances, and angles. To see this, note that if α, β are vectors in V, we have

$$\| L(\alpha) \| = (L(\alpha) \,|\, L(\alpha))^{1/2} = (\alpha \,|\, \alpha)^{1/2} = \| \alpha \|$$

$$d(L(\alpha), L(\beta)) = \| L(\beta) - L(\alpha) \| = \| L(\beta - \alpha) \|$$

$$= \| \beta - \alpha \| = d(\alpha, \beta)$$

and if ϕ' is the angle between $L(\alpha)$ and $L(\beta)$, then

$$\cos \phi' = \frac{(L(\alpha) \,|\, L(\beta))}{\| L(\alpha) \| \| L(\beta) \|} = \frac{(\alpha \,|\, \beta)}{\| \alpha \| \| \beta \|} = \cos \phi$$

where ϕ is the angle between α and β. As a consequence of this last relationship, we see that $L(\alpha)$ and $L(\beta)$ are orthogonal vectors if and only if α and β are orthogonal.

We have just observed that an orthogonal transformation preserves length, and it is of interest to note that a linear transformation which preserves length has to be orthogonal. These two observations yield the following theorem, whose proof is outlined in exercise 10.

Theorem 4.18.   The linear transformation $L: V \to V$ on the real inner product space V is orthogonal if and only if $\| L(\alpha) \| = \| \alpha \|$ for each vector α in V.

Theoretically, in order to use definition 4.8 to determine whether or not a given linear transformation $L: V \rightarrow V$ is orthogonal, we would have to test each pair of vectors α, β in V to see if $(L(\alpha)|L(\beta)) = (\alpha|\beta)$. Fortunately, however, if V is finite-dimensional, our next theorem shows that we really need to check only those vectors in any one fixed basis for V.

Theorem 4.19. Let $S = \{\alpha_1, \ldots, \alpha_n\}$ be a basis for the real inner product space V. Then the linear transformation $L: V \rightarrow V$ is orthogonal if and only if $(L(\alpha_i)|L(\alpha_j)) = (\alpha_i|\alpha_j)$ for each pair of vectors α_i, α_j in S.

Proof: The "only if" part follows immediately from definition 4.8. To prove the converse, assume that $(L(\alpha_i)|L(\alpha_j)) = (\alpha_i|\alpha_j)$ for each pair of vectors α_i, α_j in S, and let $\gamma = \sum_{k=1}^{n} c_k \alpha_k$ be a vector in V. Then we have

$$\|\gamma\|^2 = (\gamma|\gamma) = \left(\sum_{i=1}^{n} c_i \alpha_i \,\middle|\, \sum_{j=1}^{n} c_j \alpha_j\right) = \sum_{i=1}^{n} \sum_{j=1}^{n} c_i c_j (\alpha_i|\alpha_j)$$

Moreover, since L is linear, it follows that $L(\gamma) = \sum_{k=1}^{n} c_k L(\alpha_i)$, and we obtain

$$\|L(\gamma)\|^2 = (L(\gamma)|L(\gamma)) = \left(\sum_{i=1}^{n} c_i L(\alpha_i) \,\middle|\, \sum_{j=1}^{n} c_j L(\alpha_j)\right)$$

$$= \sum_{i=1}^{n} \sum_{j=1}^{n} c_i c_j (L(\alpha_i)|L(\alpha_j)) = \sum_{i=1}^{n} \sum_{j=1}^{n} c_i c_j (\alpha_i|\alpha_j)$$

We have shown that $\|L(\gamma)\| = \|\gamma\|$ for each vector γ in V, and theorem 4.18 enables us to conclude that L is orthogonal. ∎

Our next objective is to study the matrix representatives of a given orthogonal transformation $L: V \rightarrow V$. To be specific, let $P = (p_{ij})$ be the matrix which represents L with respect to the orthonormal basis $S = \{\gamma_1, \gamma_2, \ldots, \gamma_n\}$. This means that

$$L(\gamma_j) = \sum_{l=1}^{n} p_{lj}\gamma_l \quad \text{for } j = 1, 2, \ldots, n$$

and according to definition 4.8, we have

$$(\gamma_i|\gamma_j) = (L(\gamma_i)|L(\gamma_j)) = \left(\sum_{k=1}^{n} p_{ki}\gamma_k \,\middle|\, \sum_{l=1}^{n} p_{lj}\gamma_l\right)$$

$$= \sum_{k=1}^{n} \sum_{l=1}^{n} p_{ki}p_{lj}(\gamma_k|\gamma_l) = \sum_{k=1}^{n} p_{ki}p_{kj}$$

where the last equality is a consequence of the fact that $(\gamma_k|\gamma_l)$ is 1 if $k = l$

and 0 if $k \neq l$. Thus, for each pair of indices i, j we have

$$\sum_{k=1}^{n} p_{ki} p_{kj} = (\gamma_i \mid \gamma_j) = \begin{cases} 1 & \text{if } i = j \\ 0 & \text{if } i \neq j \end{cases}$$

We recognize the sum on the left as the (i, j) entry of the product matrix $P^t P$, and the above result tells us that P must satisfy the matrix equation $P^t P = I_n$. Therefore, P is a non-singular matrix whose inverse is its transpose. Matrices with this property play an important role in certain areas of linear algebra, and it is convenient to have the following terminology.

Definition 4.9. The $n \times n$ real matrix P is said to be *orthogonal* if $PP^t = I_n = P^t P$.

EXAMPLE 2. The matrices

$$\begin{bmatrix} 0 & 1 \\ 1 & 0 \end{bmatrix}, \quad \begin{bmatrix} \dfrac{\sqrt{2}}{2} & -\dfrac{\sqrt{2}}{2} \\ \dfrac{\sqrt{2}}{2} & \dfrac{\sqrt{2}}{2} \end{bmatrix}, \quad \begin{bmatrix} \dfrac{1}{2} & \dfrac{\sqrt{3}}{2} \\ \dfrac{\sqrt{3}}{2} & -\dfrac{1}{2} \end{bmatrix}$$

are all orthogonal, as are

$$\begin{bmatrix} 0 & 1 & 0 \\ 0 & 0 & 1 \\ 1 & 0 & 0 \end{bmatrix} \quad \text{and} \quad \begin{bmatrix} \dfrac{1}{\sqrt{2}} & \dfrac{1}{\sqrt{2}} & 0 \\ -\dfrac{1}{\sqrt{3}} & \dfrac{1}{\sqrt{3}} & \dfrac{1}{\sqrt{3}} \\ \dfrac{1}{\sqrt{6}} & -\dfrac{1}{\sqrt{6}} & \dfrac{2}{\sqrt{6}} \end{bmatrix}$$

The equation $P^t P = I_n$ means that the column vectors of P are mutually orthonormal with respect to the standard inner product on R_n, and a similar statement may be made regarding the equation $PP^t = I_n$ and the row vectors of P. Moreover, if either of the equations $P^t P = I_n$ and $PP^t = I_n$ is valid, then P is necessarily orthogonal, and the other equation is also valid. By combining these facts, we can make the following remarkable observation:

> The column vectors of an $n \times n$ real matrix P are mutually orthonormal if and only if the row vectors of P also have this property.

For instance, if the column vectors of the matrix P are known to be mutually orthonormal, then P must be orthogonal and its row vectors must also be mutually orthonormal. This criterion often simplifies the task of determining whether or not a given matrix is orthogonal.

We have just shown that the matrix which represents an orthogonal transformation with respect to an orthonormal basis must be orthogonal. In our next theorem, we turn things around by showing that any linear transformation which is represented by an orthogonal matrix with respect to an orthonormal basis is necessarily orthogonal.

Theorem 4.20. Let $S = \{\gamma_1, \ldots, \gamma_n\}$ be an orthonormal basis for the real inner product space V. Then the linear transformation $L: V \longrightarrow V$ is orthogonal if and only if it is represented by an orthogonal matrix with respect to the basis S.

Proof: We have already established the "only if" part. To prove the converse, let $P = (p_{ij})$ be the matrix which represents L with respect to the basis S, and assume that P is orthogonal. Then, for each pair of indices i, j we have

$$((L(\gamma_i) \,|\, L(\gamma_j)) = \left(\sum_{k=1}^{n} p_{ki}\gamma_k \,\Big|\, \sum_{l=1}^{n} p_{lj}\gamma_l \right)$$

$$= \sum_{k=1}^{n} \sum_{l=1}^{n} p_{ki}p_{lj}(\gamma_k \,|\, \gamma_l) = \sum_{k=1}^{n} p_{ki}p_{kj} = \begin{cases} 1 & \text{if } i = j \\ 0 & \text{if } i \neq j \end{cases}$$

Therefore, $(L(\gamma_i) \,|\, L(\gamma_j)) = (\gamma_i \,|\, \gamma_j)$, and theorem 4.19 enables us to conclude that L is orthogonal. ∎

EXAMPLE 3.   To see how the orthogonality criterion stated in theorem 4.20 can be used, let $L_1: R_3 \longrightarrow R_3$ and $L_2: R_3 \longrightarrow R_3$ denote the linear transformations defined by

$$L_1[(a_1, a_2, a_3)] = (-2a_1 + a_2, -a_3, a_1 + a_2 + a_3)$$

$$L_2[(a_1, a_2, a_3)] = (a_2, a_3, a_1)$$

When the standard basis is used in R_3, we find that L_1 and L_2 are represented by the matrices P and Q, respectively, where

$$P = \begin{bmatrix} -2 & 1 & 0 \\ 0 & 0 & -1 \\ 1 & 1 & 1 \end{bmatrix} \quad \text{and} \quad Q = \begin{bmatrix} 0 & 1 & 0 \\ 0 & 0 & 1 \\ 1 & 0 & 0 \end{bmatrix}$$

Since the standard basis is orthonormal with respect to the standard inner product on R_3 and since we have

$$P^t P = \begin{bmatrix} 5 & -1 & 1 \\ -1 & 2 & 1 \\ 1 & 1 & 2 \end{bmatrix} \quad \text{and} \quad Q^t Q = \begin{bmatrix} 1 & 0 & 0 \\ 0 & 1 & 0 \\ 0 & 0 & 1 \end{bmatrix}$$

theorem 4.20 enables us to conclude that L_2 is orthogonal, but L_1 is not.

It is of interest to note that a 2×2 orthogonal matrix $P = (p_{ij})$ must have one of the following forms:

$$P_{\text{rot}} = \begin{bmatrix} \cos \phi & -\sin \phi \\ \sin \phi & \cos \phi \end{bmatrix}, \quad P_{\text{ref}} = \begin{bmatrix} \cos \phi & \sin \phi \\ \sin \phi & -\cos \phi \end{bmatrix} \quad (4.14)$$

To show this, we compare entries in the matrix equation $P^t P = I_2 = PP^t$ to obtain

$$p_{11}^2 + p_{21}^2 = 1 \qquad\qquad p_{11}^2 + p_{12}^2 = 1$$

$$p_{11}p_{12} + p_{21}p_{22} = 0 \qquad p_{11}p_{21} + p_{12}p_{22} = 0 \qquad (4.15)$$

$$p_{12}^2 + p_{22}^2 = 1 \qquad\qquad p_{21}^2 + p_{22}^2 = 1$$

Since $|p_{11}| \leq 1$, there exists a real number $0 \leq \phi \leq 2\pi$ such that $p_{11} = \cos \phi$. The top equations in (4.15) tell us that both p_{21} and p_{12} are either $\sin \phi$ or $-\sin \phi$, and the bottom equations then imply that p_{22} is either $\cos \phi$ or $-\cos \phi$. There are really only two essentially different cases to consider (why not more?):

Case I	*Case II*
$p_{22} = p_{11} = \cos \phi$	$p_{11} = \cos \varphi$
$p_{12} = -\sin \phi$	$p_{12} = p_{21} = \sin \phi$
$p_{21} = \sin \phi$	$p_{22} = -\cos \varphi$

The first case leads to the form labeled P_{rot} in (4.14), while the second leads to P_{ref}.

Geometrically, the orthogonal transformation which is represented by P_{rot} with respect to the standard basis is a rotation of the plane through φ degrees about the origin (see example 1, section 4.3). Similarly, using the result of exercise 11, section 4.3, it can be shown that P_{ref} is a reflection in the line through the origin whose angle of inclination is $\frac{1}{2}\varphi$.

In general, a function which preserves the distance between each pair of points in the space V is called a *rigid motion* of V, and the study of those features of V which are preserved by all rigid motions is called the *Euclidean geometry* of V. We have just shown that any rigid motion of the plane which is also linear must be either a rotation about the origin or a reflection in a line through the origin. However, there are also rigid motions of the plane which are not linear. For instance, if ξ is a fixed vector, the function T_ξ which maps each vector α in the plane into the vector sum $\xi + \alpha$ is a rigid motion, but it is not linear unless ξ happens to be the zero vector. Such a function is said to be a *translation* of the plane in the direction of ξ, and it can be shown that any rigid motion of the plane may be expressed as a composition of a rotation or a reflection with a translation. Thus, the subject known as plane Euclidean geometry is really the study of those features of the plane which are preserved by functions composed of rotations, reflections, and translations.

This concludes our introduction to the theory of linear transformations, and our development of the fundamental ideas of linear algebra is now complete. At this point, the reader has been exposed to most of the basic principles of linear algebra, but there are many more important concepts and applications which have not yet been discussed. We shall examine some of these topics in Chapters 6 and 7, but first, we pause to develop a few important properties of determinant theory which will be used later in our work.

Exercise Set 4.7

1. Which (if any) of the following matrices is orthogonal?

$$A_1 = \begin{bmatrix} \dfrac{1}{\sqrt{2}} & \dfrac{1}{\sqrt{2}} \\ \dfrac{1}{\sqrt{2}} & -\dfrac{1}{\sqrt{2}} \end{bmatrix}, \quad A_2 = \begin{bmatrix} \dfrac{5}{13} & \dfrac{12}{13} \\ \dfrac{12}{13} & -\dfrac{5}{13} \end{bmatrix}, \quad A_3 = \begin{bmatrix} \dfrac{1}{\sqrt{3}} & -\dfrac{2}{\sqrt{6}} \\ \dfrac{2}{\sqrt{6}} & \dfrac{1}{\sqrt{3}} \end{bmatrix}$$

$$A_4 = \begin{bmatrix} \dfrac{1}{\sqrt{3}} & -\dfrac{1}{\sqrt{3}} & 0 \\ \dfrac{1}{\sqrt{3}} & 0 & -\dfrac{1}{\sqrt{3}} \\ 0 & \dfrac{1}{\sqrt{3}} & \dfrac{1}{\sqrt{3}} \end{bmatrix}, \quad A_5 = \begin{bmatrix} \dfrac{1}{\sqrt{2}} & \dfrac{1}{\sqrt{2}} & 0 \\ \dfrac{1}{\sqrt{3}} & -\dfrac{1}{\sqrt{3}} & \dfrac{1}{\sqrt{3}} \\ -\dfrac{1}{\sqrt{6}} & \dfrac{1}{\sqrt{6}} & \dfrac{2}{\sqrt{6}} \end{bmatrix}$$

2. In each of the following cases, determine whether or not the given linear transformation is orthogonal with respect to the standard inner product on R_3.
   (i) $L[(a_1, a_2, a_3)] = (a_3, (\sqrt{3}/2)a_1 + \tfrac{1}{2}a_2, \tfrac{1}{2}a_1 - (\sqrt{3}/2)a_2)$
   (ii) $L[(a_1, a_2, a_3)] = (\tfrac{3}{5}a_1 - \tfrac{4}{5}a_2, a_3, \tfrac{4}{5}a_2 + \tfrac{3}{5}a_3)$
   (iii) $L[(a_1, a_2, a_3)] = (\tfrac{5}{13}a_2 + \tfrac{12}{13}a_3, \tfrac{12}{13}a_2 - \tfrac{5}{13}a_3, a_1)$

3. Let $L: P_1 \longrightarrow P_1$ denote the linear transformation defined by

$$L[at + b] = (\sqrt{3}\,b)t + \left(\frac{\sqrt{3}}{3}a\right)$$

(i) Find the matrix which represents L with respect to the basis $\{1, \sqrt{3}\,(2t - 1)\}$.

(ii) Determine whether or not L is orthogonal with respect to the integral inner product.

Hint: Note that the basis given in part (i) is orthonormal with respect to the integral inner product.

4. For what values of a, b, c, d is the following matrix orthogonal?

$$\begin{bmatrix} \dfrac{1}{\sqrt{3}} & \dfrac{1}{\sqrt{3}} & a & \dfrac{1}{\sqrt{3}} \\[2ex] -\dfrac{2}{\sqrt{7}} & b & \dfrac{1}{\sqrt{7}} & \dfrac{1}{\sqrt{7}} \\[2ex] \dfrac{2}{\sqrt{42}} & -\dfrac{1}{\sqrt{42}} & \dfrac{6}{\sqrt{42}} & c \\[2ex] d & -\dfrac{1}{\sqrt{2}} & 0 & \dfrac{1}{\sqrt{2}} \end{bmatrix}$$

5. If P_1, P_2 are orthogonal $n \times n$ matrices, show that P_1^t and $P_1 P_2$ are also orthogonal.

6. Let V be an n-dimensional real inner product space. If $L_1: V \longrightarrow V$ and $L_2: V \longrightarrow V$ are orthogonal transformations, show that $L_2 \circ L_1$ is also orthogonal.

7. Let R_l denote the orthogonal transformation which is represented by the following matrix with respect to the standard basis for the plane:

$$P_{\text{ref}} = \begin{bmatrix} \cos\varphi & \sin\varphi \\ \sin\varphi & -\cos\varphi \end{bmatrix}$$

Show that R_l is a reflection of the plane in the line whose equation is $(1 + \cos\varphi)y = (\sin\varphi)x$.

8. In each of the following cases, determine whether the given orthogonal matrix represents a rotation or a reflection of the plane with respect to the standard basis. For the rotations, give the angle of rotation φ, and for the reflections, give the equation of the reflecting line (see exercise 7).

(i) $\begin{bmatrix} \dfrac{1}{2} & -\dfrac{\sqrt{3}}{2} \\[2ex] \dfrac{\sqrt{3}}{2} & \dfrac{1}{2} \end{bmatrix}$ (ii) $\begin{bmatrix} \dfrac{3}{5} & \dfrac{4}{5} \\[2ex] \dfrac{4}{5} & -\dfrac{3}{5} \end{bmatrix}$ (iii) $\begin{bmatrix} \dfrac{12}{13} & -\dfrac{5}{13} \\[2ex] -\dfrac{5}{13} & -\dfrac{12}{13} \end{bmatrix}$

(iv) $\begin{bmatrix} -\dfrac{\sqrt{2}}{2} & \dfrac{\sqrt{2}}{2} \\[2ex] -\dfrac{\sqrt{2}}{2} & -\dfrac{\sqrt{2}}{2} \end{bmatrix}$

9. Solve the following linear system

$$\frac{1}{\sqrt{3}}x_1 + \frac{1}{\sqrt{3}}x_2 + \qquad + \frac{1}{\sqrt{3}}x_4 = 5\sqrt{3}$$

$$\frac{1}{\sqrt{2}}x_1 + \qquad\qquad - \frac{1}{\sqrt{2}}x_4 = 9$$

$$\frac{1}{\sqrt{7}}x_1 - \frac{2}{\sqrt{7}}x_2 + \frac{1}{\sqrt{7}}x_3 + \frac{1}{\sqrt{7}}x_4 = -2\sqrt{14}$$

$$-\frac{1}{\sqrt{42}}x_1 + \frac{2}{\sqrt{42}}x_2 + \frac{6}{\sqrt{42}}x_3 - \frac{1}{\sqrt{42}}x_4 = 0$$

Hint: Take a close look at the coefficient matrix.

10. Let $L: V \longrightarrow V$ be a linear transformation such that $\| L(\gamma) \| = \| \gamma \|$ for each vector γ in V.

   (i) If α and β are vectors in V, show that

$$2\,(\alpha \,|\, \beta) = \| \alpha + \beta \|^2 - \| \alpha \|^2 - \| \beta \|^2$$

   and

$$2(L(\alpha)\,|\,L(\beta)) = \| L(\alpha) + L(\beta) \|^2 - \| L(\alpha) \|^2 - \| L(\beta) \|^2$$

   (ii) Use part (i) and the hypotheses to show that $(L(\alpha)\,|\,L(\beta)) = (\alpha\,|\,\beta)$, and conclude that L is an orthogonal transformation.

11. Let $\{\gamma_1, \ldots, \gamma_n\}$ be an orthonormal basis for the real inner product space V. Show that the linear transformation $L: V \longrightarrow V$ is orthogonal if and only if $\{L(\gamma_1), \ldots, L(\gamma_n)\}$ is an orthonormal basis for V.

   Hint: If L is represented by $A = (a_{ij})$ with respect to the given basis, show that

$$((L(\gamma_i)\,|\,L(\gamma_j)) = \sum_{k=1}^{n} a_{ki}a_{kj}$$

   and then determine conditions under which A is orthogonal.

12. Let $S = \{\alpha_1, \ldots, \alpha_n\}$ and $T = \{\beta_1, \ldots, \beta_n\}$ be orthonormal bases for the real inner product space V.

   (i) Show that the T to S change of basis matrix P is orthogonal.
   Hint: First show that $\sum_{k=1}^{n} p_{ki}p_{kj} = (\beta_i\,|\,\beta_j)$.

   (ii) If the linear transformation $L: V \longrightarrow V$ is represented by A with respect to the basis S and by B with respect to T, show that $B = P^t A P$.

SUPPLEMENTARY EXERCISES

1. Let $P_1(-2, 1, 1)$, $P_2(3, 0, 1)$, and $P_3(0, 1, 1)$ be the vertices of a triangle in 3-space, and let $L: R_3 \longrightarrow R_3$ denote the linear transformation

$$L[(a_1, a_2, a_3)] = (a_1 + 2a_2, 3a_1 - a_2 + a_3, a_1 + 3a_2)$$

(i) Find the coordinates of the center Q of triangle $P_1P_2P_3$ (see exercise 2, section 2.2).

(ii) Find $L(Q)$ and explain why this point must be the center of the triangle whose vertices are $L(P_1)$, $L(P_2)$, and $L(P_3)$.

2. A certain linear transformation $L: R_4 \longrightarrow R_3$ is known to have these two properties:

(1) $\{(1, -2, 1, 0), (1, 0, 0, 1)\}$ is a basis for $\mathfrak{N}_L$.

(2) $L[(1, 0, 0, 0)] = (5, -3, 1)$ and $L[(0, 1, 0, 0)] = (7, 1, 1)$.

Use this information to answer the questions which follow:

(i) Show that $\beta = (4, 34, -2)$ belongs to $\mathfrak{R}_L$ and find a vector α which satisfies $L(\alpha) = \beta$.

(ii) Find $L[(0, 0, 1, 0)]$ and $L[(0, 0, 0, 1]$.

(iii) Find the matrix which represents L with respect to the standard bases for R_4 and R_3.

3. Let A, B, C be $n \times n$ matrices. Assume that B and C are non-singular and that

$$B^{-1}AB = C \quad \text{and} \quad C^{-1}BC = A$$

(i) Show that A is also non-singular.

(ii) Show that B is similar to C.

4. Let $S = \{\alpha_1, \alpha_2, \alpha_3\}$ be a basis for the vector space V, and let

$$\beta_1 = 2\alpha_1 + \alpha_2 - \alpha_3$$

$$\beta_2 = \alpha_1 + 3\alpha_2 + \alpha_3$$

$$\beta_3 = -\alpha_1 \quad\quad + \alpha_3$$

(i) Show that $T = \{\beta_1, \beta_2, \beta_3\}$ is a basis for V.

(ii) Let $L: V \longrightarrow V$ be the linear transformation which is represented by the matrix

$$A = \begin{bmatrix} -2 & 0 & 0 \\ 0 & 0 & 0 \\ 0 & 0 & 1 \end{bmatrix}$$

when the basis S is used in both the domain and codomain. What matrix represents L with respect to the basis T?

5. A linear transformation $L: V \longrightarrow V$ is said to be *idempotent* if $L \circ L = L$.

(i) Show that L is idempotent if and only if $L(\beta) = \beta$ for each vector β in the range of L.

(ii) If L is idempotent, show that there exists a basis $S = \{\alpha_1, \ldots, \alpha_n\}$ for V such that $L(\alpha_i) = \alpha_i$ for $i = 1, \ldots, r$ and $L(\alpha_j) = \theta_V$ for $j = r + 1, \ldots, n$, where $r = \rho(L)$.

(iii) Describe the matrix which represents L with respect to the basis S found in part (ii).

6. The linear transformation $N: V \longrightarrow V$ is said to be *nilpotent* of index k if $N^k = 0$ but N^{k-1} is not the zero transformation. Let γ be a vector such that $N^{k-1}(\gamma) \neq \theta$.

(i) Show that the vectors γ, $N(\gamma)$, $N^2(\gamma)$, ..., $N^{k-1}(\gamma)$ are linearly independent.

(ii) Assume V has dimension n. If N is nilpotent of index n, show that the set $S = \{\gamma, N(\gamma), \ldots, N^{n-1}(\gamma)\}$ is a basis for V. Describe the matrix which represents N with respect to the basis S.

7. Describe a function $F: R_3 \longrightarrow R_1$ which is not linear but satisfies $F(t\alpha) = tF(\alpha)$ for each real number t and 3-tuple $\alpha = (x, y, z)$.

8. Let S be a finite-dimensional subspace of the real inner product space V, and define the function $L: V \longrightarrow V$ as follows:

$$L(\alpha) = \alpha_S - \alpha_{S\perp}$$

where $\alpha_S + \alpha_{S\perp}$ is the orthogonal decomposition of α (recall the notation established in section 3.4).

(i) Show that L is a linear transformation.

(ii) Show that L is orthogonal.

Hint: Use theorem 4.18.

9. Let A be a fixed $n \times n$ matrix and let $L_A: {}_nR_n \longrightarrow {}_nR_n$ denote the linear transformation defined by

$$L(X) = AX - XA \quad \text{for each } n \times n \text{ matrix } X$$

(See exercise 8, section 4.1.)

(i) Show that L_A must be singular.

(ii) Find the rank and nullity of L_A when

$$A = \begin{bmatrix} -5 & 7 \\ 2 & 3 \end{bmatrix}$$

10. Let V be a finite-dimensional vector space, and let $L_1: V \longrightarrow V$ and $L_2: V \longrightarrow V$ be linear transformations.

(i) Show that $\rho(L_1 + L_2) \leq \rho(L_1) + \rho(L_2)$.

*(ii) Show that $\nu(L_1) + \nu(L_2) \geq \nu(L_2 \circ L_1)$.

11. Let $S = \{\gamma_1, \gamma_2, \ldots, \gamma_n\}$ be an orthonormal basis for the real inner product space V, and let $L: V \longrightarrow V$ be a linear transformation which satisfies

$$(L(\alpha)\,|\,\beta) = (\alpha\,|\,L(\beta)) \quad \text{for all } \alpha, \beta \text{ in } V$$

Show that the matrix which represents L with respect to the S-basis is symmetric.

Definition. A linear transformation L from the vector space V to its field of scalars R is called a *linear functional*. The collection of all such transformations—that is, the vector space $\mathfrak{L}(V, R)$—is called the *dual space* of V and is denoted by V^*.

12. Let $S = \{\alpha_1, \ldots, \alpha_n\}$ be a basis for V, and for $j = 1, 2, \ldots, n$, let f_j be the linear functional such that

$$f_j(\alpha_i) = \begin{cases} 1 & \text{if } i = j \\ 0 & \text{if } i \neq j \end{cases} \quad \text{for } i = 1, 2, \ldots, n$$

(i) Explain why this description is enough to completely determine each f_j.

(ii) Show that $S^* = \{f_1, \ldots, f_n\}$ is a basis for V^*. This collection of functionals is called the *dual basis* of S.

(iii) Show that $\dim V = \dim V^*$.

(iv) Describe the linear functionals which constitute the dual basis of the basis $S = \{(1, 1, -1), (0, 1, 1), (1, -1, -2)\}$ for R_3.

Definition.   Let V be a finite-dimensional real inner product space, and let $L: V \longrightarrow V$ be a linear transformation. Then a function $L^*: V \longrightarrow V$ which satisfies

$$(L(\alpha)\,|\,\beta) = (\alpha\,|\,L^*(\beta)) \quad \text{for all } \alpha, \beta \text{ in } V$$

is called an *adjoint* of L.

It can be shown that L has exactly one adjoint and that L^* is a linear transformation.

*13. Let V be an n-dimensional real inner product space, and let $L: V \longrightarrow V$ be a linear transformation.

(i) Show that $\mathfrak{N}_{L^*} = \mathfrak{R}_L^{\perp}$ (see exercise 8, section 3.4).

(ii) Show that $\rho(L) = \rho(L^*)$.

(iii) If $\nu(L) > 0$, show that the equation $L(\alpha) = \beta$ is solvable if and only if β is orthogonal to every vector in $\mathfrak{N}_{L^*}$. This is another way of stating alternative (b) of the Fredholm Alternative (see exercise 11, section 4.2).

*14. (i) Show that f satisfies the first order homogeneous linear differential equation $f' + af = 0$ if and only if $f(x) = Ce^{-ax}$ for some constant C. What does this fact enable us to say about the dimension of the solution space of this equation?

(ii) Show that f satisfies the non-homogeneous differential equation $f' + af = g$ if and only if

$$f(x) = Ce^{-ax} + e^{-ax} \int e^{at}g(t)\, dt$$

for some constant C.

Hint: Multiply the given equation by e^{ax} and observe that $(e^{ax}f)' = e^{ax}g$.

*15. Let S denote the solution space of the homogeneous second order differential equation $f'' + af' + bf = 0$, and let λ_1, λ_2 be the zeroes of the polynomial $p(t) = t^2 + at + b$. Assume that λ_1 and λ_2 are both real numbers. Obtain a basis for S by completing the following steps:

(i) Let f be a fixed solution of the given differential equation, and define the function g by the equation $g = f' - \lambda_2 f$. Show that $g' - \lambda_1 g = 0$. *Hint:* Note that $\lambda_1 + \lambda_2 = -a$ and $\lambda_1 \lambda_2 = b$ (why?).

(ii) Show that $g(x) = C_1 e^{\lambda_1 x}$ for some constant C_1, and conclude that f must satisfy the non-homogeneous first order differential equation $f' - \lambda_2 f = C_1 e^{\lambda_1 x}$.

(iii) Show that f satisfies the given second order equation if and only if

$$f(x) = C_2 e^{\lambda_2 x} + C_1 e^{\lambda_2 x} \int e^{(\lambda_1 - \lambda_2)t} \, dt$$

for constants C_1, C_2.
Hint: See exercise 14, part (ii).

(iv) If $\lambda_1 \neq \lambda_2$, show that $\{e^{\lambda_1 x}, e^{\lambda_2 x}\}$ is a basis of S, and in the case where $\lambda_1 = \lambda_2$, show that $\{e^{\lambda_1 x}, x e^{\lambda_1 x}\}$ is a basis.

CHAPTER

5

Determinants

Determinants were invented in the nineteenth century by mathematicians who observed that many questions regarding the solvability of a general $n \times n$ linear system are related in one way or another to a certain type of algebraic expression. For instance, it can be shown that the 2×2 linear system

$$a_{11}x_1 + a_{12}x_2 = b_1$$

$$a_{21}x_1 + a_{22}x_2 = b_2$$

$$(5.1)$$

does not have a unique solution if the expression $\Delta_2 = a_{11}a_{22} - a_{12}a_{21}$ is zero, but if Δ_2 is different from zero, then the system has the unique solution given by

$$x_1 = \frac{b_1 a_{22} - b_2 a_{12}}{\Delta_2}, \qquad x_2 = \frac{b_2 a_{11} - b_1 a_{21}}{\Delta_2}$$

In this sense, the solvability of the system is determined by Δ_2, and so this expression came to be known as the *determinant* of the system. Since Δ_2 is formed from the entries of the coefficient matrix A of system (5.1), we shall also say that Δ_2 is the determinant of A and write

$$\Delta_2 = \det \begin{bmatrix} a_{11} & a_{12} \\ a_{21} & a_{22} \end{bmatrix} = a_{11}a_{22} - a_{12}a_{21}$$

Similarly, it can be shown that the solvability of the general 3×3 linear system

$$a_{11}x_1 + a_{12}x_2 + a_{13}x_3 = b_1$$

$$a_{21}x_1 + a_{22}x_2 + a_{23}x_3 = b_2$$

$$a_{31}x_1 + a_{32}x_2 + a_{33}x_3 = b_3$$

is determined by the expression

$$\Delta_3 = \det \begin{bmatrix} a_{11} & a_{12} & a_{13} \\ a_{21} & a_{22} & a_{23} \\ a_{31} & a_{32} & a_{33} \end{bmatrix}$$

$$= a_{11}a_{22}a_{33} - a_{11}a_{23}a_{32} - a_{12}a_{21}a_{33}$$

$$+ a_{12}a_{23}a_{31} + a_{13}a_{21}a_{32} - a_{13}a_{22}a_{31}$$

Note that each term in the above expression is a signed product of three entries of the matrix $A = (a_{ij})$, one from each row and column. Eventually, we shall define the determinant of an $n \times n$ matrix in an analogous fashion, but first, we must develop some additional terminology and notation to be used in describing the pattern of signs which appear in such expressions.

Recall that a rearrangement of the numbers in the sequence $1, 2, \ldots, n$ is called a *permutation* of this sequence. We shall denote the permutation in which the number i is replaced by the number j_i by writing

$$\begin{Bmatrix} 1 & 2 & 3 & \ldots & n \\ j_1 & j_2 & j_3 & \ldots & j_n \end{Bmatrix}$$

For example, the following are permutations of the sequence $1, 2, 3, 4, 5$:

$$\begin{Bmatrix} 1 & 2 & 3 & 4 & 5 \\ 1 & 3 & 4 & 5 & 2 \end{Bmatrix} \quad \text{and} \quad \begin{Bmatrix} 1 & 2 & 3 & 4 & 5 \\ 5 & 2 & 1 & 3 & 4 \end{Bmatrix}$$

In the first these numbers are rearranged to appear in the order $1, 3, 4, 5, 2$, while in the second the rearrangement is $5, 2, 1, 3, 4$.

Returning to the general case, for each i, let τ_i denote the number of integers less than j_i which appear to the right of j_i in the permutation

$$\pi = \begin{Bmatrix} 1 & 2 & \dots & n \\ j_1 & j_2 & \dots & j_n \end{Bmatrix}$$

Then the *index* $\sigma(\pi)$ is defined to be the sum

$$\sigma(\pi) = \sum_{i=1}^{n} \tau_i$$

and π is said to be an *even permutation* if $\sigma(\pi)$ is an even number and an *odd permutation* if $\sigma(\pi)$ is odd. For example, for the permutation

$$\pi = \begin{Bmatrix} 1 & 2 & 3 & 4 & 5 \\ 3 & 5 & 1 & 2 & 4 \end{Bmatrix}$$

we have

$\tau_1 = 2$ since both 1 and 2 appear to the right of $j_1 = 3$
$\tau_2 = 3$ since 1, 2, 4 appear to the right of $j_2 = 5$
$\tau_3 = 0$ since there is no positive integer smaller than $j_3 = 1$
$\tau_4 = 0$ since only 4 appears to the right of $j_4 = 2$
$\tau_5 = 0$ since $j_5 = 4$ is the right-most number

and it follows that π is an odd permutation since

$$\sigma(\pi) = 2 + 3 + 0 + 0 + 0 = 5$$

It can be shown that there are $n!$ permutations of the number sequence $1, 2, \dots, n$, of which half are even and half are odd. Now we are ready to make the following definition.

Definition 5.1.   The determinant of the $n \times n$ matrix $A = (a_{ij})$ is the expression

$$\det A = \sum_{\pi} (-1)^{\sigma(\pi)} a_{1j_1} a_{2j_2} \dots a_{nj_n}$$

where the sum is taken over all $n!$ permutations of the number sequence $1, 2, \dots, n$.

We shall refer to the term $a_{1j_1} a_{2j_2} \dots a_{nj_n}$ as the *generalized diagonal product* of A which corresponds to the permutation

$$\pi = \begin{Bmatrix} 1 & 2 & \dots & n \\ j_1 & j_2 & \dots & j_n \end{Bmatrix}.$$

For instance, as indicated below, $a_{13}a_{21}a_{32}$ is the generalized diagonal product which corresponds to $\pi = \left\{\begin{matrix} 1 & 2 & 3 \\ 3 & 1 & 2 \end{matrix}\right\}$:

$$\begin{bmatrix} a_{11} & a_{12} & \boxed{a_{13}} \\ \boxed{a_{21}} & a_{22} & a_{23} \\ a_{31} & \boxed{a_{32}} & a_{33} \end{bmatrix}$$

Note that since the sequence $j_1, j_2, \ldots, j_n$ is just a rearrangement of $1, 2, \ldots, n$, the generalized diagonal product $a_{1j_1}a_{2j_2}\ldots a_{nj_n}$ contains exactly one entry from each row and one entry from each column of A. Moreover, the sign of this term in det A is positive if π is an even permutation and negative if π is odd.

EXAMPLE 1.   The generalized diagonal products of the matrix

$$A = \begin{bmatrix} 3 & -2 & 1 \\ -5 & 4 & 0 \\ 2 & 1 & 6 \end{bmatrix}$$

may be computed as follows:

General form of the product	Parity of the associated permutation	Numerical value in A
$a_{11}a_{22}a_{33}$	even	$(3)(4)(6) = 72$
$a_{11}a_{23}a_{32}$	odd	$(3)(0)(1) = 0$
$a_{12}a_{21}a_{33}$	odd	$(-2)(-5)(6) = 60$
$a_{12}a_{23}a_{31}$	even	$(-2)(0)(2) = 0$
$a_{13}a_{21}a_{32}$	even	$(1)(-5)(1) = -5$
$a_{13}a_{22}a_{31}$	odd	$(1)(4)(2) = 8$

Therefore, we have

det $A = (+1)(72) + (-1)(0) + (-1)(60) + (+1)(0) + (+1)(-5) + (-1)(8)$

$= -1$

Even when n is as small as 10 or 11, the number $n!$ is very large, and for this reason, it is impractical to evaluate the determinant of an $n \times n$ matrix A by finding its generalized diagonal products. In order to develop an alternate method for evaluating det A, we need to know more about the general properties of determinants, and several of these properties are described in the theorems which follow. Since the main purpose of this chapter is to develop determinant theory not so much for its own sake as for its application to the study of matrices, we shall omit certain details in the proofs of these results. Nevertheless, in each case, we shall describe the logical strategy behind the proof of the theorem in question and shall provide examples to illustrate key features in the general argument. The reader who wants to know more about the technical aspects of determinant theory can find this material in any of the intermediate or advanced level texts listed in the bibliography.

Theorem 5.1.   If A is an $n \times n$ matrix, then $\det A = \det A^t$.

Proof: This result follows immediately from the fact that each generalized diagonal product of A^t is the same as a product in A taken in a different order. For instance, in the 3×3 case, we have

$$A = \begin{bmatrix} a_{11} & a_{12} & \textcircled{a_{13}} \\ \textcircled{a_{21}} & a_{22} & a_{23} \\ a_{31} & \textcircled{a_{32}} & a_{33} \end{bmatrix}, \quad A^t = \begin{bmatrix} a_{11} & \textcircled{a_{21}} & a_{31} \\ a_{12} & a_{22} & \textcircled{a_{32}} \\ \textcircled{a_{13}} & a_{23} & x_{33} \end{bmatrix}$$

$$+ \, a_{13}a_{21}a_{32} \qquad\qquad\qquad + \, a_{21}a_{32}a_{13}$$

It is important to note that any result which involves a relationship between the rows of A and $\det A$ can be converted by theorem 5.1 into an analogous result relating the determinant to the columns of A, and vice versa. The next three theorems show how the determinant of a given matrix is affected by each elementary row operation and the corresponding elementary column operation.

Theorem 5.2.   If the matrix A' is derived from A by exchanging two rows (or two columns), then $\det A' = -\det A$.

Proof: This result can be proved by using a few additional properties of permutations to show that each generalized diagonal product of A' is the same as a product in A taken in a different order and having the opposite sign. For example, in the 3×3 case, suppose A' is derived from A by exchanging the first and third rows. Then we have

$$A = \begin{bmatrix} a_{11} & \textcircled{a_{12}} & a_{13} \\ a_{21} & a_{22} & \textcircled{a_{23}} \\ \textcircled{a_{31}} & a_{32} & a_{33} \end{bmatrix}, \quad A' = \begin{bmatrix} \textcircled{a_{31}} & a_{32} & a_{33} \\ a_{21} & a_{22} & \textcircled{a_{23}} \\ a_{11} & \textcircled{a_{12}} & a_{13} \end{bmatrix}$$

$$a_{12}a_{23}a_{31} \qquad\qquad\qquad\quad - \, a_{31}a_{23}a_{12}$$

Theorem 5.3.   If the matrix A' is derived from A by multiplying a row (or a column) by the scalar s, then det $A' = s$ det A.

Proof: It is easy to see that each generalized diagonal product of A is s times the corresponding product in A, and this observation is enough to prove the result in question. For instance,

$$A = \begin{bmatrix} a_{11} & \textcircled{a_{12}} & a_{13} \\ \textcircled{a_{21}} & a_{22} & a_{23} \\ a_{31} & a_{32} & \textcircled{a_{33}} \end{bmatrix}, \quad A' = \begin{bmatrix} a_{11} & \textcircled{a_{12}} & a_{13} \\ \textcircled{sa_{21}} & sa_{22} & sa_{23} \\ a_{31} & a_{32} & \textcircled{a_{33}} \end{bmatrix}$$

$$- a_{12}a_{21}a_{33} \qquad\qquad\qquad - a_{12}(sa_{21})a_{33}$$

Theorem 5.4.   If the matrix A' is derived from A by adding a scalar multiple of one row to another (or by adding a scalar multiple of one column to another), then det $A' =$ det A.

Proof: To illustrate the main ideas behind this result, consider the case where

$$A = \begin{bmatrix} a_{11} & a_{12} \\ a_{21} & a_{22} \end{bmatrix} \quad \text{and} \quad A' = \begin{bmatrix} a_{11} & a_{12} \\ a_{21} + sa_{11} & a_{22} + sa_{12} \end{bmatrix}$$

Applying definition 5.1, we find that

$$\det A' = a_{11}(a_{22} + sa_{12}) - a_{12}(a_{21} + sa_{11})$$
$$= (a_{11}a_{22} - a_{12}a_{21}) + s(a_{11}a_{12} - a_{12}a_{11}) = \det A + s \det B$$

where B is the matrix

$$\begin{bmatrix} a_{11} & a_{12} \\ a_{11} & a_{12} \end{bmatrix}$$

Since det $B = 0$, it follows that det $A' =$ det A.

In the general case, suppose that A' is derived from the $n \times n$ matrix A by adding s times the kth row to the lth row. As in the special case analyzed above, it can be shown that det $A' =$ det $A + s$ det B, where B is the matrix formed by replacing the lth row of A by the kth row (see exercise 12). Finally, since two rows of B are identical, it can be shown that det $B = 0$ (see exercise 6), and we conclude that det $A' =$ det A. ∎

An $n \times n$ matrix T is said to be *triangular* if it has all zero entries above (or below) the main diagonal. For instance, the matrices

$$\begin{bmatrix} -3 & 0 & 0 \\ 7 & 1 & 0 \\ 4 & -2 & 9 \end{bmatrix}, \quad \begin{bmatrix} 9 & -5 & 1 \\ 0 & 1 & 3 \\ 0 & 0 & 2 \end{bmatrix}, \quad \begin{bmatrix} 1 & 0 & -3 & 7 \\ 0 & 2 & 5 & -2 \\ 0 & 0 & 0 & 1 \\ 0 & 0 & 0 & 4 \end{bmatrix}$$

are all in triangular form. In general, it is not easy to compute the determinant of a given $n \times n$ matrix, but our next theorem shows that the determinant of a triangular matrix is just the product of its diagonal entries.

Theorem 5.5. If $T = (t_{ij})$ is a triangular matrix, then det $T = t_{11} t_{22} \ldots t_{nn}$.

Proof: Since T is triangular, each generalized diagonal product except $t_{11} t_{22} \ldots t_{nn}$ necessarily contains a zero entry. Therefore, det $T = t_{11} t_{22} \ldots t_{nn}$ as claimed. ∎

EXAMPLE 2.

$$\det \begin{bmatrix} -5 & 0 & 0 \\ 3 & 7 & 0 \\ 2 & -4 & 1 \end{bmatrix} = -5 \cdot 7 \cdot 1 - (-5) \cdot 0 \cdot (-4) + 0 \cdot 0 \cdot 2 - 0 \cdot 3 \cdot 1$$
$$+ 0 \cdot 3 \cdot (-4) - 0 \cdot 7 \cdot 2 = -5 \cdot 7 \cdot 1 = -35$$

By following the same procedure used in the reduction algorithm, a given $n \times n$ matrix A can be placed in triangular form T by a finite sequence of elementary row operations. Theorem 5.5 tells us how to compute det T, and theorems 5.2, 5.3, and 5.4 can be used to determine the relationship between det T and det A. This procedure for computing determinants is illustrated in the following example.

EXAMPLE 3.   To evaluate the determinant of the matrix

$$A = \begin{bmatrix} 0 & 4 & 1 & 3 \\ 3 & -5 & 2 & 1 \\ 7 & 1 & 4 & 3 \\ 1 & 1 & -2 & 1 \end{bmatrix}$$

we shall use elementary row operations to place A in triangular form. At the end of the first stage of this reduction, we have

$$
A = \begin{bmatrix} 0 & 4 & 1 & 3 \\ 3 & -5 & 2 & 1 \\ 7 & 1 & 4 & 3 \\ 1 & 1 & -2 & 1 \end{bmatrix} \xrightarrow[\substack{-3R_1+R_2\to R_2 \\ -7R_1+R_3\to R_3}]{R_4\longleftrightarrow R_1} \begin{bmatrix} 1 & 1 & -2 & 1 \\ 0 & -8 & 8 & -2 \\ 0 & -6 & 18 & -4 \\ 0 & 4 & 1 & 3 \end{bmatrix} = A_1
$$

According to theorem 5.2, the type I operation $R_4 \longleftrightarrow R_1$ changes the sign of det A, but the other two operations are of type III and have no effect on the determinant. Thus, we have det $A_1 = -\det A$. Continuing, we find that

$$
A_1 = \begin{bmatrix} 1 & 1 & -2 & 1 \\ 0 & -8 & 8 & -2 \\ 0 & -6 & 18 & -4 \\ 0 & 4 & 1 & 3 \end{bmatrix} \xrightarrow[\substack{-5/12\ R_3+R_4\to R_4}]{\substack{-3/4\ R_2+R_3\to R_3 \\ 1/2\ R_2+R_4\to R_4}} \begin{bmatrix} 1 & 1 & -2 & 1 \\ 0 & -8 & 8 & -2 \\ 0 & 0 & 12 & -\frac{5}{2} \\ 0 & 0 & 0 & \frac{73}{24} \end{bmatrix} = T
$$

and since only type III operations have been used in this last stage, we have det $T = \det A_1 = -\det A$. Finally, since T is triangular, theorem 5.5 tells us that det $T = (1)(-8)(12)(\frac{73}{24}) = -292$, and it follows that det $A = -\det T = 292$.

Now that we know how to evaluate the determinant of a given matrix, our next objective is to show how determinant theory is related to the theory of matrices. Several important features of this relationship are discussed in the next section.

Exercise Set 5.1

1. In each of the following cases, use definition 5.1 to evaluate the determinant of the given matrix.

(i) $\begin{bmatrix} 1 & 1 & 0 \\ 0 & 1 & 1 \\ 1 & 0 & 1 \end{bmatrix}$

(ii) $\begin{bmatrix} 7 & 1 & 9 \\ -2 & 9 & 0 \\ 3 & 1 & 4 \end{bmatrix}$

(iii) $\begin{bmatrix} 5 & -4 & 3 \\ 3 & 1 & 7 \\ 2 & -5 & -4 \end{bmatrix}$

(iv) $\begin{bmatrix} 1 & -3 & 7 \\ 7 & 1 & 4 \\ 9 & 5 & -3 \end{bmatrix}$

2. Evaluate the determinant of each of the following matrices by means of the method illustrated in example 3.

(i) $\begin{bmatrix} 3 & -5 & -2 \\ 1 & 7 & 0 \\ 2 & 1 & -1 \end{bmatrix}$
(ii) $\begin{bmatrix} 5 & 1 & 3 \\ -2 & 1 & 1 \\ 1 & 1 & 2 \end{bmatrix}$

(iii) $\begin{bmatrix} -2 & 1 & 1 & 0 \\ 1 & -3 & 1 & 5 \\ 6 & -2 & 0 & 3 \\ 7 & 1 & 1 & 1 \end{bmatrix}$
(iv) $\begin{bmatrix} 3 & 5 & -2 & 1 \\ 0 & 2 & 3 & 1 \\ 1 & 0 & 4 & 0 \\ -2 & 1 & 1 & 1 \end{bmatrix}$

3. In each of the following cases, find all values of x which satisfy the given determinantal equation.

(i) $\det \begin{bmatrix} 3 - x & 6 \\ 1 & 4 + x \end{bmatrix} = 0$
(ii) $\det \begin{bmatrix} 2 - x & 3x \\ -2 & 7 + x \end{bmatrix} = -6$

(iii) $\det \begin{bmatrix} 5 - x & 9 & -13 \\ 0 & x & 4 \\ 0 & 0 & x + 3 \end{bmatrix} = 0$
(iv) $\det \begin{bmatrix} x + 5 & 4 & 4 \\ -4 & x - 3 & -4 \\ -4 & -4 & x - 3 \end{bmatrix} = 0$

4. If A is an $n \times n$ matrix and s is a scalar, show that $\det (sA) = s^n \det A$.
5. If the $n \times n$ matrix A has a row or a column of zeroes, show that $\det A = 0$.
6.   (i) Verify that the following matrix has determinant 0:

$$\begin{bmatrix} a_{11} & a_{12} & a_{13} \\ a_{21} & a_{22} & a_{23} \\ a_{11} & a_{12} & a_{13} \end{bmatrix}$$

*(ii) Generalize the result of part (i) by showing that an $n \times n$ matrix A which has two identical rows or two identical columns must have determinant 0. *Hint:* Show that for each generalized diagonal product with a positive sign, there is a second product which has the same terms but a negative sign.

7. If a row of the matrix A can be expressed as a linear combination of other rows (or a column, as a linear combination of other columns), show that $\det A = 0$.
8. If A is an $n \times n$ matrix whose rank is less than n, show that $\det A = 0$. *Hint:* Interpret the result of exercise 7.
9. Show that $\det I_n = 1$. Also, if E is any $n \times n$ reduced row echelon matrix other than I_n, show that $\det E = 0$.
10. Show that

$$\det \begin{bmatrix} a & b & c \\ b & c & a \\ c & a & b \end{bmatrix} = 3abc - (a^3 + b^3 + c^3)$$

*11. Show that

$$\det \begin{bmatrix} 1 & a & a^2 \\ 1 & b & b^2 \\ 1 & c & c^2 \end{bmatrix} = (b - a)(c - a)(c - b)$$

for all scalars a, b, c. This expression is known as the *Vandermonde determinant* of order 3.

12. (i) If c_1, c_2, c_3 are scalars, verify that

$$\det \begin{bmatrix} a_{11} & a_{12} & a_{13} \\ a_{21} + c_1 & a_{22} + c_2 & a_{23} + c_3 \\ a_{31} & a_{32} & a_{33} \end{bmatrix}$$

$$= \det \begin{bmatrix} a_{11} & a_{12} & a_{13} \\ a_{21} & a_{22} & a_{23} \\ a_{31} & a_{32} & a_{33} \end{bmatrix} + \det \begin{bmatrix} a_{11} & a_{12} & a_{13} \\ c_1 & c_2 & c_3 \\ a_{31} & a_{32} & a_{33} \end{bmatrix}$$

*(ii) State and prove a generalization of the result in part (i) that applies to $n \times n$ matrices.

13. Given that

$$\det \begin{bmatrix} -3 & 2 & 5 \\ 4 & 1 & 1 \\ 3 & 2 & 1 \end{bmatrix} = 26$$

and

$$\det \begin{bmatrix} -3 & 2 & 5 \\ 1 & 3 & 0 \\ 3 & 2 & 1 \end{bmatrix} = -46$$

what is

$$\det \begin{bmatrix} -3 & 2 & 5 \\ 4 + 1 & 1 + 3 & 1 + 0 \\ 3 & 2 & 1 \end{bmatrix}$$

14. Let Q be the $n \times n$ matrix whose block form is

$$Q = \begin{bmatrix} A & B \\ O & C \end{bmatrix}$$

where A and C are $p \times p$ and $(n - p) \times (n - p)$ square submatrices, respectively, and O is an $(n - p) \times p$ block of zeroes. Show that det $Q = $ (det A)(det C).

15. Evaluate the following determinants:

(i) det $\begin{bmatrix} -3 & 7 & 8 & 9 & -2 \\ 4 & -1 & -3 & 6 & 1 \\ 0 & 0 & 0 & -3 & 1 \\ 0 & 0 & 2 & 1 & 0 \\ 0 & 0 & 1 & -3 & 1 \end{bmatrix}$ (ii) det $\begin{bmatrix} 7 & 1 & 0 & 0 \\ 3 & 5 & 0 & 0 \\ 2 & -1 & 1 & 1 \\ 4 & 0 & 1 & 0 \end{bmatrix}$

16. Let A be a 2×2 orthogonal matrix. Show that det $A = 1$ if A represents a rotation and det $A = -1$ if A represents a reflection.
Hint: Examine the forms displayed in (4.14) of section 4.7.

5.2 THE PRODUCT RULE FOR DETERMINANTS

The determinantal equation det $(AB) = (\det A)(\det B)$ is used for many different purposes in linear algebra. In this section, we shall first prove that this "product rule" for determinants is valid and then examine some of its applications. We begin by dealing with the special case where A is an $n \times n$ reduced row echelon matrix.

Lemma 5.1. If R and B are $n \times n$ matrices and R is in reduced row echelon form, then det $(RB) = (\det R)(\det B)$.

Proof: If $R = I_n$, we have det $R = 1$ and $RB = B$, and the lemma is clearly true. If R is not I_n, it must have at least one row of zeroes (why?), and RB must also have a row of zeroes (why?). Therefore, in this case, we have det $R = 0 = \det (RB)$ (see exercise 5, section 5.1), and it follows that det $(RB) = 0 = (\det R)(\det B)$. ∎

Next, recall from section 1.7 that if R is the reduced row echelon form of the matrix A, there exist elementary matrices $E_1, E_2, \ldots, E_p$ such that $A = E_p E_{p-1} \ldots E_1 R$. Therefore, we can write $AB = (E_p \ldots E_1 R)B$, and since we already know that det $(RB) = (\det R)(\det B)$, the general product rule may be obtained by showing that

$$\det (AB) = (\det E_p) \ldots (\det E_1) \det (RB)$$

and that

$$\det A = (\det E_p) \ldots (\det E_1)(\det R)$$

These relationships may be derived quite easily once we have shown that det $(EC) = (\det E)(\det C)$, where E is an elementary matrix. To this end, recall that the three different types of elementary matrices may be described as follows:

P_{ij} is obtained by exchanging the ith and jth rows of I_n.

$S_k(s)$ is obtained by multiplying the kth row of I_n by the scalar s.
$M_k(i; s)$ is obtained by multiplying the ith row of I_n by s and adding the result to the kth row.
Therefore, according to theorems 5.2, 5.3, and 5.4, we have

$$\det P_{ij} = -\det I_n = -1$$

$$\det S_k(s) = s \det I_n = s$$

$$\det M_k(i; s) = \det I_n = 1$$

We can now prove our next lemma.

Lemma 5.2. If E is an $n \times n$ elementary matrix, then $\det(EC) = (\det E)(\det C)$ for each $n \times n$ matrix C.

Proof: To verify the product rule in the case where $E = P_{ij}$, recall that the product matrix $P_{ij}C$ is formed by exchanging the ith row of C with the jth row. Therefore, according to theorem 5.2, we have $\det(P_{ij}C) = -\det C = (\det P_{ij})(\det C)$, since $\det P_{ij} = -1$. Similarly, using theorems 5.3 and 5.4, it can be shown that

$$\det S_k(s)C = s \det C = \det S_k(s) \det C$$

and

$$\det M_k(i; s)C = \det C = \det M_k(i; s) \det C$$

and the proof is complete. ∎

Using mathematical induction, we can extend the result of lemma 5.2 as follows.

Corollary 5.1. If $E_1, E_2, \ldots, E_p$ are elementary matrices, then $\det(E_p E_{p-1} \ldots E_1 C) = (\det E_p) \ldots (\det E_1)(\det C)$.

Proof: Exercise.

We now have the tools to prove the product rule for determinants in full generality.

Theorem 5.6. If A and B are $n \times n$ matrices, then

$$\det AB = (\det A)(\det B)$$

Proof: Let R be the reduced row echelon form of A, and let $E_1, \ldots, E_p$ be elementary matrices such that $A = E_p \ldots E_1 R$. Then, according to corollary 5.1, we have

$$\det A = (\det E_p) \dots (\det E_1)(\det R)$$

and

$$\det AB = \det (E_p \dots E_1(RB)) = (\det E_p) \dots (\det E_1)(\det RB)$$

Finally, since R is a reduced row echelon matrix, lemma 5.1 tells us that $\det RB = (\det R)(\det B)$, and it follows that

$$\det AB = (\det E_p) \dots (\det E_1)(\det R)(\det B)$$
$$= [(\det E_p) \dots (\det E_1) \det R](\det B) = (\det A)(\det B)$$

as desired. ∎

EXAMPLE 1.   If

$$A = \begin{bmatrix} 1 & 0 & -2 \\ 0 & 1 & 4 \\ -1 & 1 & 0 \end{bmatrix} \quad \text{and} \quad B = \begin{bmatrix} -2 & 1 & -3 \\ 1 & 0 & 1 \\ 5 & 2 & 0 \end{bmatrix}$$

we find that $\det A = -6$, $\det B = 3$, and

$$\det AB = \det \begin{bmatrix} -12 & -3 & -3 \\ 21 & 8 & 1 \\ 3 & -1 & 4 \end{bmatrix} = -18 = (\det A)(\det B)$$

Next, we shall demonstrate the value of the product rule by using it to prove three important facts about matrices. The first of these results provides us with a useful determinantal criterion for singularity.

Corollary 5.2.   The $n \times n$ matrix A is singular if and only if $\det A = 0$.

Proof: If A is non-singular, we have $AA^{-1} = I_n$ and $(\det A)(\det A^{-1}) = \det (AA^{-1}) = \det I_n = 1$. Thus, it is clear that the determinant of a non-singular matrix cannot be zero.

Conversely, suppose A is singular and let R be its reduced row echelon form. Then R must have a row of zeroes since its rank is less than n, and it follows that $\det R = 0$. Therefore, if $E_1, E_2, \dots, E_p$ are elementary matrices which satisfy $A = E_p \dots E_1 R$, we find that

$$\det A = (\det E_p) \dots (\det E_1)(\det R) = 0$$

as claimed. ∎

PROBLEM 1. For what values of t does the matrix

$$A(t) = \begin{bmatrix} 3-t & -1 & 0 \\ 1 & 2-t & -1 \\ 5 & -2 & -1-t \end{bmatrix}$$

have an inverse?

Solution: We find that

$$\det A = t^3 - 4t^2 + 8 = (t-2)(t^2 - 2t - 4)$$

and according to corollary 5.2, $A(t)$ has an inverse if and only if this polynomial is not zero. Therefore, $[A(t)]^{-1}$ exists whenever t is any number other than 2, $1 + \sqrt{5}$, or $1 - \sqrt{5}$.

PROBLEM 2. Determine whether or not the linear transformation $L: R_3 \to R_3$ defined by

$$L[(a_1, a_2, a_3)] = (-2a_1 + 3a_2 + a_3, 4a_1 - a_2 + 3a_3, -a_1 + 2a_2 + a_3)$$

is singular.

Solution: When the standard basis is used in both the domain and codomain of L, this linear transformation is represented by the matrix

$$A = \begin{bmatrix} -2 & 3 & 1 \\ 4 & -1 & 3 \\ -1 & 2 & 1 \end{bmatrix}$$

and we find that $\det A = 0$. Therefore, A is singular, and L must also be singular.

Thanks to corollary 5.2, we know that the determinant of a non-singular matrix A cannot be zero. In our next corollary, we show that $\det A$ and $\det A^{-1}$ are reciprocals of one another.

Corollary 5.3. If A is an $n \times n$ non-singular matrix, then $\det A^{-1} = (\det A)^{-1}$.

Proof: Since $AA^{-1} = I_n$, we have $(\det A)(\det A^{-1}) = \det I_n = 1$, and it follows that

$$\det A^{-1} = \frac{1}{\det A}$$

EXAMPLE 2.   If

$$A = \begin{bmatrix} -1 & 3 & -5 \\ 2 & 0 & 1 \\ 0 & -3 & 4 \end{bmatrix}$$

then

$$A^{-1} = \begin{bmatrix} 1 & 1 & 1 \\ -\frac{8}{3} & -\frac{4}{3} & -3 \\ -2 & -1 & -2 \end{bmatrix}$$

and we have $\det A = 3$ and $\det A^{-1} = \frac{1}{3}$, as predicted by corollary 5.3.

As the third and perhaps most important consequence of theorem 5.6, we shall now show that similar matrices have the same determinant.

Corollary 5.4.   If the $n \times n$ matrices A and B are similar, then $\det A = \det B$.

Proof: If P is a non-singular matrix such that $B = P^{-1}AP$, then we have

$$\det B = \det(P^{-1}AP) = (\det P^{-1})(\det A)(\det P)$$
$$= (\det P^{-1})(\det P)(\det A) = \det A$$

since $(\det P^{-1})(\det P) = 1$. ∎

Unfortunately, the converse of this corollary is not generally true. For instance, the matrices

$$A = \begin{bmatrix} -1 & 0 & 0 \\ 0 & -1 & 0 \\ 0 & 0 & 1 \end{bmatrix} \quad \text{and} \quad I_3 = \begin{bmatrix} 1 & 0 & 0 \\ 0 & 1 & 0 \\ 0 & 0 & 1 \end{bmatrix}$$

both have determinant 1, but A cannot be similar to I_3, for if it were, we could find a non-singular matrix P such that $A = P^{-1}I_3P$, and this is clearly impossible (why?).

Note that since two matrices represent the same linear transformation $L: V \to V$ if and only if they are similar, it follows from corollary 5.4 that all matrices which represent L must have the same determinant. This observation provides a quick way of showing that a given matrix cannot possibly represent L.

PROBLEM 3. Can the linear transformation $L: R_3 \rightarrow R_3$ defined by

$$L[(a_1, a_2, a_3)] = (-3a_1 + a_2 + 4a_3, \quad 2a_1 - 3a_3, \quad a_2 + 5a_3)$$

be represented by the matrix

$$A = \begin{bmatrix} 1 & -2 & 2 \\ 3 & 1 & 5 \\ 2 & 1 & 4 \end{bmatrix}$$

with respect to a suitable basis for R_3?

Solution: When the standard basis is used in the domain and codomain, L is represented by the matrix

$$B = \begin{bmatrix} -3 & 1 & 4 \\ 2 & 0 & -3 \\ 0 & 1 & 5 \end{bmatrix}$$

Since det $B = -11$ while det $A = 5$, it follows that A and B are not similar, and hence, A cannot represent L.

By now, the reader should have a fairly good idea of just what can be done with the product rule for determinants, and several additional applications of this relationship are provided in the exercises. In the next section, we continue our development of the theory of determinants by deriving a formula for the inverse of a non-singular matrix A which involves only det A and the determinant of each $(n - 1) \times (n - 1)$ submatrix of A.

Exercise Set 5.2

1. Verify that det $(AB) = (\det A)(\det B)$ in the case where

$$A = \begin{bmatrix} -1 & 3 & 1 \\ 4 & -2 & 7 \\ 1 & 1 & 5 \end{bmatrix} \quad \text{and} \quad B = \begin{bmatrix} 1 & 1 & -1 \\ 2 & 1 & 3 \\ 5 & 1 & -2 \end{bmatrix}$$

2. In each of the following cases, determine the values of t for which the given matrix is singular.

(i) $\begin{bmatrix} 8 & -3 & 1 \\ 3 & t & -2 \\ 2 & -5 & 5 \end{bmatrix}$

(ii) $\begin{bmatrix} t & 1 & 0 \\ 0 & t & 1 \\ 15 & 17 & t+1 \end{bmatrix}$

(iii) $\begin{bmatrix} t-6 & 1 & 5 \\ 0 & t+1 & -1 \\ -9 & 1 & t+8 \end{bmatrix}$ (iv) $\begin{bmatrix} 3 & t & 1 \\ 3 & 1 & t \\ t+5 & 2 & 2 \end{bmatrix}$

3. In each of the following cases, determine whether or not the given linear transformation is singular.
   (i) $L[(a_1, a_2, a_3)] = (-2a_1 + a_2 - 5a_3, 3a_1 + a_3, 5a_2 + 4a_3)$
   (ii) $L[(a_1, a_2, a_3)] = (7a_1 + 3a_2 - 2a_3, 2a_1 + a_2 + 5a_3, a_1 - 17a_3)$
   (iii) $L[(a_1, a_2, a_3)] = (5a_1 - 8a_2, 7a_2 - a_3, a_1 + 3a_2 - a_3)$

4. Let K be an $n \times n$ skew-symmetric matrix (that is, $K = -K^t$). If n is odd, show that K must be singular.

5. Show that an $n \times n$ nilpotent matrix N must be singular. (See supplementary exercise 9 of chapter 2.)

6. If A is an $n \times n$ idempotent matrix (that is, $A^2 = A$), show that det A is either 0 or 1.

7. If P is an $n \times n$ orthogonal matrix (that is, $P^t = P^{-1}$), show that det P is either 1 or -1.

8. Let A be an $n \times n$ matrix, and let λ be a scalar such that $AX = \lambda X$ for some real n-tuple X. It is necessary that X be non zero; otherwise, A could be 0 but $\lambda \neq 0$.
   (i) Show that det $(\lambda I_n - A) = 0$.
   (ii) If $B = P^{-1}AP$, show that det $(\lambda I_n - B) = 0$ also.

9. If A is a 3×3 matrix whose determinant is 5 and if $B = QAP$, where

$$Q = \begin{bmatrix} -3 & 1 & -2 \\ 1 & 0 & 4 \\ 2 & 1 & 1 \end{bmatrix} \quad \text{and} \quad P = \begin{bmatrix} 1 & 0 & 1 \\ 1 & -1 & 1 \\ -2 & 1 & 3 \end{bmatrix}$$

what is det B?

10. If A is a matrix which represents the linear transformation $L: P_2 \longrightarrow P_2$ defined by

$$L[at^2 + bt + c] = (-3a + b)t^2 + (a - 2c)t + (a - b + c)$$

what is det A?

5.3 COFACTOR EXPANSION OF THE DETERMINANT FUNCTION

In this section, we shall show that the determinant of a given $n \times n$ matrix A can be written as a linear expression involving the determinants of certain $(n - 1) \times (n - 1)$ submatrices of A. This so-called *cofactor expansion* of the determinant function will then be used to derive several important theoretical results, including an explicit form for the inverse of a non-singular matrix and Cramer's rule for solving an $n \times n$ system of linear equations. We begin with the following definition.

Definition 5.2.   Let $A = (a_{ij})$ be an $n \times n$ matrix, and let A_{pq} denote the submatrix formed by excluding the pth row and qth column of A. Then the number det A_{pq} is called the *minor* associated with the (p, q) entry of A, and the signed minor $(-1)^{p+q}$ det A_{pq} is called the *cofactor* associated with that entry.

To simplify the notation in the work which follows, we shall denote the cofactor $(-1)^{p+q}$ det A_{pq} by the symbol α_{pq}.

EXAMPLE 1.   If

$$A = \begin{bmatrix} 4 & -3 & 2 \\ 7 & 5 & 1 \\ 0 & 2 & 9 \end{bmatrix}$$

we have

$$A_{12} = \begin{bmatrix} 7 & 1 \\ 0 & 9 \end{bmatrix} \quad \text{and} \quad A_{23} = \begin{bmatrix} 4 & -3 \\ 0 & 2 \end{bmatrix}$$

Thus, the minor associated with the $(1, 2)$ entry of A is det $A_{12} = 7 \cdot 9 - 1 \cdot 0 = 63$, and the cofactor associated with the $(2, 3)$ entry is given by

$$\alpha_{23} = (-1)^{2+3} \det A_{23} = (-1)[4 \cdot 2 - (-3) \cdot 0] = -8.$$

In section 5.1, we defined the determinant of the 3×3 matrix

$$A = \begin{bmatrix} a_{11} & a_{12} & a_{13} \\ a_{21} & a_{22} & a_{23} \\ a_{31} & a_{32} & a_{33} \end{bmatrix}$$

to be the following sum of signed generalized diagonal products:

$$\det A = a_{11}a_{22}a_{33} - a_{11}a_{23}a_{32} + a_{12}a_{23}a_{31} \qquad (5.2)$$
$$- a_{12}a_{21}a_{33} + a_{13}a_{21}a_{32} - a_{13}a_{22}a_{31}$$

This expression can be rewritten as

$$a_{11}(a_{22}a_{33} - a_{23}a_{32}) + a_{12}(a_{23}a_{31} - a_{21}a_{33}) + a_{13}(a_{21}a_{32} - a_{22}a_{31})$$

which means that

$$\det A = a_{11}\alpha_{11} + a_{12}\alpha_{12} + a_{13}\alpha_{13} \qquad (5.3)$$

since we have

$$\mathcal{Q}_{11} = (-1)^2 \det A_{11} = a_{22}a_{33} - a_{13}a_{32}$$

$$\mathcal{Q}_{12} = (-1)^3 \det A_{12} = -(a_{21}a_{33} - a_{23}a_{31})$$

$$\mathcal{Q}_{13} = (-1)^4 \det A_{13} = a_{21}a_{32} - a_{22}a_{31}$$

Moreover, the sum in (5.2) can also be written in the form

$$a_{13}(a_{21}a_{32} - a_{22}a_{31}) + a_{23}(a_{12}a_{31} - a_{11}a_{32}) + a_{33}(a_{11}a_{22} - a_{12}a_{21})$$

from which we conclude that

$$\det A = a_{13}\mathcal{Q}_{13} + a_{23}\mathcal{Q}_{23} + a_{33}\mathcal{Q}_{33} \tag{5.4}$$

Formula (5.3) is said to be the *cofactor expansion* of det A with respect to the first row, while (5.4) is the expansion with respect to the third column. In general, the determinant of an $n \times n$ matrix can be expanded in a similar fashion with respect to any one of its rows or columns. This observation is given precise formulation in the following lemma.

Lemma 5.3.   If A is an $n \times n$ matrix and p, q are fixed indices, then we have

$$\det A = a_{p1}\mathcal{Q}_{p1} + a_{p2}\mathcal{Q}_{p2} + \ldots + a_{pn}\mathcal{Q}_{pn} \tag{5.5}$$

and

$$\det A = a_{1q}\mathcal{Q}_{1q} + a_{2q}\mathcal{Q}_{2q} + \ldots + a_{nq}\mathcal{Q}_{nq} \tag{5.6}$$

Proof: First, note that (5.6) follows from (5.5) and theorem 5.1. To prove (5.5), it suffices to show that each signed generalized diagonal product of A which includes the entry a_{pj} appears as a term in the expression $a_{pj}\mathcal{Q}_{pj}$, and vice versa. It is not especially difficult to verify this fact, but the proof is a bit tedious and involves several steps which are not really relevant to the rest of our work. For this reason, we omit the details.

EXAMPLE 2.   Let

$$A = \begin{bmatrix} 3 & -7 & 5 \\ -2 & 1 & 0 \\ 4 & 8 & -9 \end{bmatrix}$$

The cofactor expansion of det A with respect to the entries in the third row is as follows:

$$\det A = 4(-1)^{3+1} \det \begin{bmatrix} -7 & 5 \\ 1 & 0 \end{bmatrix} + 8(-1)^{3+2} \det \begin{bmatrix} 3 & 5 \\ -2 & 0 \end{bmatrix}$$

$$+ (-9)(-1)^{3+3} \det \begin{bmatrix} 3 & -7 \\ -2 & 1 \end{bmatrix}$$

$$= 4(1)(-5) + 8(-1)(10) + (-9)(1)(-11) = -1$$

Similarly, expanding $\det A$ by the entries and cofactors of the second column, we find that

$$\det A = (-7)(-1)^{1+2} \det \begin{bmatrix} -2 & 0 \\ 4 & -9 \end{bmatrix} + 1(-1)^{2+2} \det \begin{bmatrix} 3 & 5 \\ 4 & -9 \end{bmatrix}$$

$$+ 8(-1)^{3+2} \det \begin{bmatrix} 3 & 5 \\ -2 & 0 \end{bmatrix}$$

$$= (-7)(-1)(18) + (1)(1)(-47) + (8)(-1)(10) = -1$$

Lemma 5.3 tells us that $\det A$ can be computed by expanding by the entries and corresponding cofactors of any row or column of A. However, our next lemma shows that an expansion formed by using the entries of one row (or column) and the cofactors of the entries in another must always be zero.

Lemma 5.4. Let A be an $n \times n$ matrix, and let p and q be fixed indices. If i and j are indices such that $i \neq p$ and $j \neq q$, then

$$a_{p1}\alpha_{i1} + a_{p2}\alpha_{i2} + \ldots + a_{pn}\alpha_{in} = 0 \tag{5.7}$$

and

$$a_{1q}\alpha_{1j} + a_{2q}\alpha_{2j} + \ldots + a_{nq}\alpha_{nj} = 0 \tag{5.8}$$

Proof: To prove (5.7), let $A' = (a'_{ij})$ denote the $n \times n$ matrix whose ith row is the same as the pth row of A and whose other $n - 1$ rows are identically the same as the corresponding rows of A. Then we have $a'_{ij} = a_{pj}$ and $\alpha'_{ij} = \alpha_{ij}$ for $j = 1, 2, \ldots, n$, and since the ith and pth rows of A' are the same, it follows that $\det A' = 0$. Therefore, if the determinant of A' is expanded by the entries and cofactors in its ith row, we obtain

$$0 = \det A' = a'_{i1}\alpha'_{i1} + a'_{i2}\alpha'_{i2} + \ldots + a'_{in}\alpha'_{in}$$

$$= a_{p1}\alpha_{i1} + a_{p2}\alpha_{i2} + \ldots + a_{pn}\alpha_{in}$$

as desired. This proves formula (5.7), and (5.8) follows by applying theorem 5.1. ∎

EXAMPLE 3.   Let A be the matrix discussed in example 2. Expanding by the entries of the first row and the cofactors of the entries in the third row, we find that

$$3(-1)^{3+1} \det \begin{bmatrix} -7 & 5 \\ 1 & 0 \end{bmatrix} + (-7)(-1)^{3+2} \det \begin{bmatrix} 3 & 5 \\ -2 & 0 \end{bmatrix}$$

$$+ 5(-1)^{3+3} \det \begin{bmatrix} 3 & -7 \\ -2 & 1 \end{bmatrix}$$

$$= (3)(1)(-5) + (-7)(-1)(10) + (5)(1)(-11) = 0$$

as predicted by lemma 5.4. Note that this is what we would obtain by expanding the determinant of the singular matrix

$$A' = \begin{bmatrix} 3 & -7 & 5 \\ -2 & 1 & 0 \\ 3 & -7 & 5 \end{bmatrix}$$

by the entries and corresponding cofactors of the third row.

Before continuing our discussion of the results contained in lemmas 5.3 and 5.4, we pause to make the following definition.

Definition 5.3.   The *adjugate* of the $n \times n$ matrix A is the $n \times n$ matrix whose (i, j) entry is $\mathcal{C}_{ji}$, the cofactor of the entry in the jth row and ith column of A. We shall denote this matrix by writing adj A.

EXAMPLE 4.   If

$$A = \begin{bmatrix} -3 & 1 & 2 \\ 4 & 5 & -6 \\ 0 & 3 & -1 \end{bmatrix}$$

then

$$\text{adj } A = \begin{bmatrix} \mathcal{C}_{11} & \mathcal{C}_{21} & \mathcal{C}_{31} \\ \mathcal{C}_{12} & \mathcal{C}_{22} & \mathcal{C}_{32} \\ \mathcal{C}_{13} & \mathcal{C}_{23} & \mathcal{C}_{33} \end{bmatrix} = \begin{bmatrix} 13 & 7 & -16 \\ 4 & 3 & -10 \\ 12 & 9 & -19 \end{bmatrix}$$

For many years, the matrix defined above was referred to as either the *adjoint* or the *classical adjoint* in the mathematical literature. However, the term "adjoint" is also applied to a concept in operator theory, and recently, there has been a movement among mathematicians to change the name of the matrix described in definition 5.3 to either "adjugate" or "matrix of cofactors." Our decision to use the former is a matter of personal taste. The main reason for defining the adjugate matrix is contained in the following theorem.

Theorem 5.7. If A is an $n \times n$ matrix, then

$$A(\text{adj } A) = (\det A)I_n = (\text{adj } A)A$$

Proof: Let $B = \text{adj } A$. Then $b_{ij} = \alpha_{ji}$, and the (i, j) entry of the product matrix $A(\text{adj } A) = AB$ is given by the sum

$$a_{i1}b_{1j} + a_{i2}b_{2j} + \ldots + a_{in}b_{nj} = a_{i1}\alpha_{j1} + a_{i2}\alpha_{j2} + \ldots + a_{in}\alpha_{jn}$$

According to lemmas 5.3 and 5.4, the sum on the right is $\det A$ if $i = j$ and 0 if $i \neq j$, and it follows that $AB = (\det A)I_n$. A similar argument can be used to show that $(\text{adj } A)A = (\det A)I_n$. ∎

EXAMPLE 5. If

$$A = \begin{bmatrix} -3 & 1 & 2 \\ 4 & 5 & -6 \\ 0 & 3 & -1 \end{bmatrix}$$

we have

$$\text{adj } A = \begin{bmatrix} 13 & 7 & -16 \\ 4 & 3 & -10 \\ 12 & 9 & -19 \end{bmatrix} \quad \text{and} \quad \det A = -11$$

and we find that

$$A(\text{adj } A) = \begin{bmatrix} -3 & 1 & 2 \\ 4 & 5 & -6 \\ 0 & 3 & -1 \end{bmatrix}\begin{bmatrix} 13 & 7 & -16 \\ 4 & 3 & -10 \\ 12 & 9 & -19 \end{bmatrix} = \begin{bmatrix} -11 & 0 & 0 \\ 0 & -11 & 0 \\ 0 & 0 & -11 \end{bmatrix}$$

$$= -11I_3 = (\det A)I_3$$

as predicted by theorem 5.7.

In the case where A is non-singular, we have $\det A \neq 0$, and theorem 5.7 provides the following formula for A^{-1}.

Corollary 5.5 If A is a non-singular $n \times n$ matrix, then

$$A^{-1} = \frac{1}{\det A}(\text{adj } A) \tag{5.9}$$

EXAMPLE 6.   If

$$A = \begin{bmatrix} -3 & 1 & 5 \\ 2 & 0 & 7 \\ 4 & -1 & -1 \end{bmatrix}$$

we find that $\det A = -1$ and that

$$\operatorname{adj} A = \begin{bmatrix} 7 & -4 & 7 \\ 30 & -17 & 31 \\ -2 & 1 & -2 \end{bmatrix}$$

Therefore, according to corollary 5.5, we have

$$A^{-1} = \frac{1}{(-1)} \operatorname{adj} A = \begin{bmatrix} -7 & 4 & -7 \\ -30 & 17 & -31 \\ 2 & -1 & 2 \end{bmatrix}$$

From the computational standpoint, the formula contained in corollary 5.5 is not a very efficient way to find A^{-1}, but it does have great value as a theoretical tool. One of its applications is the following result, known as Cramer's rule, which provides an explicit formula for the solution of a system of n linear equations in n unknowns.

Theorem 5.8. (Cramer's rule.) If $A = (a_{ij})$ is a non-singular $n \times n$ matrix, then the linear system $AX = \beta$ has a unique solution $X = (x_k)$ with

$$x_1 = \frac{\det \hat{A}_1}{\det A}, \qquad x_2 = \frac{\det \hat{A}_2}{\det A}, \qquad \ldots, \qquad x_n = \frac{\det \hat{A}_n}{\det A}$$

where $\hat{A}_k$ denotes the matrix formed by replacing the kth column of A by the n-tuple $\beta = (b_k)$.

Proof: Since A is non-singular, the system $AX = \beta$ has the unique solution $X = A^{-1}\beta$. Therefore, with the formula for A^{-1} displayed in (5.9), we can prove that

$$x_k = \frac{\det \hat{A}_k}{\det A}$$

by showing that the kth component of the vector $(\operatorname{adj} A)\beta$ is $\det \hat{A}_k$. Because α_{jk} is the (k, j) entry of $\operatorname{adj} A$, the component in question must be

$$\sum_{j=1}^{n} \alpha_{jk} b_j = b_1 \alpha_{1k} + \ldots + b_n \alpha_{nk} \tag{5.10}$$

Finally, expanding by the entries and cofactors of the kth column of $\hat{A}_k$, we find that the expression on the right of (5.10) is det $\hat{A}_k$, and the proof is complete. ∎

EXAMPLE 7. The 3×3 linear system

$$3x_1 - 7x_2 + 2x_3 = -1$$

$$4x_1 + x_2 = 12$$

$$2x_1 - 4x_2 + x_3 = 1$$

can be written in the form $AX = \beta$, where

$$A = \begin{bmatrix} 3 & -7 & 2 \\ 4 & 1 & 0 \\ 2 & -4 & 1 \end{bmatrix}, \quad X = \begin{bmatrix} x_1 \\ x_2 \\ x_3 \end{bmatrix}, \quad \text{and} \quad \beta = \begin{bmatrix} -1 \\ 12 \\ 1 \end{bmatrix}$$

To solve the system by Cramer's rule, we first form the matrices

$$\hat{A}_1 = \begin{bmatrix} \overset{\beta}{-1} & -7 & 2 \\ 12 & 1 & 0 \\ 1 & -4 & 1 \end{bmatrix}, \quad \hat{A}_2 = \begin{bmatrix} 3 & \overset{\beta}{-1} & 2 \\ 4 & 12 & 0 \\ 2 & 1 & 1 \end{bmatrix}, \quad \hat{A}_3 = \begin{bmatrix} 3 & -7 & \overset{\beta}{-1} \\ 4 & 1 & 12 \\ 2 & -4 & 1 \end{bmatrix}$$

and then compute the unique solution $X = (x_1, x_2, x_3)$ as follows:

$$x_1 = \frac{\det \hat{A}_1}{\det A} = \frac{-15}{-5} = 3$$

$$x_2 = \frac{\det \hat{A}_2}{\det A} = \frac{0}{-5} = 0$$

$$x_3 = \frac{\det \hat{A}_3}{\det A} = \frac{25}{-5} = -5$$

Cramer's rule is a useful theoretical result, but it is ridiculously inadequate for practical computations. For instance, in order to solve a system of 25 equations in 25 unknowns, a very fast digital computer would take just a few seconds using the reduction algorithm but several thousand years using Cramer's rule. A researcher would have to be quite patient to put up with such a delay, and besides, who could afford to pay for the electricity?

Exercise Set 5.3

1. Let

$$A = \begin{bmatrix} -7 & 3 & 2 & -1 \\ 0 & 5 & 8 & 3 \\ 2 & 1 & 1 & 4 \\ 1 & 0 & 2 & -8 \end{bmatrix}$$

(i) What is the minor associated with the $(3, 4)$ position in A?

(ii) Evaluate the cofactor associated with the $(2, 3)$ position in A.

(iii) Find a_{42}.

2. In each case, evaluate the determinant of the given matrix by expanding with respect to the entries and cofactors of (1) the second row and (2) the third column.

(i) $A = \begin{bmatrix} -1 & 2 & 7 \\ 3 & 1 & 1 \\ 4 & -2 & 9 \end{bmatrix}$ (ii) $B = \begin{bmatrix} 2 & 1 & -9 \\ 5 & 3 & 0 \\ -7 & -1 & 2 \end{bmatrix}$

(iii) $C = \begin{bmatrix} -3 & 2 & 1 & 0 \\ 5 & 3 & -2 & 1 \\ 0 & 1 & 0 & 1 \\ 1 & 8 & 0 & 2 \end{bmatrix}$

3. Compute the adjugate of each of the following matrices.

(i) $\begin{bmatrix} 7 & 4 \\ -2 & 3 \end{bmatrix}$ (ii) $\begin{bmatrix} -3 & 5 & 1 \\ 2 & 0 & 3 \\ 1 & 1 & 3 \end{bmatrix}$ (iii) $\begin{bmatrix} 7 & -3 & 4 \\ 8 & 1 & 1 \\ 1 & -1 & 1 \end{bmatrix}$

(iv) $\begin{bmatrix} 4 & 2 & -5 \\ 1 & 3 & 1 \\ 2 & -4 & -7 \end{bmatrix}$ (v) $\begin{bmatrix} -3 & -9 & 6 \\ 1 & 3 & -2 \\ 2 & 6 & -4 \end{bmatrix}$

4. Use the formula established in corollary 5.5 to find the inverse of each of the following non-singular matrices.

(i) $\begin{bmatrix} 7 & 4 \\ -2 & 3 \end{bmatrix}$ (ii) $\begin{bmatrix} -3 & 5 & 1 \\ 2 & 0 & 3 \\ 1 & 1 & 3 \end{bmatrix}$ (iii) $\begin{bmatrix} 7 & -3 & 4 \\ 8 & 1 & 1 \\ 1 & -1 & 1 \end{bmatrix}$

(iv) $\begin{bmatrix} 0 & 3 & 0 \\ 0 & 0 & -7 \\ 5 & 0 & 0 \end{bmatrix}$ *(v) $\begin{bmatrix} 0 & 3 & -1 & 0 \\ 0 & 0 & 5 & 13 \\ 2 & 0 & 0 & 1 \\ 7 & -4 & 0 & 0 \end{bmatrix}$

5. Solve each of the following linear systems by Cramer's rule.

(i) $7x_1 + 4x_2 = -10$

     $-2x_1 + 3x_2 = 7$

(ii) $-3x_1 + 5x_2 + x_3 = 3$

     $2x_1 \qquad + 3x_3 = 6$

     $x_1 + x_2 + 3x_3 = 7$

(iii) $7x_1 - 3x_2 + 4x_3 = 0$

     $8x_1 + x_2 + x_3 = 15$

     $x_1 - x_2 + x_3 = -2$

(iv) $3x_2 - x_3 \qquad = 0$

     $5x_3 + 13x_4 = 2$

     $2x_1 \qquad + x_4 = 3$

     $7x_1 - 4x_2 \qquad = 10$

6. If A is an $n \times n$ matrix, show that
   (i) adj $(kA) = k^{n-1}(\text{adj } A)$ for each scalar k
   (ii) adj $A^t = (\text{adj } A)^t$

7. If A is a non-singular $n \times n$ matrix, show that adj A is also non-singular and that

$$(\text{adj } A)^{-1} = \left(\frac{1}{\det A}\right) A = \text{adj } A^{-1}$$

8. If A is singular, show that adj A is also singular.
   Hint: Note that $A(\text{adj } A) = 0$, and explain why this means adj A cannot have an inverse.

9. The adjugate of a certain matrix A is

$$\text{adj } A = \begin{bmatrix} 3 & 2 & 5 \\ 1 & 1 & 8 \\ 4 & 0 & 3 \end{bmatrix}$$

Is A singular or non-singular? Explain your reasoning.

10. If A is an $n \times n$ matrix, show that

$$\det (\text{adj } A) = (\det A)^{n-1}$$

11. Show that the matrix

$$B = \begin{bmatrix} 3 & -2 & 1 \\ 4 & 1 & 1 \\ 0 & 1 & -1 \end{bmatrix}$$

cannot be the adjugate of another real matrix.
   Hint: Use the result of exercise 10.

12. The inverse of a certain 4×4 matrix A is the matrix

$$A^{-1} = \begin{bmatrix} -2 & 1 & 3 & 1 \\ 4 & 1 & 0 & 2 \\ 1 & -3 & -2 & 5 \\ 12 & 0 & -7 & 0 \end{bmatrix}$$

Use this information to find det A and then compute adj A.

13. Let A be a non-singular $n \times n$ matrix whose entries are all integers.

   (i) Show that adj A is also a matrix of integers.

   (ii) Show that A^{-1} is a matrix of integers if and only if det A is either 1 or -1.

   (iii) Find non-zero integers x, y, z so that the inverse of the matrix

$$A = \begin{bmatrix} 3 & -1 & 2 \\ 1 & 1 & 5 \\ x & y & z \end{bmatrix}$$

will contain only integers.

SUPPLEMENTARY EXERCISES

1.   (i) Show that the equation of the line through the distinct points $P_1(a_1, b_1)$ and $P_2(a_2, b_2)$ is given by

$$\det \begin{bmatrix} x & y & 1 \\ a_1 & b_1 & 1 \\ a_2 & b_2 & 1 \end{bmatrix} = 0$$

   (ii) Derive a similar determinantal equation for the plane which contains the non-collinear points $P_1(a_1, b_1, c_1)$, $P_2(a_2, b_2, c_2)$, and $P_3(a_3, b_3, c_3)$.

2. Show that the area of the triangle whose vertices are the non-collinear points $P_1(a_1, b_1)$, $P_2(a_2, b_2)$, and $P_3(a_3, b_3)$ is given by $\frac{1}{2}|D|$, where

$$D = \det \begin{bmatrix} a_1 & b_1 & 1 \\ a_2 & b_2 & 1 \\ a_3 & b_3 & 1 \end{bmatrix}$$

3. Let α, β be vectors in 3-space, and let $(\,|\,)$ be the standard inner product on R_3. Show that the area of the triangle determined by the vectors α and β is given by $\frac{1}{2}\sqrt{G_2}$, where

$$G_2 = \det \begin{bmatrix} (\alpha\,|\,\alpha) & (\alpha\,|\,\beta) \\ (\beta\,|\,\alpha) & (\beta\,\beta) \end{bmatrix}$$

(See exercise 14, section 3.1 and supplementary exercise 7, in chapter 3.)

Definition.   The *determinantal rank* of an $m \times n$ matrix $A = (a_{ij})$ is the dimension of the largest non-singular square submatrix of A.

---

4. Verify that the matrix

$$A = \begin{bmatrix} -2 & 1 & 3 & 1 & -3 \\ 5 & -3 & 1 & 4 & 2 \\ 3 & -2 & 4 & 5 & -1 \\ 1 & -1 & 7 & 6 & -4 \end{bmatrix}$$

has determinantal rank 2 by showing that its 3×3 and 4×4 submatrices are all singular but it has at least one non-singular 2×2 submatrix.

Hint: If $\rho_1, \rho_2, \rho_3, \rho_4$ are the row vectors of A, note that $\rho_3 = \rho_1 + \rho_2$ and $\rho_4 = 2\rho_1 + \rho_2$.

*5. Show that the rank of a matrix and its determinantal rank are the same.

Hint: First show that row equivalent matrices have the same determinantal rank.

6. Let A be an $n \times n$ matrix with rank r.

(i) If $r = n$, show that adj A has rank n.

(ii) If $r \leq n - 2$, show that adj $A = O$.

Hint: Use the result of exercise 5, above.

*(iii) If $r = n - 1$, show that adj A has rank 1.

Hint: Since A adj $A = O$ (why?), the column vectors of adj A all belong to the solution space of the homogeneous system $AX = O$.

7. If A is an $n \times n$ matrix, show that

$$\text{adj (adj } A) = (\det A)^{n-2} A$$

8. If A and B are non-singular $n \times n$ matrices, show that

$$\text{adj } (AB) = (\text{adj } B)(\text{adj } A)$$

Is this formula necessarily valid if either A or B is singular?

9. An $n \times n$ array $H_n = (h_{ij})$ is said to be a *Jacobi matrix* if $h_{ij} = 0$ whenever $|i - j| \geq 2$. Suppose H_n also has the property that for each index i, $h_{ii} = a$, $h_{i, i+1} = b$, and $h_{i, i-1} = c$. For instance,

$$H_4 = \begin{bmatrix} a & b & 0 & 0 \\ c & a & b & 0 \\ 0 & c & a & b \\ 0 & 0 & c & a \end{bmatrix}$$

(i) Show that $\det H_n = a \det H_{n-1} - bc \det H_{n-2}$ for $n = 3, 4, \ldots$.

(ii) Find $\det H_6$.

*10. Let L be a non-singular linear transformation of the plane onto itself, and suppose that

$$L(x, y) = (c_{11}x + c_{12}y, c_{21}x + c_{22}y)$$

Let $P_1(a_1, b_1)$, $P_2(a_2, b_2)$, and $P_3(a_3, b_3)$ be points in the plane which are not collinear.

(i) Find the coordinates of the points Q_1, Q_2, Q_3 which satisfy $L(P_j) = Q_j$ for $j = 1, 2, 3$. Explain why these points cannot be collinear.

(ii) Show that the area of the triangle whose vertices are Q_1, Q_2, Q_3 is $|m|$ times that of triangle $P_1 P_2 P_3$, where

$$m = \det \begin{bmatrix} c_{11} & c_{12} \\ c_{21} & c_{22} \end{bmatrix}$$

The number $|m|$ is called the *mapping factor* of L.
Hint: Use the result of exercise 2, above.

(iii) Show that L preserves the ratio of triangular areas. In other words, if Δ_1 and Δ_2 are triangles, then the ratio of their areas is the same as the ratio of the areas of the image triangles $L(\Delta_1)$ and $L(\Delta_2)$.

11. Let $P_1(0, 0)$, $P_2(7, 0)$, $P_3(3, 5)$ be points in the plane, and let T_1, T_2, T_3 and S_1, S_2, S_3 be points located as indicated in the following figure:

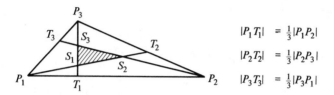

$$|P_1 T_1| = \tfrac{1}{3}|P_1 P_2|$$
$$|P_2 T_2| = \tfrac{1}{3}|P_2 P_3|$$
$$|P_3 T_3| = \tfrac{1}{3}|P_3 P_1|$$

(i) Describe a non-singular linear transformation L such that $L(P_j) = Q_j$ for $j = 1, 2, 3$, where $Q_1(0, 0)$, $Q_2(1, 0)$, and $Q_3(\tfrac{1}{2}, \sqrt{3}/2)$ are the vertices of an equilateral triangle. Is there only one such transformation?

(ii) Verify that $L(T_1)$, the image of T_1 under L, is located $\tfrac{1}{3}$ the distance from Q_1 to Q_2. Show that similar statements can be made for $L(T_2)$ and $L(T_3)$.

(iii) Use the results of exercise 10 above and supplementary exercise 2 of chapter 2 to prove that the area of triangle $S_1 S_2 S_3$ is $\tfrac{1}{7}$ that of triangle $P_1 P_2 P_3$.

12. Let

$$F = \begin{bmatrix} f_{11} & f_{12} \\ f_{21} & f_{22} \end{bmatrix}$$

be the 2×2 matrix whose (i, j) entry is the differentiable function $f_{ij}(t)$, and let $D = \det F$. Show that

$$\frac{dD}{dt} = \det \begin{bmatrix} f'_{11} & f_{12} \\ f'_{21} & f_{22} \end{bmatrix} + \det \begin{bmatrix} f_{11} & f'_{12} \\ f_{21} & f'_{22} \end{bmatrix}$$

where, as usual, f'_{ij} denotes the derivative of f_{ij}. State and prove a generalization of this result which applies to any $n \times n$ variable matrix F.

13. Let $T_n = (t_{ij})$ denote the $n \times n$ matrix such that $t_{ii} = a$ for each index i and $t_{ij} = b$ for all $j \neq i$. For example,

$$T_4 = \begin{bmatrix} a & b & b & b \\ b & a & b & b \\ b & b & a & b \\ b & b & b & a \end{bmatrix}$$

Show that $\det T_n = (a - b)^{n-1}(a + (n - 1)b)$ for $n = 2, 3, \ldots$

CHAPTER

6

Eigenvalues, Eigenvectors, and Diagonability

6.1 EIGENVALUES AND EIGENVECTORS

When analyzing a given linear transformation $L: V \to V$, it is desirable to know as much as possible about the relationship between each vector v and its image $L(v)$. For most vectors, this relationship is too complicated to be of use, but if $L(v)$ happens to be a scalar multiple of v, then L may be regarded as a simple elongation or contraction in the direction determined by v. In this chapter, we shall first develop a method for determining which vectors (if any) have this property, and then we shall show that L has a diagonal matrix representative whenever V has a basis comprised of such vectors. We begin by providing names for certain features of the relationship we plan to study.

Definition 6.1. Let $L: V \to V$ be a linear transformation on the vector space V, and let λ be a scalar. If v is a non-zero vector in V such that $L(v) = \lambda v$, then v is said to be an *eigenvector* of L associated with the *eigenvalue* λ.

The mathematical literature contains many other names for the two terms defined above. In fact, the word "eigenvalue" is an unhappy marriage of two

284

other names—namely, the German "eigenwert" and the English "character-istic value." There have been many attempts to replace "eigenvalue" by terms such as "latent root" and "autovalue," but to no avail, for even though "eigenvalue" offends the esthetic sensibility of practically everyone, it does have a catchy sound and is easy to remember.

EXAMPLE 1.   Let $L: R_3 \to R_3$ be the linear transformation defined by

$$L[(a_1, a_2, a_3)] = (2a_1 - a_2 + 3a_3, a_2, 4a_2 + 3a_3)$$

Then if $\alpha = (3, 0, 1)$, we have

$$L(\alpha) = (9, 0, 3) = 3(3, 0, 1) = 3\alpha$$

and it follows that α is an eigenvector of L associated with the eigenvalue 3.

EXAMPLE 2.   Let L_m denote the reflection of the plane in the line $y = mx$. Since each vector γ is mapped by L_m into its mirror image with respect to the reflecting line (see figure 6.1), it follows that each vector α which lies in this line must be mapped into itself. In other words, $L_m(\alpha) = \alpha$ and α is an eigen-vector of L_m associated with the eigenvalue 1.

On the other hand, if β lies in the line $x = -my$, which is normal to $y = mx$, then the reflection process guarantees that $L_m(\beta) = -\beta$, and β is an eigenvector associated with the eigenvalue -1. If γ is a vector which lies on neither the reflecting line $y = mx$ nor the normal line $x = -my$, then it is clear that $L_m(\gamma)$ is not a multiple of γ. These statements are illustrated in figure 6.1.

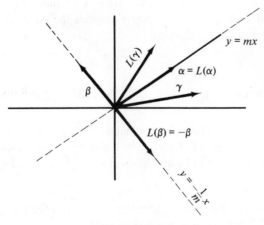

FIGURE 6.1

EXAMPLE 3.   Let L_ϕ denote the rotation of the plane through an angle of ϕ radians about the origin. If $\phi = 0$, every vector is mapped into itself, and if $\phi = \pi$, each vector α is mapped into $-\alpha$. However, if $0 < \phi < \pi$, only the zero vector will be mapped by L_ϕ into a multiple of itself, and in this case, L_ϕ has no eigenvalues.

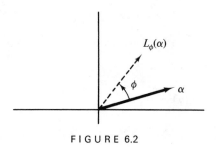

FIGURE 6.2

EXAMPLE 4.   Let D denote the differentiation operator [that is, $D(f) = f'$]. Then the function f is an eigenvector of D associated with the eigenvalue λ if f satisfies the differential equation $D(f) = f' = \lambda f$. Solving, we find that for each fixed real number λ, the corresponding eigenvectors are functions of the form $f(x) = Ce^{\lambda x}$. Note that D is an example of an operator with infinitely many eigenvalues.

Incidentally, in situations where V is a function space, the eigenvectors of a given linear transformation $L: V \longrightarrow V$ are often referred to as *eigenfunctions*.

Next, suppose that V is a finite-dimensional vector space, and let A be the matrix which represents the linear transformation $L: V \longrightarrow V$ with respect to the basis S for V. According to theorem 4.5, we have $[L(v)]_S = A[v]_S$ for each vector v in V, and it follows that v satisfies the equation $L(v) = \lambda v$ if and only if the n-tuple $[v]_S$ satisfies

$$A[v]_S = [L(v)]_S = \lambda [v]_S$$

This observation encourages us to make the following definition.

Definition 6.2.   Let A be an $n \times n$ matrix. If λ is a number from the same scalar field as the entries of A and X is a non-zero scalar n-tuple such that $AX = \lambda X$, then X is said to be an *eigenvector* of A associated with the *eigenvalue* λ.

With this terminology, the conclusions reached in the paragraph preceding definition 6.2 may be summarized as follows.

Theorem 6.1. Let A be the matrix which represents the linear transformation $L: V \rightarrow V$ with respect to the basis $S = \{\alpha_1, \ldots, \alpha_n\}$. Then λ is an eigenvalue of L associated with the eigenvector v if and only if λ is an eigenvalue of A associated with the eigenvector $[v]_S$.

According to theorem 6.1, the eigenvalues and eigenvectors of L may be determined by solving the vector equation $AX = \lambda X$, where A is a particular matrix representative of L. Rewriting this equation in the form $(\lambda I_n - A)X = O$, we find that it has a non-zero solution whenever $\lambda I_n - A$ is a singular matrix. Combining this observation with the fact that a matrix is singular if and only if its determinant is zero, we have the following important result.

Theorem 6.2. If λ is a number in the same scalar field as the entries of the $n \times n$ matrix A, then it is an eigenvalue of A if and only if it satisfies the determinantal equation $\det (\lambda I_n - A) = 0$.

EXAMPLE 5. For the 3×3 matrix

$$A = \begin{bmatrix} -15 & 16 & 32 \\ -4 & 5 & 8 \\ -4 & 4 & 9 \end{bmatrix}$$

we find that

$$\det (\lambda I_3 - A) = \det \begin{bmatrix} \lambda + 15 & -16 & -32 \\ 4 & \lambda - 5 & -8 \\ 4 & -4 & \lambda - 9 \end{bmatrix}$$

$$= \lambda^3 + \lambda^2 - 5\lambda + 3 = (\lambda - 1)^2(\lambda + 3)$$

According to theorem 6.2, the eigenvalues of A are 1 and -3, since these numbers are the solutions of the determinantal equation $\det (\lambda I_3 - A) = 0$.

If A is an $n \times n$ matrix, it is easy to see that $\det (xI_n - A)$ is a polynomial of the form $x^n + c_{n-1}x^{n-1} + \ldots + c_1 x + c_0$. Because of theorem 6.2, this polynomial plays such an important role in our work it is convenient to make the following definition.

Definition 6.3. If A is an $n \times n$ matrix, the polynomial $p_A(x) = \det (xI_n - A)$ is called the *characteristic polynomial* of A, and the equation $p_A(x) = 0$ is referred to as the *characteristic equation* of A.

Using the terminology introduced in this definition, theorem 6.2 may be restated as follows.

Theorem 6.2. Let A be an $n \times n$ matrix, and let λ be a number in the same scalar field as the entries of A. Then λ is an eigenvalue of A if and only if it is a zero of the characteristic polynomial $p_A(x)$.

It is important to remember that the zeroes of a polynomial with real coefficients are not necessarily real numbers. For instance, the characteristic polynomial of the matrix

$$A = \begin{bmatrix} -7 & 0 & 0 \\ 3 & 6 & -4 \\ -5 & 4 & 0 \end{bmatrix}$$

is $p_A(x) = (x + 7)(x^2 - 6x + 16)$, and we find that the zeroes of this polynomial are -7, $3 + \sqrt{7}\,i$, and $3 - \sqrt{7}\,i$. Thus, -7 is the only *real* eigenvalue of A, but if A is regarded as a complex matrix, then it has eigenvalues -7, $3 + \sqrt{7}\,i$, and $3 - \sqrt{7}\,i$. In other words, it is not possible to find a *real* 3-tuple X such that $AX = (3 + \sqrt{7}\,i)X$, but this equation can be satisfied if we allow X to have *complex* entries; for example,

$$\begin{bmatrix} -7 & 0 & 0 \\ 3 & 6 & -4 \\ -5 & 4 & 0 \end{bmatrix} \begin{bmatrix} 0 \\ 3 + \sqrt{7}\,i \\ 4 \end{bmatrix} = (3 + \sqrt{7}\,i) \begin{bmatrix} 0 \\ 3 + \sqrt{7}\,i \\ 4 \end{bmatrix}$$

In theorem 6.1, we observed that the eigenvalues of the linear transformation L are the same as those of any matrix representative of L. Since the matrices A and B both represent the same linear transformation if and only if they are similar, it is not surprising to find that similar matrices have the same eigenvalues. In fact, our next result shows that they have the same characteristic polynomial.

Theorem 6.3. If the $n \times n$ matrices A and B are similar, they have the same characteristic polynomial and hence, the same eigenvalues.

Proof: If $B = P^{-1}AP$, we have

$$xI_n - B = xI_n - P^{-1}AP = x(P^{-1}I_nP) - (P^{-1}AP) = P^{-1}[xI_n - A]P$$

and it follows that

$$p_B(x) = \det(xI_n - B) = \det[P^{-1}(xI_n - A)P]$$
$$= (\det P^{-1})[\det(xI_n - A)](\det P) = \det(xI_n - A) = p_A(x)$$

as claimed. The statement regarding the eigenvalues of A and B is a consequence of theorem 6.2. ∎

Unfortunately, the converse of this theorem is not generally true. For example, the matrices

$$\begin{bmatrix} 0 & 1 \\ 0 & 0 \end{bmatrix} \text{ and } \begin{bmatrix} 0 & 0 \\ 0 & 0 \end{bmatrix}$$

both have the characteristic polynomial x^2, but they are not similar (why not?).

According to the so-called Fundamental Theorem of Algebra, a polynomial of the form $x^n + c_{n-1}x^{n-1} + \ldots + c_1 x + c_0$, where the coefficients c_j are complex numbers, can be expressed as the product of n factors of the form $x - r$, where r is a zero of the polynomial. Therefore, since the eigenvalues of a complex $n \times n$ matrix C are the zeroes of its characteristic polynomial, we can write

$$p_C(x) = (x - \lambda_1)^{k_1}(x - \lambda_2)^{k_2} \ldots (x - \lambda_j)^{k_j} \qquad (6.1)$$

where $\lambda_1, \lambda_2, \ldots, \lambda_j$ are the distinct eigenvalues of C and $k_1 + k_2 + \ldots + k_j = n$. For each index i, the integer k_i is known as the *algebraic multiplicity* of λ_i. For example, the characteristic polynomial of the matrix

$$C = \begin{bmatrix} 2+i & 7i & 3-2i & 5i \\ 0 & 8 & 0 & 0 \\ 0 & 2-9i & 0 & 1 \\ 0 & 2+3i & -5 & 4 \end{bmatrix}$$

is $p_C(x) = [x - (2 + i)]^2[x - (2 - i)](x - 8)$. Thus, C has three distinct eigenvalues, and we see that $\lambda_1 = 2 + i$ has multiplicity 2, while $\lambda_2 = 2 - i$ and $\lambda_3 = 8$ both have multiplicity 1.

From the decomposition displayed in (6.1), it is obvious that an $n \times n$ complex (or real) matrix can have at most n distinct eigenvalues. There are several additional results which may be derived from (6.1). For instance, since $p_C(x) = \det(xI_n - C)$, we have

$$\det(xI_n - C) = (x - \lambda_1)^{k_1}(x - \lambda_2)^{k_2} \ldots (x - \lambda_j)^{k_j}$$

In particular, when $x = 0$, the equation becomes

$$\det(0I_n - C) = \det(-C) = (-1)^n \det C$$
$$= (0 - \lambda_1)^{k_1}(0 - \lambda_2)^{k_2} \ldots (0 - \lambda_j)^{k_j}$$

and since $k_1 + k_2 + \ldots + k_j = n$, it follows that

$$(-1)^n \det C = (-1)^n(\lambda_1)^{k_1}(\lambda_2)^{k_2} \ldots (\lambda_j)^{k_j}$$

Thus, we have the following useful theorem.

Theorem 6.4.   The determinant of an $n \times n$ matrix equals the product of the n zeroes of its characteristic polynomial.

EXAMPLE 6.   In example 5, we observed that the characteristic polynomial of the matrix

$$A = \begin{bmatrix} -15 & 16 & 32 \\ -4 & 5 & 8 \\ -4 & 4 & 9 \end{bmatrix}$$

is $p_A(x) = (x - 1)^2(x + 3)$. Therefore, the zeroes of $p_A(x)$ are 1, 1, and -3, and theorem 6.4 tells us that

$$\det A = (1)(1)(-3) = -3$$

In chapter 5, we showed that an $n \times n$ matrix is singular if and only if its determinant is zero, and this result can now be combined with theorem 6.4 to yield the following criterion for singularity.

Corollary 6.1.   An $n \times n$ matrix is singular if and only if at least one of its eigenvalues is zero.

Proof: Since $\det A = \lambda_1^{k_1} \lambda_2^{k_2} \ldots \lambda_j^{k_j}$, it follows that $\det A = 0$ if and only if one of the λ's is zero. ∎

This section has been devoted to developing the basic ideas and terminology associated with the study of eigenvalues and eigenvectors. In the next section, we continue this study by investigating the collection of all eigenvectors associated with a given eigenvalue.

Exercise Set 6.1

1. Find the characteristic polynomial and the eigenvalues of each of the following matrices.

(i) $\begin{bmatrix} 1 & 1 \\ -3 & 5 \end{bmatrix}$ (ii) $\begin{bmatrix} 2 & 1 \\ -16 & 10 \end{bmatrix}$ (iii) $\begin{bmatrix} -2 & 1 \\ 7 & 4 \end{bmatrix}$

(iv) $\begin{bmatrix} 3 & 1 & 0 \\ 0 & 3 & 1 \\ 3 & -7 & 8 \end{bmatrix}$ (v) $\begin{bmatrix} 1 & 1 & 0 \\ 0 & 1 & 1 \\ 5 & 1 & -4 \end{bmatrix}$

Hint: One eigenvalue is 4.

2. Let $L: R_3 \longrightarrow R_3$ denote the linear transformation defined by

$$L((a_1, a_2, a_3)) = (2a_1 + a_2, 2a_2 + a_3, 6a_1 - a_2 - 2a_3).$$

   (i) Find the matrix A which represents L with respect to the standard basis for R_3.

   (ii) Compute the eigenvalues of A.

3. Find the eigenvalues of the linear transformation $L: P_2 \longrightarrow P_2$ defined by

$$L[at^2 + bt + c] = (2a + 5b - 2c)t^2 + (a - c)t + (4a + 6b - 4c)$$

4. Let V denote the vector space for which the ordered set $S = \{e^{-t}, te^{-t}, \sin t, \cos t\}$ is a basis, and let $L: V \longrightarrow V$ denote the linear transformation defined by $L(f) = f'' + f$.

   (i) Find the matrix which represents L with respect to the S-basis.

   (ii) Find the eigenvalues of L.

5. Let A be an $n \times n$ matrix, and suppose that $AX = \lambda X$.

   (i) For each positive integer k, show that $A^k X = \lambda^k X$.

   (ii) If s is a scalar, show that $(sA)X = (s\lambda)X$.

   (iii) If $B = a_k A^k + a_{k-1} A^{k-1} + \ldots + a_1 A + a_0 I_n$, show that

$$BX = (a_k \lambda^k + a_{k-1} \lambda^{k-1} + \ldots + a_1 \lambda + a_0)X.$$

   (iv) If A is non-singular, show that $A^{-1} X = (1/\lambda)X$.

6. (i) If A is an $n \times n$ matrix, show that $\det (xI_n - A) = \det (xI_n - A^t)$

   (ii) Show that A and A^t have the same eigenvalues.

   (iii) If $AX = \lambda X$, can we claim that $A^t X = \lambda X$? Justify your answer with a proof or a counterexample.

7. Let A be an $n \times n$ matrix which satisfies the equation $A^2 - 3A + 2I_n = 0$.

   (i) If $AX = \lambda X$, show that $(\lambda^2 - 3\lambda + 2)X = 0$.

   Hint: See exercise 6, above.

   (ii) If λ is an eigenvalue of A, show that λ must be either 1 or 2.

   (iii) Show that A must be non-singular.

8. Let A and B be $n \times n$ matrices and assume that A is non-singular.

   (i) Show that AB is similar to BA.

   (ii) Show that AB and BA have the same eigenvalues.

9. If V is a finite-dimensional vector space, show that the linear transformation $L: V \longrightarrow V$ is singular if and only if it has a zero eigenvalue.

10. Let L be an orthogonal linear transformation on the real inner product space V. If λ is an eigenvalue of L, show that $|\lambda| = 1$.

   Hint: Recall that $\|L(v)\| = \|v\|$ for each vector v in V.

11. If λ is an eigenvalue of the $n \times n$ real orthogonal matrix P, show that $|\lambda| = 1$.

12. What are the eigenvalues and eigenvectors of the diagonal matrix $D = \operatorname{diag} (-3, 7, 5, 0)$? State and prove a general result regarding the eigenvalues and eigenvectors of the $n \times n$ diagonal matrix $D = \operatorname{diag}(d_1, \ldots, d_n)$.

13. Recall that an $n \times n$ matrix N is *nilpotent* of index k if $N^k = 0$ but N^{k-1} has at least one non-zero entry. If λ is an eigenvalue of such a matrix, show that λ must be 0.

14. The $n \times n$ matrix A is said to be idempotent if $A^2 = A$. If λ is an eigenvalue of such a matrix, show that λ is either 0 or 1. What can be said about a non-singular idempotent matrix?

15. An $n \times n$ *Markov* (or probability) *matrix* $A = (a_{ij})$ has the property that $0 \le a_{ij} \le 1$ for all i, j and

$$\sum_{j=1}^{n} a_{ij} = 1 \quad \text{for } i = 1, 2, \ldots, n$$

Show that $\lambda = 1$ is an eigenvalue of each such matrix. (See section 1.8.)

16. Let

$$A = \begin{bmatrix} 0 & 1 & 0 \\ 0 & 0 & 1 \\ 2 & 0 & -1 \end{bmatrix}$$

 (i) Compute $p_A(x)$, the characteristic polynomial of A.
 (ii) Verify that $\lambda_1 = 1$ is a zero of $p_A(x)$, and find the other two zeroes, λ_2 and λ_3, of this polynomial.
 (iii) Does there exist a real column vector X such that $AX = \lambda_2 X$? Explain.
 (iv) Does there exist a complex column vector Z such that $AZ = \lambda_2 Z$?

17. Let A be an $n \times n$ real matrix. If n is odd, show that A must have at least one real eigenvalue.

18. Let A, B be $n \times n$ matrices, and suppose that $AX = \lambda X$ and $BX = \mu X$.
 (i) Show that $(A + B)X = (\lambda + \mu)X$.
 (ii) Show that $(AB)X = (\lambda \mu)X$.

19. If $\lambda_1, \lambda_2, \ldots, \lambda_n$ are the eigenvalues of the $n \times n$ matrix A and c is a scalar, show that the matrix $A + cI_n$ has eigenvalues $\lambda_1 + c, \lambda_2 + c, \ldots, \lambda_n + c$.

20. Let $p_A(x)$ be the characteristic polynomial of the $n \times n$ matrix A. Show that $(-1)^n \det A = p_A(0)$.

6.2 EIGENSPACES

According to theorem 6.1, the eigenvalues of a given linear transformation $L: V \longrightarrow V$ can be computed by finding the zeroes of the characteristic polynomial of any matrix representative of L. Of course, this is somewhat easier said than done, for in practice, it is often difficult to carry out the algebraic steps involved with finding the characteristic polynomial and even more difficult to compute or even approximate its zeroes. Such computational questions are important, but they fall mainly in the area of numerical analysis and for the most part, shall be ignored throughout this text. Instead, we shall assume that the eigenvalues of L are known so that we can focus attention on the algebraic structure of the collection of all eigenvectors associated with a given eigenvalue.

First of all, note that whenever v_1 and v_2 satisfy the functional equation $L(v) = \lambda v$, then so does the linear expression $c_1 v_1 + c_2 v_2$, for we have

$$L(c_1 v_1 + c_2 v_2) = c_1 L(v_1) + c_2 L(v_2)$$
$$= c_1(\lambda v_1) + c_2(\lambda v_2) = \lambda(c_1 v_1 + c_2 v_2)$$

Therefore, the collection of all solutions of this equation is a subspace of V, and we are justified in using the language of the following definition.

Definition 6.4. Let $L: V \rightarrow V$ be a linear transformation, and let λ be a fixed scalar. Then the subspace of V comprised of all vectors v which satisfy $L(v) = \lambda v$ is called an *eigenspace* of L and is denoted by $\mathcal{C}_\lambda$.

If V happens to be finite-dimensional and A is a matrix representative of L, then the eigenspace $\mathcal{C}_\lambda$ is closely associated with the solution space of the linear system $AX = \lambda X$. This observation provides motivation for the following definition.

Definition 6.5. Let A be an $n \times n$ matrix, and let λ be a fixed scalar. Then the collection of all n-tuples X which satisfy $AX = \lambda X$ is called the *eigenspace* of A associated with λ and is denoted by $\mathcal{S}_\lambda$.

When λ is not an eigenvalue of A, the set $\mathcal{S}_\lambda$ contains only the zero n-tuple (why?). For this reason, we shall be primarily interested in studying eigenspaces associated with eigenvalues, and in this case, a basis for $\mathcal{S}_\lambda$ may be found by computing the fundamental solutions of the homogeneous linear system $(\lambda I_n - A)X = O$. To carry out these computations, we can use the methods of chapter 2, as illustrated in the solved problems which follow.

PROBLEM 1. First find the eigenvalues of the matrix

$$A = \begin{bmatrix} 2 & -1 & -1 \\ 1 & 0 & -1 \\ -1 & 1 & 2 \end{bmatrix}$$

and then obtain a basis for the eigenspace associated with each distinct eigenvalue.

Solution: Expanding $\det(xI_3 - A)$, we obtain

$$p_A(x) = x^3 - 4x^2 + 5x - 2 = (x - 1)^2(x - 2)$$

and thus, the distinct eigenvalues of A are $\lambda_1 = 1$ and $\lambda_2 = 2$.

To find a basis for the eigenspace $\mathcal{S}_{\lambda_1}$, we form the matrix $I_3 - A$, perform the reduction

$$\begin{bmatrix} -1 & 1 & 1 \\ -1 & 1 & 1 \\ 1 & -1 & -1 \end{bmatrix} \longrightarrow \begin{bmatrix} 1 & -1 & -1 \\ 0 & 0 & 0 \\ 0 & 0 & 0 \end{bmatrix}$$

$$(I_3 - A)$$

and observe that the vector $X = (x_i)$ satisfies the system $(I_3 - A)X = 0$ if and only if its components satisfy the equation $x_1 - x_2 - x_3 = 0$. Since the fundamental solutions of this system are

$$\eta_1 = \begin{bmatrix} 1 \\ 1 \\ 0 \end{bmatrix} \quad \text{and} \quad \eta_2 = \begin{bmatrix} 1 \\ 0 \\ 1 \end{bmatrix} \tag{6.2}$$

we conclude that $\{\eta_1, \eta_2\}$ is a basis for $\mathcal{S}_{\lambda_1}$.

Similarly, from the reduction

$$\begin{bmatrix} 0 & 1 & 1 \\ -1 & 2 & 1 \\ 1 & -1 & 0 \end{bmatrix} \longrightarrow \begin{bmatrix} 1 & 0 & 1 \\ 0 & 1 & 1 \\ 0 & 0 & 0 \end{bmatrix}$$
$$(2I_3 - A)$$

it follows that the vector

$$\eta_3 = \begin{bmatrix} -1 \\ -1 \\ 1 \end{bmatrix} \tag{6.3}$$

is the only fundamental solution of the system $(2I_3 - A)X = O$. Therefore, $\{\eta_3\}$ is a basis for $\mathcal{S}_{\lambda_2}$.

The next solved problem shows how the eigenvalues and eigenvectors of a given linear transformation may be found once we know the corresponding features of one of its matrix representatives.

PROBLEM 2. Find the eigenvalues and eigenvectors of the linear transformation $L: P_2 \to P_2$ defined by

$$L(at^2 + bt + c) = (2a - b - c)t^2 + (a - c)t + (-a + b + 2c)$$

Solution: When the basis $S = \{t^2, t, 1\}$ is used in P_2, we find that L is represented by the matrix

$$A = \begin{bmatrix} 2 & -1 & -1 \\ 1 & 0 & -1 \\ -1 & 1 & 2 \end{bmatrix}$$

which was analyzed in problem 1. Therefore, the distinct eigenvalues of L are $\lambda_1 = 1$ and $\lambda_2 = 2$, and bases for the eigenspaces $\mathcal{C}_{\lambda_1}$ and $\mathcal{C}_{\lambda_2}$ may be constructed by finding polynomials p_1, p_2, p_3 which satisfy $[p_j]_S = \eta_j$ for $j =$

1, 2, 3, where η_1, η_2, η_3 are the 3-tuples displayed in (6.2) and (6.3). Specifically we have

$$p_1(t) = (1)t^2 + (1)t + (0)1 = t^2 + t$$
$$p_2(t) = (1)t^2 + (0)t + (1)1 = t^2 + 1$$
$$p_3(t) = (-1)t^2 + (-1)t + (1)1 = -t^2 - t + 1$$

and it follows that $\{p_1, p_2\}$ and $\{p_3\}$ are bases for $\mathcal{C}_{\lambda_1}$ and $\mathcal{C}_{\lambda_2}$, respectively.

If $L(v) = \lambda v$, then L may be regarded geometrically as a contraction or elongation with scaling factor λ when viewed in the direction of v, as indicated in figure 6.3. Since it makes no sense to have two different scaling factors for

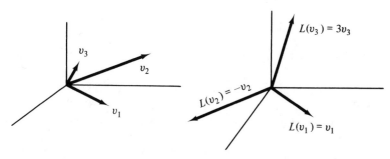

FIGURE 6.3

the same direction, we would expect to find that v cannot also be an eigenvector associated with another eigenvalue μ which is different from λ. In fact, as our next theorem shows, eigenvectors which correspond to distinct eigenvalues are not only different but linearly independent.

Theorem 6.5. Let $\lambda_1, \ldots, \lambda_k$ be distinct eigenvalues of the linear transformation $L: V \longrightarrow V$, and for each index j, let v_j be an eigenvector associated with λ_j. Then the vectors $v_1, v_2, \ldots, v_k$ are linearly independent.

Proof: We shall use mathematical induction on k. Our argument has a starting point since the theorem is clearly true when $k = 1$. For the inductive step, we assume that the vectors $v_1, \ldots, v_{k-1}$ are linearly independent and then use this fact to prove that $v_1, v_2, \ldots, v_{k-1}, v_k$ are independent. To this end, let $c_1, \ldots, c_k$ be scalars which satisfy

$$c_1 v_1 + c_2 v_2 + \ldots + c_k v_k = \theta_V \tag{6.4}$$

multiplying both sides of this equation by λ_k, we obtain

$$\theta_V = \lambda_k \theta_V = \lambda_k \left[\sum_{j=1}^{k} c_j v_j \right] = \sum_{j=1}^{k} c_j (\lambda_k v_j)$$

Moreover, since $L(v_j) = \lambda_j v_j$ for $j = 1, 2, \ldots, k$, we have

$$\theta_V = L(\theta_V) = L\left[\sum_{j=1}^{k} c_j v_j\right] = \sum_{j=1}^{k} c_j L(v_j) = \sum_{j=1}^{k} c_j(\lambda_j v_j)$$

Equating these two representations of θ_V, we find that

$$\sum_{j=1}^{k} c_j(\lambda_k v_j) = \theta_V = \sum_{j=1}^{k} c_j(\lambda_j v_j)$$

and by subtracting the sum on the right from that on the left, we obtain

$$\sum_{j=1}^{k} c_j(\lambda_k - \lambda_j)v_j = \theta_V$$

Since $\lambda_k - \lambda_k = 0$, this equation becomes

$$c_1(\lambda_k - \lambda_1)v_1 + \ldots + c_{k-1}(\lambda_k - \lambda_{k-1})v_{k-1} = \theta_V$$

and the inductive hypothesis (namely, the assumption that $v_1, \ldots, v_{k-1}$ are independent) enables us to conclude that $c_j(\lambda_k - \lambda_j) = 0$ for $j = 1, 2, \ldots,$ $k - 1$. Finally, since the λ's are distinct, we have $\lambda_k - \lambda_j \neq 0$ for each index j, and it follows that $c_1 = c_2 = \ldots = c_{k-1} = 0$. Referring back to equation (6.4), we see that c_k must also be 0, and we have shown that the vectors $v_1, \ldots, v_k$ are linearly independent, as desired. ∎

Corollary 6.2. Let $\lambda_1, \ldots, \lambda_k$ be distinct eigenvalues of the $n \times n$ matrix A, and for each index j, let X_j be an eigenvector associated with λ_j. Then the n-tuples $X_1, X_2, \ldots, X_k$ are linearly independent.

According to theorem 6.5, a given linear transformation $L: V \to V$ has at least as many linearly independent eigenvectors as it has distinct eigenvalues. In fact, it may happen that there are enough linearly independent eigenvectors of L to form a basis for V, and when this occurs, it is possible to obtain a simple description of L. To be specific, suppose that $L(v_j) = \lambda_j v_j$ for $j = 1, 2, \ldots, n$ and that $T = \{v_1, v_2, \ldots, v_n\}$ is a basis for V. Then, since we have

$$L(v_1) = \lambda_1 v_1 = \lambda_1 v_1 + 0v_2 + \ldots + 0v_n$$
$$L(v_2) = \lambda_2 v_2 = 0v_1 + \lambda_2 v_2 + \ldots + 0v_n$$
$$\vdots \qquad \vdots \qquad \vdots \qquad \qquad \vdots$$
$$L(v_n) = \lambda_n v_n = 0v_1 + 0v_2 + \ldots + \lambda_n v_n$$

it follows that L is represented by the diagonal matrix

$$D = \begin{bmatrix} \lambda_1 & & & 0 \\ & \lambda_2 & & \\ & & \ddots & \\ 0 & & & \lambda_n \end{bmatrix} = \text{diag} (\lambda_1, \ldots, \lambda_n)$$

with respect to the basis T.

We shall find that not every linear transformation can be represented by a diagonal matrix, but those which do have such a representation are so important that we provide them with a special name.

Definition 6.6   The linear transformation $L: V \longrightarrow V$ is said to be *diagonable* if it has a diagonal matrix representative.

For instance, the reflection R_l of the plane in the line $y = mx$ is diagonable. To prove this statement, let α_1 and α_2 be non-zero vectors which point along $y = mx$ and its normal, $x = -my$, respectively (see example 2, section 6.1). Then we have $R_l(\alpha_1) = \alpha_1$ and $R_l(\alpha_2) = -\alpha_2$ and it follows that R_l is represented by the diagonal matrix

$$\begin{bmatrix} 1 & 0 \\ 0 & -1 \end{bmatrix}$$

with respect to the basis $S = \{\alpha_1, \alpha_2\}$.

As a second example, recall that

$$L(v_1) = v_1$$
$$L(v_2) = v_2$$
$$L(v_3) = 2v_3$$

where $L: P_2 \longrightarrow P_2$ is the linear transformation analyzed in problem 2, and $v_1 = t^2 + t$, $v_2 = t^2 + 1$, $v_3 = -t^2 - t + 1$. Therefore, L is represented by the diagonal matrix

$$\begin{bmatrix} 1 & 0 & 0 \\ 0 & 1 & 0 \\ 0 & 0 & 2 \end{bmatrix}$$

with respect to the basis $S = \{v_1, v_2, v_3\}$.

We have just observed that $L: V \longrightarrow V$ is diagonable if V has a basis comprised entirely of eigenvectors of L. Conversely, if L is represented by $D = \text{diag}(\lambda_1, \ldots, \lambda_n)$ with respect to the basis $\{w_1, \ldots, w_n\}$, then it is easy to show that $L(w_j) = \lambda_j w_j$ for each index j (see exercise 8). Thus, we have the following result.

Theorem 6.6.   The linear transformation $L: V \rightarrow V$ is diagonable if and only if V has a basis comprised entirely of eigenvectors of L. Moreover, in the case where L is diagonable, its eigenvalues appear on the main diagonal of its diagonal matrix representative.

In chapter 4, we showed that two matrices represent the same linear transformation if and only if they are similar. Consequently, if L is diagonable, each of its matrix representatives must be similar to a diagonal matrix, and conversely, if a given matrix representative of L is similar to a diagonal matrix, then L is diagonable. These observations yield the following corollary.

Corollary 6.3.   The linear transformation $L: V \rightarrow V$ is diagonable if and only if each of its matrix representatives is similar to the diagonal matrix $D = \text{diag}(\lambda_1, \ldots, \lambda_n)$, where the λ's are the eigenvalues of L.

Because of the result contained in corollary 6.3, it is convenient to have the terminology introduced in the following definition.

Definition 6.7.   An $n \times n$ matrix is said to be *diagonable* if it is similar to a diagonal matrix.

Earlier, we observed that not every linear transformation is diagonable. To prove this assertion, it suffices to show that not all matrices are diagonable (why?). Accordingly, consider the matrix

$$A = \begin{bmatrix} 0 & 1 \\ 0 & 0 \end{bmatrix}$$

The eigenvalues of A are clearly $\lambda_1 = \lambda_2 = 0$, and if A were diagonable, we could find a non-singular matrix P such that $P^{-1}AP = \text{diag}(0, 0) = O$. Obviously, no such matrix exists (why not?), and A cannot be diagonable.

In general, it is not easy to determine whether or not a given linear transformation or matrix is diagonable. However, in our next section, we shall show that there are a few special cases where such a determination can be made without too much trouble.

Exercise Set 6.2

1. In each of the following cases, find the eigenvalues of the given matrix and then obtain a basis for the eigenspace associated with each distinct eigenvalue.

(i) $\begin{bmatrix} 6 & -3 & -6 \\ -1 & 3 & 1 \\ 4 & -3 & -4 \end{bmatrix}$ (ii) $\begin{bmatrix} -4 & 6 & 5 \\ 6 & -5 & -6 \\ -5 & 6 & 6 \end{bmatrix}$

(iii) $\begin{bmatrix} 4 & 0 & -6 \\ -3 & -2 & 3 \\ 3 & 0 & -5 \end{bmatrix}$ (iv) $\begin{bmatrix} 5 & -3 & -7 \\ -2 & 1 & 2 \\ 2 & -3 & -4 \end{bmatrix}$

2. Let $L: P_2 \longrightarrow P_2$ denote the linear transformation defined by $L[at^2 + bt + c] = (7a - 10c)t^2 + (-5a - 3b + 5c)t + (5a - 8c)$.
   (i) Find the matrix A which represents L with respect to the basis $\{t^2, t, 1\}$.
   (ii) Verify that $\lambda_1 = 2$ is a zero of the characteristic polynomial of A, and find the other two zeroes of this polynomial.
   (iii) Find a basis for the eigenspace associated with each distinct eigenvalue of A.
   (iv) Find a basis T for P_2 which is comprised entirely of eigenvectors of L, and then compute the matrix D which represents L with respect to the basis T.

3. In each of the following cases, find the eigenvalues of the given linear transformation and obtain a basis for the eigenspace associated with each distinct eigenvalue.
   (i) $L[at^2 + bt + c] = (a - b - 2c)t^2 + (b + c)t + (a - b - 2c)$
   (ii) $L[(a_1, a_2, a_3)] = (8a_1 + 3a_2 - 4a_3, -3a_1 + a_2 + 3a_3, 4a_1 + 3a_2)$
   (iii) $L[(a_1, a_2, a_3, a_4)] = (a_2, a_3, a_4, 2a_1 - 5a_2 + 3a_3 + a_4)$

4. The matrix

$$A = \begin{bmatrix} -1 & -1 & 1 & -1 \\ -2 & 0 & 1 & -1 \\ -2 & 3 & -2 & -1 \\ 4 & -10 & 10 & 3 \end{bmatrix}$$

is known to have two distinct eigenvalues, one of which is $\lambda_1 = 1$. Moreover, it is known that

$$\eta = \begin{bmatrix} 1 \\ 1 \\ 0 \\ 1 \end{bmatrix}$$

is an eigenvector of A. Use this information to answer the following questions:
   (i) What is the other eigenvalue of A?
   (ii) Find a basis for the eigenspace associated with each distinct eigenvalue of A.
   (iii) Is A diagonable? Explain your reasoning.

5. If A is an $n \times n$ diagonable matrix, show that each of the following matrices is also diagonable.
   (i) sA for each scalar s
   (ii) A^t
   (iii) A^{-1} if A is non-singular

6. Suppose the $n \times n$ matrices A and B are jointly diagonable in the sense that $P^{-1}AP = \text{diag}(\lambda_1, \ldots, \lambda_n)$ and $P^{-1}BP = \text{diag}(\mu_1, \ldots, \mu_n)$. Show that the matrices $A + B$ and AB are also diagonable.

7. Let A be an $n \times n$ diagonable matrix whose eigenvalues are $\lambda_1, \lambda_2, \ldots, \lambda_n$. Show that

$$\text{tr } A = \sum_{j=1}^{n} \lambda_j$$

Note: It can be shown that this formula for the trace of A is valid even when A is not diagonable.

8. Suppose that the linear transformation $L: V \longrightarrow V$ is represented by the diagonal matrix $D = \text{diag}(d_1, \ldots, d_n)$ with respect to the basis $S = \{\alpha_1, \ldots, \alpha_n\}$.
   (i) Show that $L(\alpha_j) = d_j\alpha_j$ for $j = 1, 2, \ldots, n$.
   (ii) If γ is the vector in V whose S-coordinates are $[\gamma]_S = (c_1, c_2, \ldots, c_n)$, what are the S-coordinates of its image, $L(\gamma)$?

9. Let V denote the vector space for which the ordered set $S = \{1, x, x^2, e^{-x}, \sin x, \cos x\}$ is a basis.
   (i) Find the matrix A which represents the linear transformation $L: V \longrightarrow V$ defined by

$$L(f) = f'' - 2f' + f$$

   with respect to the S-basis.
   (ii) For what real numbers λ and functions f in V does the differential equation

$$f'' - 2f' + (1 - \lambda)f = 0$$

   have a solution? Note that the differential equation is equivalent to $L(f) = \lambda f$.

6.3 DIAGONABLE LINEAR TRANSFORMATIONS AND MATRICES

In the last section, we observed that the linear transformation $L: V \longrightarrow V$ can be represented by a diagonal matrix whenever V has a basis comprised entirely of eigenvectors. In this section, we shall first discuss some implications of this diagonability criterion and then examine a few useful features of diagonable matrices.

Suppose $\lambda_1, \lambda_2, \ldots, \lambda_n$ are distinct eigenvalues of the linear transformation $L: V \longrightarrow V$, and for each index j, let v_j be an eigenvector of L which is associated with λ_j. Then, according to theorem 6.5, the vectors $v_1, v_2, \ldots, v_n$ are linearly independent, and if V has dimension n, it follows that $\{v_1, v_2, \ldots, v_n\}$ is a basis for V. Combining this fact with the diagonability criterion established in theorem 6.6, we obtain the following useful result.

Theorem 6.7. If V is an n-dimensional vector space and the linear transformation $L: V \longrightarrow V$ has n distinct eigenvalues, then L is diagonable.

In practice, it is often more convenient to use the matrix version of theorem 6.7, which may be stated as follows.

Theorem 6.8.   An $n \times n$ matrix with n distinct eigenvalues is diagonable.

EXAMPLE 1.   The matrix

$$A = \begin{bmatrix} -3 & 2 & 1 \\ -7 & 6 & 5 \\ 2 & -2 & -2 \end{bmatrix}$$

has three distinct eigenvalues, $\lambda_1 = 0$, $\lambda_2 = 2$, and $\lambda_3 = -1$. Therefore, according to theorem 6.8, A is diagonable.

EXAMPLE 2.   Since the matrix

$$A = \begin{bmatrix} 1 & 1 \\ -1 & 1 \end{bmatrix}$$

has eigenvalues $\lambda_1 = 1 + i$ and $\lambda_2 = 1 - i$, theorem 6.8 tells us that there exists a non-singular *complex* matrix Q such that $Q^{-1}AQ = \text{diag}\,(1 + i, 1 - i)$.

It is important to note that even though A itself is real, there is no real non-singular matrix P such that $P^{-1}AP$ is a diagonal matrix.

We have just seen that a given linear transformation $L: V \to V$ is diagonable if all its eigenvalues are distinct, but even in the case where some of its eigenvalues are repeated, L will still be diagonable if V has a basis comprised of eigenvectors. In our next theorem, we show that this occurs whenever the sum of the dimensions of the eigenspaces associated with distinct eigenvalues of L is the same as the dimension of V.

Theorem 6.9.   Let L be a linear transformation on the finite-dimensional vector space V, and let $\lambda_1, \lambda_2, \ldots, \lambda_k$ be the distinct eigenvalues of L. Then L is diagonable if and only if

$$\dim \mathcal{C}_{\lambda_1} + \dim \mathcal{C}_{\lambda_2} + \ldots + \dim \mathcal{C}_{\lambda_k} = \dim V$$

Proof: For each index j, let S_j be a basis for the eigenspace $\mathcal{C}_{\lambda_j}$, and let S denote the collection of all vectors in all these bases, that is,

$$S = S_1 \cup S_2 \cup \ldots \cup S_k$$

Using theorem 6.5, it can be shown that S contains

$$\sigma = \dim \mathbb{C}_{\lambda_1} + \dim \mathbb{C}_{\lambda_2} + \ldots + \dim \mathbb{C}_{\lambda_k}$$

members and that these eigenvectors are linearly independent (verify this statement). Furthermore, since S is contained in V, we have $\sigma \leq \dim V$.

Therefore, if $\sigma = \dim V$, it follows that S is a basis for V, and since S contains only eigenvectors of L, theorem 6.6 assures us that L is diagonable. Conversely, suppose L is diagonable, and let $T = \{v_1, \ldots, v_n\}$ be a basis for V comprised of eigenvectors of L. Since each v_i is contained in one of the eigenspaces $\mathbb{C}_{\lambda_j}$, it follows that S contains at least as many members as T (why?). In other words, $\sigma \geq n = \dim V$, and since we already know that $\sigma \leq n$, we conclude that $\sigma = \dim V$, as claimed. ∎

The matrix form of theorem 6.9 may be stated as follows.

Theorem 6.10. The $n \times n$ matrix A with distinct eigenvalues $\lambda_1, \lambda_2, \ldots,$ λ_k is diagonable if and only if

$$\dim \mathcal{S}_{\lambda_1} + \dim \mathcal{S}_{\lambda_2} + \ldots + \dim \mathcal{S}_{\lambda_k} = n$$

The following solved problem illustrates how these results may be used to determine whether or not a given linear transformation or matrix is diagonable.

PROBLEM 1. Determine whether or not the linear transformation $L: R_3 \to R_3$ defined by

$$L[(a_1, a_2, a_3)] = (-11a_1 - 8a_2 - 8a_3, 8a_1 + 5a_2 + 8a_3, 4a_1 + 4a_2 + a_3)$$

is diagonable.

Solution: We find that L is represented by the matrix

$$A = \begin{bmatrix} -11 & -8 & -8 \\ 8 & 5 & 8 \\ 4 & 4 & 1 \end{bmatrix} \tag{6.5}$$

with respect to the standard basis for R_3, and we know that L is diagonable if and only if A is diagonable. Computing the characteristic polynomial of A, we obtain

$$p_A(x) = x^3 + 5x^2 + 3x - 9 = (x - 1)(x + 3)^2$$

and it follows that the distinct eigenvalues of A (and L) are $\lambda_1 = 1$ and $\lambda_2 = -3$. Following the procedure established in problem 1 of section 6.2, we find that $\{\eta_1\}$ and $\{\eta_2, \eta_3\}$ are bases for the eigenspaces S_{λ_1} and S_{λ_2}, respectively, where

$$\eta_1 = \begin{bmatrix} -2 \\ 2 \\ 1 \end{bmatrix} \quad \text{and} \quad \eta_2 = \begin{bmatrix} -1 \\ 1 \\ 0 \end{bmatrix}, \eta_3 = \begin{bmatrix} -1 \\ 0 \\ 1 \end{bmatrix}$$

Thus, we have $\dim S_{\lambda_1} = 1$ and $\dim S_{\lambda_2} = 2$, and since

$$\dim S_{\lambda_1} + \dim S_{\lambda_2} = 1 + 2 = 3$$

it follows from theorem 6.10 that A is diagonable. Consequently, L is also diagonable.

In the problem analyzed above, we were able to show that the matrix A displayed in (6.5) is diagonable without actually constructing the non-singular matrix P which satisfies $P^{-1}AP = \text{diag}(1, -3, -3)$. However, there is a simple way to form P using the eigenvectors of A. To be specific, let P be the 3×3 matrix whose jth column vector is the eigenvector η_j:

$$\begin{matrix} \quad\quad \eta_1 \quad \eta_2 \quad \eta_3 \end{matrix}$$
$$P = \begin{bmatrix} -2 & -1 & -1 \\ 2 & 1 & 0 \\ 1 & 0 & 1 \end{bmatrix} \tag{6.6}$$

Then, if D is the diagonal matrix $\text{diag}(1, -3, -3)$, we have

$$AP = \begin{bmatrix} -11 & -8 & -8 \\ 8 & 5 & 8 \\ 4 & 4 & 1 \end{bmatrix}\begin{bmatrix} -2 & -1 & -1 \\ 2 & 1 & 0 \\ 1 & 0 & 1 \end{bmatrix} = \begin{bmatrix} -2 & 3 & 3 \\ 2 & -3 & 0 \\ 1 & 0 & -3 \end{bmatrix}$$

and

$$PD = \begin{bmatrix} -2 & -1 & -1 \\ 2 & 1 & 0 \\ 1 & 0 & 1 \end{bmatrix}\begin{bmatrix} 1 & 0 & 0 \\ 0 & -3 & 0 \\ 0 & 0 & -3 \end{bmatrix} = \begin{bmatrix} -2 & 3 & 3 \\ 2 & -3 & 0 \\ 1 & 0 & -3 \end{bmatrix}$$

and it follows that $AP = PD$. Finally, P is non-singular since its column vectors, η_1, η_2, η_3, are linearly independent, and we have $P^{-1}AP = D$. This result is a special case of our next theorem.

Theorem 6.11. Let $\lambda_1, \lambda_2, \ldots, \lambda_n$ be the eigenvalues (not necessarily distinct) of the diagonable $n \times n$ matrix A, and let $\eta_1, \eta_2, \ldots \eta_n$, be linearly independent eigenvectors arranged so that η_j is associated with λ_j. Then the matrix P whose jth column vector is η_j is non-singular and

$$P^{-1}AP = \text{diag}\,(\lambda_1, \lambda_2, \ldots, \lambda_n)$$

Proof: The matrix P is non-singular since its column vectors are linearly independent. Therefore, to show that $P^{-1}AP = D$, where $D = \text{diag}\,(\lambda_1, \ldots, \lambda_n)$, it suffices to prove that $AP = PD$. To this end, note that the jth column vector of AP is $A\eta_j$, while that of PD is $\lambda_j\eta_j$ (see exercise 12, section 1.4). Therefore, since $A\eta_j = \lambda_j\eta_j$ for $j = 1, 2, \ldots, n$, it follows that the jth column vector of AP is the same as that of PD, and we conclude that $AP = PD$, as desired. ∎

We are interested in diagonable matrices mainly because if $P^{-1}AP = D$, then A and D have many properties in common, and the properties of D are extremely easy to analyze. In particular, if $D = \text{diag}\,(d_1, \ldots, d_n)$, we find that:

1. $D^k = \text{diag}\,(d_1^k, d_2^k, \ldots, d_n^k)$ for each positive integer k
2. D is non-singular if and only if each d_j is a non-zero number, and in this case,

$$D^{-1} = \text{diag}\,(d_1^{-1}, d_2^{-1}, \ldots, d_n^{-1})$$

3. If $q(x) = c_m x^m + c_{m-1}x^{m-1} + \ldots + c_1 x + c_0$, then

$$q(D) = c_m D^m + c_{m-1}D^{m-1} + \ldots + c_1 D + c_0 I_n$$
$$= \text{diag}\,(q(d_1), q(d_2), \ldots, q(d_n))$$

For example, if $D = \text{diag}\,(-3, 5, 2)$, then

1. $D^3 = \text{diag}\,(-27, 125, 8)$
2. D is non-singular and $D^{-1} = \text{diag}\,(-\frac{1}{3}, \frac{1}{5}, \frac{1}{2})$.
3. If $q(x) = 2x^2 - 3x + 1$, we have
   $q(D) = 2D^2 - 3D + I_3 = \text{diag}\,(q(-3), q(5), q(2)) = \text{diag}\,(28, 36, 3)$

The following solved problem shows how these three formulas can be used to carry out computations involving a diagonable matrix.

PROBLEM 2.   If A is the matrix displayed in (6.5), show that $A^3 + 5A^2 + 3A - 9I_3 = O$.

Solution: We have already shown that $P^{-1}AP = D$, where P is the non-singular matrix displayed in (6.6) and $D = \text{diag}\,(1, -3, -3)$. Let $q(x) =$

$x^3 + 5x^2 + 3x - 9$. Then, according to formula 3 above, we have

$$q(D) = D^3 + 5D^2 + 3D - 9I_3 = \text{diag}\,[q(1), q(-3), q(-3)]$$
$$= \text{diag}\,(0, 0, 0) = O$$

and since $q(A) = Pq(D)P^{-1}$ (see exercise 5, section 4.6), we conclude that $q(A) = O$, as desired.

In addition to the computational advantages already noted, it is often easier to prove theoretical results for matrices which are diagonal than for those which are not. For instance, the characteristic polynomial of the matrix A displayed in (6.5) is $p_A(x) = x^3 + 5x^2 + 3x - 9$, and in problem 2, we showed that A satisfies its characteristic equation, in the sense that $p_A(A) = O$. Our next result, which is known as the Cayley-Hamilton theorem, states that *every* matrix satisfies its characteristic equation.

Theorem 6.12. (Cayley-Hamilton.)   If A is an $n \times n$ matrix whose characteristic polynomial is

$$p_A(x) = x^n + c_{n-1}x^{n-1} + \ldots + c_1x + c_0$$

then

$$p_A(A) = A^n + c_{n-1}A^{n-1} + \ldots + c_1A + c_0I_n = O$$

Proof: A general proof of this theorem is beyond the scope of this text. However, we do have all the tools needed to handle the important special case where A is diagonable.

Accordingly, let $\lambda_1, \lambda_2, \ldots, \lambda_n$ be the eigenvalues of A, and let P be a non-singular matrix such that $P^{-1}AP = D$, where $D = \text{diag}\,(\lambda_1, \lambda_2, \ldots, \lambda_n)$. Since each λ_j is a zero of the characteristic polynomial $p_A(x)$, we have $p_A(\lambda_j) = 0$ for $j = 1, 2, \ldots, n$, and it follows that

$$p_A(D) = \text{diag}\,[p_A(\lambda_1), \ldots, p_A(\lambda_n)] = O$$

Therefore, since $A = PDP^{-1}$, we conclude that

$$p_A(A) = P[p_A(D)]P^{-1} = O$$

which means that $A^n + c_{n-1}A^{n-1} + \ldots + c_0I_n = O$, as claimed. ∎

According to the Cayley-Hamilton theorem, if A is an $n \times n$ matrix, then A^n can be expressed as a linear combination of $A^{n-1}, \ldots, A, I_n$, and this in turn enables us to express all higher powers of A in the same form.

For instance, since the characteristic polynomial of the matrix

$$A = \begin{bmatrix} -3 & 1 & 5 \\ 7 & 2 & -6 \\ 1 & 1 & 1 \end{bmatrix}$$

is $p_A(x) = x^3 - 13x + 12$, the Cayley-Hamilton theorem tells us that $A^3 - 13A + 12I_3 = 0$. It follows that $A^3 = 13A - 12I_3$, and we have

$$A^4 = AA^3 = A(13A - 12I_3) = 13A^2 - 12A$$

$$A^5 = AA^4 = A(13A^2 - 12A) = 13A^3 - 12A^2$$

$$= 13(13A - 12I_3) - 12A^2 = -12A^2 + 169A - 156I_3$$

and so on. A few additional applications of the Cayley-Hamilton theorem are outlined in the exercises.

Since there are great advantages to knowing that a matrix is diagonable, it would be nice to be able to determine whether or not a given matrix A has this property by simply examining its entries. Usually, this is not possible, but in one important special case, it is: namely, real symmetric matrices are always diagonable. We shall verify this fact in chapter 7 as part of our development of the properties of symmetric matrices.

In the last three sections, we have discussed a few of the basic ideas associated with the study of eigenvalues, eigenvectors, and diagonability. Most topics which involve these concepts lie beyond the scope of this text, but even the introductory material of this chapter has interesting applications. In the next section, we illustrate this fact by using our results to investigate the solution set of a system of linear differential equations.

Exercise Set 6.3

1. In each of the following cases, determine whether or not the given matrix is diagonable.

(i) $\begin{bmatrix} 7 & 4 & 0 \\ -8 & -5 & 0 \\ 8 & 4 & -1 \end{bmatrix}$ (ii) $\begin{bmatrix} 3 & 0 & -1 \\ 14 & -5 & -7 \\ 2 & 0 & 0 \end{bmatrix}$

(iii) $\begin{bmatrix} 3 & 0 & -1 \\ -4 & 3 & 2 \\ 4 & 0 & -1 \end{bmatrix}$ (iv) $\begin{bmatrix} -9 & 1 & 5 \\ 2 & 3 & -1 \\ -10 & 2 & 6 \end{bmatrix}$

2. Which (if any) of the following linear transformations is diagonable?

(i) $L[(a_1, a_2, a_3)] = (7a_1 + 4a_2, -8a_1 - 5a_2, 8a_1 + 4a_2 - a_3)$

(ii) $L[at^2 + bt + c] = (3a - c)t^2 + (-4a + 3b + 2c)t + (4a - c)$

(iii) $L[at^2 + bt + c] = (4a - c)t^2 + (21a - 5b - 7c)t + (6a - c)$

3. (i) Find the characteristic polynomial of the matrix

$$K = \begin{bmatrix} 0 & 3 & 4 \\ -3 & 0 & -12 \\ -4 & 12 & 0 \end{bmatrix}$$

(ii) Explain why there exists a complex matrix Q such that $Q^{-1}KQ$ is diagonal. Does there exist a real matrix P such that $P^{-1}KP$ is diagonal?

4. (i) Find the eigenvalues of the matrix

$$A = \begin{bmatrix} -5 & 1 & 5 \\ -7 & 4 & 4 \\ -1 & 1 & 1 \end{bmatrix}$$

(ii) Construct a basis for the eigenspace associated with each distinct eigenvalue of A.

(iii) Obtain a non-singular matrix P such that $P^{-1}AP$ is diagonal.

5. Let $L: P_2 \longrightarrow P_2$ denote the linear transformation defined by

$$L(at^2 + bt + c) = (b + c)t^2 + (-a - 2b - c)t + (a + b)$$

Find a basis for P_2 which is comprised of eigenvectors of L.

6. Show that a non-zero nilpotent matrix N cannot be diagonable (see exercise 13, section 6.1).

7. Let A be a diagonable 3×3 matrix whose characteristic polynomial is $p_A(x) = x^3 - 3x^2 + 4$.

(i) If $q(x)$ is a polynomial which satisfies $q(-1) = q(2) = 0$, show that $q(A) = O$.

(ii) Find the polynomial $m(x)$ of smallest degree for which $m(A) = O$.

(iii) Express each of the matrices A^3, A^4, A^5, A^{-1} as a linear combination of A and I_3.

8. Let A be a diagonable $n \times n$ matrix, and let $p(x) = (x - \lambda_1)(x - \lambda_2) \ldots (x - \lambda_k)$, where $\lambda_1, \ldots, \lambda_k$ are the distinct eigenvalues of A.

(i) If D is a diagonal matrix which satisfies $P^{-1}AP = D$, show that $p(D) = (D - \lambda_1 I_n) \ldots (D - \lambda_k I_n) = O$.

(ii) Show that $p(A) = O$.

(iii) If $q(x)$ is a polynomial such that $q(A) = O$, show that $p(x)$ must divide $q(x)$. *Hint:* Observe that $q(\lambda_1) = q(\lambda_2) = \ldots = q(\lambda_k) = 0$.

9. Let A be an $n \times n$ diagonable matrix, and suppose $P^{-1}AP = D$ where D is diagonable.

(i) Show that A^t is also similar to D.

(ii) Show that A^t is similar to A and that there exists a symmetric matrix Q such that $Q^{-1}AQ = A^t$. *Hint:* Consider the matrix PP^t.

10. Let A be an $n \times n$ matrix whose characteristic polynomial is $p_A(x) = x^m + c_{m-1}x^{m-1} + \ldots + c_1 x + c_0$.

(i) Show that $c_0 = (-1)^m \det A$.

  Hint: Note that $p_A(0) = \det(0I_n - A)$.

(ii) If A is non-singular, show that $c_0 \neq 0$ and that

$$A^{-1} = -\frac{1}{c_0}[A^{m-1} + c_{m-1}A^{m-2} + \ldots + c_2 A + c_1 I_n]$$

6.4 (OPTIONAL) SYSTEMS OF LINEAR DIFFERENTIAL EQUATIONS

Many problems in applied mathematics can be described in terms of systems of differential equations, and matrix methods play an important role in the analysis of such systems. For instance, in supplementary exercise 9 of chapter 1, we outlined a method for solving a system of the form $A(dX/dt) = \beta$. As a second example, consider the system

$$\frac{dx_1}{dt} = -11x_1 - 8x_2 - 8x_3$$

$$\frac{dx_2}{dt} = 8x_1 + 5x_2 + 8x_3 \qquad (6.7)$$

$$\frac{dx_3}{dt} = 4x_1 + 4x_2 + x_3$$

Using the notational conventions established in supplementary exercise 7 of chapter 1, we find that system (6.7) can be expressed in the form $dX/dt = AX$, where

$$\frac{dX}{dt} = \begin{bmatrix} \dfrac{dx_1}{dt} \\ \dfrac{dx_2}{dt} \\ \dfrac{dx_3}{dt} \end{bmatrix}, \qquad A = \begin{bmatrix} -11 & -8 & -8 \\ 8 & 5 & 8 \\ 4 & 4 & 1 \end{bmatrix}, \qquad X = \begin{bmatrix} x_1 \\ x_2 \\ x_3 \end{bmatrix}$$

As the first step in solving system (6.7), we recall from problem 1 of section 6.3 that $P^{-1}AP = D$, where

$$P = \begin{bmatrix} -2 & -1 & -1 \\ 2 & 1 & 0 \\ 1 & 0 & 1 \end{bmatrix} \quad \text{and} \quad D = \begin{bmatrix} 1 & 0 & 0 \\ 0 & -3 & 0 \\ 0 & 0 & -3 \end{bmatrix}$$

In other words, $A = PDP^{-1}$ and by setting $Y = P^{-1}X$, we obtain

$$\frac{dX}{dt} = AX = (PDP^{-1})X = P(DY)$$

Since

$$\frac{dY}{dt} = P^{-1}\frac{dX}{dt}$$

(see supplementary exercise 7 of chapter 1), it follows that

$$\frac{dY}{dt} = P^{-1}\frac{dX}{dt} = DY$$

and the original system has been reduced to the form

$$\frac{dy_1}{dt} = y_1$$

$$\frac{dy_2}{dt} = -3y_2$$

$$\frac{dy_3}{dt} = -3y_3$$

Solving, we find that

$$y_1 = C_1 e^t, \qquad y_2 = C_2 e^{-3t}, \qquad y_3 = C_3 e^{-3t}$$

where C_1, C_2, C_3 are arbitrary constants. Finally, since $Y = P^{-1}X$ we have $X = PY$, which means that

$$x_1 = -2y_1 - y_2 - y_3$$
$$x_2 = \quad 2y_1 + y_2$$
$$x_3 = \quad y_1 \quad + y_3$$

and we conclude that the original system has the general solution

$$x_1 = -2C_1 e^t - C_2 e^{-3t} - C_3 e^{-3t}$$
$$x_2 = 2C_1 e^t + C_2 e^{-3t} \tag{6.8}$$
$$x_3 = C_1 e^t + \quad\quad C_3 e^{-3t}$$

To evaluate the C's, we need some additional information about system (6.7). For example, suppose the system is known to satisfy the initial conditions $x_1(0) = 1$, $x_2(0) = -3$, $x_3(0) = 5$. Then we have

$$-2C_1 - C_2 - C_3 = 1$$
$$2C_1 + C_2 \quad\quad = -3$$
$$C_1 \quad\quad + C_3 = 5$$

and it follows that $C_1 = 3$, $C_2 = -9$, $C_3 = 2$, since

$$\begin{bmatrix} C_1 \\ C_2 \\ C_3 \end{bmatrix} = P^{-1}X_0 = \begin{bmatrix} 1 & 1 & 1 \\ -2 & -1 & -2 \\ -1 & -1 & 0 \end{bmatrix} \begin{bmatrix} 1 \\ -3 \\ 5 \end{bmatrix} = \begin{bmatrix} 3 \\ -9 \\ 2 \end{bmatrix}$$

By generalizing the result of the preceding example, we obtain the following theorem.

Theorem 6.13.   Let A be a diagonable $n \times n$ matrix and let P be a nonsingular matrix such that $P^{-1}AP = D = \text{diag}(\lambda_1, \ldots, \lambda_n)$. Then the system

$$\frac{dX}{dt} = AX$$

has the general solution

$$\begin{bmatrix} x_1 \\ \cdot \\ \cdot \\ \cdot \\ x_n \end{bmatrix} = P \begin{bmatrix} e^{\lambda_1 t} & & 0 \\ & \cdot & \\ & & \cdot \\ 0 & & e^{\lambda_n t} \end{bmatrix} \begin{bmatrix} C_1 \\ \cdot \\ \cdot \\ \cdot \\ C_n \end{bmatrix}$$

where the C's are arbitrary constants.

If A is not diagonable, it is more difficult to solve the system $dX/dt = AX$ since we cannot then reduce the system to the simple form $dY/dt = DY$. However, even this case can be handled, for it can be shown that every complex matrix is similar to a matrix that is "almost" diagonal. We can be a little more explicit in describing the form of such a matrix, but first, we need to introduce a few special terms. Accordingly, a matrix $T = (t_{ij})$ is said to be a *Jordan block* if it has the following features:

1. All the diagonal entries of T are the same.
2. Each "superdiagonal" entry $t_{i,i+1}$ is 1.
3. $t_{ij} = 0$ if j is not i or $i + 1$.

For instance, the matrices

$$\begin{bmatrix} 1 & 1 & 0 \\ 0 & 1 & 1 \\ 0 & 0 & 1 \end{bmatrix} \qquad \begin{bmatrix} 0 & 1 & 0 \\ 0 & 0 & 1 \\ 0 & 0 & 0 \end{bmatrix} \qquad \begin{bmatrix} -3 & 1 & 0 & 0 \\ 0 & -3 & 1 & 0 \\ 0 & 0 & -3 & 1 \\ 0 & 0 & 0 & -3 \end{bmatrix}$$

are all Jordan blocks. A matrix J is said to be in *Jordan form* if it can be expressed as a block diagonal matrix in which the diagonal submatrices are all Jordan blocks. Note the pattern of Jordan blocks on the main diagonal of each of the following Jordan form matrices:

$$\begin{bmatrix} 3 & 1 & 0 & 0 & 0 \\ 0 & 3 & 1 & 0 & 0 \\ 0 & 0 & 3 & 0 & 0 \\ 0 & 0 & 0 & 5 & 1 \\ 0 & 0 & 0 & 0 & 5 \end{bmatrix} \quad \begin{bmatrix} 4 & 1 & 0 & 0 & 0 \\ 0 & 4 & 0 & 0 & 0 \\ 0 & 0 & 4 & 1 & 0 \\ 0 & 0 & 0 & 4 & 0 \\ 0 & 0 & 0 & 0 & -1 \end{bmatrix} \quad \begin{bmatrix} 2 & 1 & 0 & 0 & 0 \\ 0 & 2 & 1 & 0 & 0 \\ 0 & 0 & 2 & 0 & 0 \\ 0 & 0 & 0 & 2 & 0 \\ 0 & 0 & 0 & 0 & 0 \end{bmatrix}$$

In general, a Jordan form matrix J has its eigenvalues on the main diagonal (why?), a pattern of 1's and 0's on the next higher diagonal (the so-called "superdiagonal"), and 0's elsewhere.

Using advanced techniques, it can be shown that each complex matrix C is similar to exactly one matrix J in Jordan form. This result, which is generally referred to as the *Jordan canonical form theorem*, is an extremely useful tool in theoretical linear algebra. Neither a proof of this theorem nor a description of how to construct a matrix P such that $P^{-1}CP = J$ lies within the scope of this text. Instead, we shall demonstrate the usefulness of this result by showing how easy it is to solve the system $dX/dt = JX$ in the case where J is a matrix in Jordan form. For instance, consider the Jordan form matrix

$$J = \begin{bmatrix} 3 & 1 & 0 \\ 0 & 3 & 0 \\ 0 & 0 & -1 \end{bmatrix}$$

Writing out the differential system $dX/dt = JX$, we have

$$\frac{dx_1}{dt} = 3x_1 + x_2$$

$$\frac{dx_2}{dt} = \qquad 3x_2 \qquad\qquad (6.9)$$

$$\frac{dx_3}{dt} = \qquad\qquad - x_3$$

The last two equations are easy to solve, and we find that $x_2 = C_2 e^{3t}$ and $x_3 = C_3 e^{-t}$. Substituting $C_2 e^{3t}$ for x_2 in the first equation of (6.9), we obtain the first order linear differential equation $dx_1/dt = 3x_1 + C_2 e^{3t}$. Solving by the method developed in supplementary exercise 14 of chapter 4, we find that

$$x_1 = e^{3t}[C_1 + C_2 \int e^{-3t}e^{3t}\, dt] = C_1 e^{3t} + C_2 t e^{3t}$$

and we now have a complete general solution of the differential system displayed in (6.9).

Although the Jordan form matrix associated with system (6.9) is rather simple, the method we used to solve the system is really quite general. In particular, the form of the Jordan matrix J guarantees that each equation in the system $dX/dt = JX$ can be dealt with either by direct integration or by solving a first order linear equation of the form

$$\frac{dx}{dt} = cx + p(t)e^{ct}$$

where $p(t)$ is a suitable polynomial. Moreover, if the complex matrix C satisfies the equation $P^{-1}CP = J$ and the non-singular matrix P and the Jordan form matrix J are both known, then the process of solving the differential system $dX/dt = CX$ boils down to handling the relatively simple system $dY/dt = JY$, where $Y = P^{-1}X$. The basic ideas involved in this procedure are illustrated in exercise 4.

Incidentally, an nth order linear differential equation may be regarded as a system and as such can be solved by the methods we have just developed. For instance, if we identify y with x_1, y' with x_2, and y'' with x_3, the differential equation

$$y''' - 3y'' + 5y' + 7y = 0 \tag{6.10}$$

may be regarded as the system

$$\frac{dx_1}{dt} = y' = \qquad\qquad x_2$$

$$\frac{dx_2}{dt} = y'' = \qquad\qquad\qquad x_3 \tag{6.11}$$

$$\frac{dx_3}{dt} = y''' = -7x_1 - 5x_2 + 3x_3$$

Note that the matrix associated with system (6.11) is

$$C = \begin{bmatrix} 0 & 1 & 0 \\ 0 & 0 & 1 \\ -7 & -5 & 3 \end{bmatrix}$$

This array is known as the *companion matrix* of the polynomial $p(x) = x^3 - 3x^2 + 5x + 7$, and it is easy to verify that p is the characteristic polynomial of C (see supplementary exercise 6). For this reason, we say that p is the *characteristic polynomial* of the differential equation (6.10).

If the characteristic polynomial p of the nth order differential equation

$$a_n y^{(n)} + a_{n-1} y^{(n-1)} + \ldots + a_1 y' + a_0 y = 0$$

has n distinct real zeroes, $\lambda_1, \ldots, \lambda_n$, then the associated matrix is diagonable. In this case, theorem 6.13 shows that $\{e^{\lambda_1 t}, \ldots, e^{\lambda_n t}\}$ is a basis for the solution space of the equation. Even when p has zeroes which are not distinct or are complex, it can be shown that the solution space of the differential equation has dimension n, but it is not easy to write down the general form of a suitable basis for this space. To illustrate what is involved in describing such a basis, consider the differential equation

$$y^{(vi)} - 5y^{(v)} + 10y^{(iv)} - 12y''' + 11y'' - 7y' + 2y = 0$$

The characteristic polynomial of this equation is

$$p(x) = x^6 - 5x^5 + 10x^4 - 12x^3 + 11x^2 - 7x + 2$$

and the zeroes of p are 1, 1, 1, 2, i, and $-i$. Then it can be shown that

$$\{e^t, te^t, t^2 e^t, e^{2t}, \sin t, \cos t\}$$

is a basis for the solution space of the given differential equation.

The reader who is interested in knowing more about the relationship between linear algebra and differential equations is encouraged to examine the material in *Linear Algebra with Differential Equations* by D. L. Bentley and K. Cooke, Holt, Rinehart, and Winston, New York, 1973.

Exercise Set 6.4

1. In each of the following cases, find the general solution of the given system of differential equations.

(i) $\dfrac{dx_1}{dt} = 2x_1 + x_2$

$\dfrac{dx_2}{dt} = x_1 + 2x_2$

(ii) $\dfrac{dx_1}{dt} = \qquad - 2x_2$

$\dfrac{dx_2}{dt} = 3x_1 + 5x_2$

(iii) $\dfrac{dx_1}{dt} = 2x_1 + x_2 + x_3$

$\dfrac{dx_2}{dt} = \qquad 3x_2 + 2x_3$

$\dfrac{dx_3}{dt} = \qquad x_2 + 2x_3$

(iv) $\dfrac{dx_1}{dt} = \qquad x_2 + 3x_3$

$\dfrac{dx_2}{dt} = 2x_1 + x_2 - 3x_3$

$\dfrac{dx_3}{dt} = 4x_1 - 2x_2 - 4x_3$

2. In each of the following cases, determine whether or not the given matrix is in Jordan form.

(i)
$$\begin{bmatrix} 2 & 1 & 0 & 0 \\ 0 & 2 & 0 & 0 \\ 0 & 0 & 3 & 0 \\ 0 & 0 & 0 & 0 \end{bmatrix}$$

(ii)
$$\begin{bmatrix} -1 & 0 & 0 & 0 \\ 0 & 5 & 1 & 0 \\ 0 & 0 & 5 & 0 \\ 0 & 0 & 0 & 5 \end{bmatrix}$$

(iii)
$$\begin{bmatrix} 0 & 0 & 0 & 0 \\ 0 & 4 & 1 & 0 \\ 0 & 0 & 4 & 1 \\ 0 & 0 & 0 & -2 \end{bmatrix}$$

(iv)
$$\begin{bmatrix} -3 & 1 & 0 & 0 \\ 0 & -3 & 1 & 0 \\ 0 & 0 & -3 & 0 \\ 0 & 0 & 0 & 7 \end{bmatrix}$$

3. Find the general solution of the system

$$\frac{dx_1}{dt} = 2x_1 + x_2$$

$$\frac{dx_2}{dt} = \qquad 2x_2 + x_3$$

$$\frac{dx_3}{dt} = \qquad\qquad 2x_3$$

Note that the matrix associated with this system is in Jordan form.

4. Consider the matrices

$$P = \begin{bmatrix} 1 & 1 & 1 \\ -2 & 0 & 1 \\ 2 & 1 & 1 \end{bmatrix} \quad\text{and}\quad A = \begin{bmatrix} 4 & 0 & -1 \\ -8 & 4 & 6 \\ 6 & -1 & -2 \end{bmatrix}$$

(i) Find P^{-1} and verify that the matrix $J = P^{-1}AP$ is in Jordan form.

(ii) Show that X satisfies the system $dX/dt = AX$ if and only if the vector $Y = P^{-1}X$ satisfies $dY/dt = JY$.

(iii) Find the solution of the system

$$\frac{dx_1}{dt} = \qquad 4x_1 \qquad - x_3$$

$$\frac{dx_2}{dt} = -8x_1 + 4x_2 + 6x_3$$

$$\frac{dx_3}{dt} = \qquad 6x_1 - x_2 - 2x_3$$

which satisfies the initial conditions $x_1(0) = x_2(0) = 0$ and $x_3(0) = 1$.

Hint: Use parts (i) and (ii) and the result of exercise 3.

5. In each of the following cases, express the given differential equation as a system, then find its characteristic polynomial and obtain a basis for its solution space.

(i) $y'' - 3y' + 2y = 0$ (ii) $y'' - 9y = 0$

(iii) $y''' - 7y' + 6y = 0$

SUPPLEMENTARY EXERCISES

1. Let $E_n = (e_{ij})$ denote the $n \times n$ matrix such that $e_{ij} = 1$ for all i, j.
   (i) Show that the distinct eigenvalues of E_n are 0 and n. What is the characteristic polynomial of E_n?
   (ii) If c is a scalar, show that $\det (E_n + cI_n) = c^{n-1}(c + n)$.
   Hint: Use theorem 6.4 and the result of exercise 19, section 6.1.
2. A certain 4×4 real matrix is known to have these properties:
   1. Two of the eigenvalues of A are $\lambda_1 = 3$ and $\lambda_2 = 2$.
   2. The number 3 is an eigenvalue of the matrix $A + 2I_4$.
   3. $\det A = 12$.
   Use this information to answer the following questions about A.
   (i) What are the other two eigenvalues of A?
   (ii) What is the characteristic polynomial of A? of A^t?
   (iii) What is the characteristic polynomial of A^{-1}?
3. (i) The characteristic polynomial of a certain 3×3 matrix A is $p_A(x) = x^3 - 7x^2 + 5x - 9$. Use this fact to express adj A as a linear combination of A^2, A, and I_3.
   (ii) If A is an $n \times n$ non-singular matrix, show that adj A can be expressed as a linear combination of A^{n-1}, A^{n-2}, ..., A, I_n.
4. Let

$$A = \begin{bmatrix} 1 & -2 \\ -2 & 1 \end{bmatrix}$$

and let $L: {}_2R_2 \longrightarrow {}_2R_2$ denote the linear transformation defined by

$$L(X) = AX - XA$$

   (i) Find the 4×4 matrix which represents L with respect to the standard basis for ${}_2R_2$.
   (ii) Find the eigenvalues of L and obtain a basis for the eigenspace associated with each distinct eigenvalue.
   (iii) Is L diagonable? Explain.
5. Let S be a finite-dimensional subspace of the real inner product space V, and let $L: V \longrightarrow V$ denote the linear transformation defined by

$$L(\alpha) = \alpha_S - \alpha_{S^\perp}$$

where $\alpha = \alpha_S + \alpha_{S^\perp}$ is the orthogonal decomposition of α along S.
   (i) If $\{\alpha_1, \ldots, \alpha_k\}$ is a basis for S and $\{\alpha_{k+1}, \ldots, \alpha_n\}$ is a basis for $S^\perp$, find the matrix which represents L with respect to the basis $\{\alpha_1, \ldots, \alpha_n\}$ for V.
   (ii) What are the eigenvalues of L?
6. (i) If a_1, a_2, a_3 are real numbers, verify that the matrix

$$C(p) = \begin{bmatrix} 0 & 1 & 0 \\ 0 & 0 & 1 \\ a_3 & a_2 & a_1 \end{bmatrix}$$

has characteristic polynomial $p(x) = x^3 - a_1x^2 - a_2x - a_3$. This matrix is said to be the *companion matrix* of $p(x)$.

   (ii) Describe an $n \times n$ matrix whose characteristic polynomial is $x^n - a_1x^{n-1} - \ldots - a_{n-1}x - a_n$.

7. Let $A = (a_{ij})$ be a 3×3 matrix, and for $k = 1, 2, 3$, let c_k denote the sum of the $k \times k$ principal minors of A. Then it can be shown that the characteristic polynomial of A is

$$p_A(x) = x^3 - c_1x^2 + c_2x - c_3$$

(Note: A minor is said to be principal if the diagonal entries of the associated submatrix are also diagonal entries of A. Thus, in particular, we must have $c_1 = \text{tr } A$ and $c_3 = \det A$.) For instance, consider the matrix

$$A = \begin{bmatrix} -9 & 3 & -7 \\ 4 & 2 & 1 \\ 8 & 0 & 5 \end{bmatrix}$$

We find that $c_1 = \text{tr } A = -9 + 2 + 5 = -2$, $c_3 = \det A = -14$, and

$$c_2 = \det \begin{bmatrix} -9 & 3 \\ 4 & 2 \end{bmatrix} + \det \begin{bmatrix} -9 & -7 \\ 8 & 5 \end{bmatrix} + \det \begin{bmatrix} 2 & 1 \\ 0 & 5 \end{bmatrix} = -9$$

and it follows that

$$p_A(x) = x^3 - (-2)x^2 + (-9)x - (-14) = x^3 + 2x^2 - 9x + 14$$

   (i) Use this formula to find the characteristic polynomial of the following matrices:

$$A_1 = \begin{bmatrix} -5 & 3 & 2 \\ -1 & 1 & 0 \\ 7 & 1 & -4 \end{bmatrix} \quad A_2 = \begin{bmatrix} 3 & -7 & 4 \\ 1 & 0 & 1 \\ 3 & 2 & -3 \end{bmatrix}$$

   (ii) State an analogous result for 4×4 matrices. Do you think it is possible to compute the characteristic polynomial of an $n \times n$ matrix in a similar fashion?

8. Let A and B be $n \times n$ matrices which commute (that is, $AB = BA$), and assume that the eigenspace associated with the eigenvalue λ of A has dimension 1.

   (i) If $AX = \lambda X$, show that $Y = BX$ is an eigenvector of A associated with the eigenvalue λ.

   (ii) Show that there exists a scalar μ such that $BX = \mu X$. Note that μ is an eigenvalue of B.

   (iii) Show that $\lambda + \mu$ is an eigenvalue of $A + B$ and $\lambda\mu$ is an eigenvalue of AB.

9. If A and B are $n \times n$ matrices, show that AB and BA have the same eigenvalues by completing the steps outlined below.

(i) Let J_m be the $n \times n$ matrix whose block form is

$$J_m = \begin{bmatrix} I_m & 0 \\ 0 & 0 \end{bmatrix}$$

If C is an $n \times n$ matrix, show that CJ_m and $J_m C$ have the same characteristic polynomial.

(ii) Let r be the rank of A, and let P and Q be non-singular matrices such that $QAP = J_r$ (see supplementary exercise 11, in chapter 2). Show that $Q(AB)Q^{-1}$ and $P^{-1}(BA)P$ have the same characteristic polynomial.

(iii) Complete the proof by showing that AB and BA have the same characteristic polynomial and hence, the same eigenvalues.

*10. Let $C = (c_{ij})$ be an $n \times n$ complex matrix and for $k = 1, 2, \ldots, n$, let $p_k = \sum_{\substack{i=1 \\ i \neq k}}^{n} |c_{kj}|$. In other words, p_k is the sum of the moduli of the off-diagonal entries in the kth row of C.

By completing the steps outlined below, prove the following result, which is known as the Gerschgorin disk theorem.

Theorem. The eigenvalues of the $n \times n$ complex matrix $C = (c_{ij})$ are all contained in the union of the n disks defined by

$$|z - c_{mm}| \leq p_m \quad m = 1, 2, \ldots, n$$

Note: The mth such disk is centered at c_{mm} in the complex plane and has radius p_m. We shall refer to this region as the mth *Gerschgorin disk*.

(i) Let λ be an eigenvalue of C, and suppose that $CX = \lambda X$. Show that

$$|(\lambda - c_{ii})x_i| \leq \sum_{\substack{j=1 \\ j \neq i}}^{n} |c_{ij}||x_j|$$

for $i = 1, 2, \ldots, n$.

(ii) Let m be chosen so that $|x_j| \leq |x_m|$ for $j = 1, 2, \ldots, n$. That is, $|x_m|$ is at least as large as any other component of X. Show that

$$|\lambda - c_{mm}| \leq p_m$$

and complete the proof of the theorem.

*11. Plot the Gerschgorin disks for the matrix

$$C = \begin{bmatrix} 4 & i & 0 & 1+i \\ 0 & -3+i & 2 & i \\ 3+4i & 0 & 0 & 1 \\ 0 & i & 1 & 3-i \end{bmatrix}$$

Show that 7 cannot be an eigenvalue of C.

*12. The $n \times n$ matrix $H = (h_{ij})$ is said to be (strictly) *diagonally dominant* if

$$\sum_{\substack{j=1 \\ j \neq i}}^{n} |h_{ij}| < |h_{ii}| \quad \text{for} \quad i = 1, 2, \ldots, n$$

For instance, the matrices

$$\begin{bmatrix} 2+i & 1 & i \\ -1+i & 3 & 1 \\ 4 & 0 & 3-4i \end{bmatrix} \quad \text{and} \quad \begin{bmatrix} 9 & 1 & -3 & 2 \\ 0 & 4 & 1 & 2 \\ 3 & -1 & 6 & 1 \\ 1 & 0 & 1 & 3 \end{bmatrix}$$

both have this property.

Show that a diagonally dominant matrix must be non-singular.

Hint: Show that 0 lies outside the union of the Gerschgorin disks.

13. Let $T_n = (t_{ij})$ denote the $n \times n$ matrix such that for each index i, $t_{ii} = a$ and $t_{ij} = b$ for $j \neq i$. (See supplementary exercise 13, chapter 5.)

(i) Verify that $T_n = (a - b)I_n + bE_n$, where E_n is the $n \times n$ matrix of all 1's.

(ii) Find the eigenvalues of T_n.

(iii) Show that $\det T_n = (a - b)^{n-1}(a + (n - 1)b)$.

Hint: Use supplementary exercise 1, above.

7

Quadratic Forms
and Symmetric Matrices

7.1 INTRODUCTION TO QUADRATIC FORMS

A quadratic form is an expression such as $x_1^2 - 2x_1x_2 + 3x_2^2$ in which the sum of the degrees of the variables in each term is exactly two. This kind of expression plays an important role in such diverse areas as probability theory, statistics, differential geometry, mechanics, operations research, and approximation theory. In this chapter, we shall develop matrix methods for investigating the properties of quadratic forms, and the first step in this development is to provide a more formal definition of such expressions.

Definition 7.1. A *real quadratic form* is a function Q of n real variables $x_1, x_2, \ldots, x_n$ which satisfies

$$Q(x_1, x_2, \ldots, x_n) = \sum_{i=1}^{n} \sum_{\substack{j=1 \\ j \geq i}}^{n} a_{ij} x_i x_j \qquad (7.1)$$

where the coefficients a_{ij} are real numbers. At times, we shall find it convenient to denote such a function by $Q(X)$, where X is the variable n-tuple $(x_1, x_2, \ldots, x_n)$.

The notation $j \geq i$ which appears in the inner summation of (7.1) guarantees that $x_i x_j$ does not appear in the quadratic expression whenever $i > j$. For instance, $2x_1^2 - 3x_1 x_2 + x_2^2$ conforms to definition 7.1, but $2x_1^2 - 3x_2 x_1 + x_2^2$ does not.

We have already mentioned a few of the many subjects in which quadratic forms appear. Each of these areas is extremely important, but we shall be especially interested in applying our results to the study of the geometric configurations described in the following example.

EXAMPLE 1. (Central conics and quadrics.) In plane geometry, it is shown that the equation of a conic section centered at the origin of the plane has the form $Ax^2 + Bxy + Cy^2 = 1$. The expression $D = 4AC - B^2$ is called the discriminant of the conic, and it can be shown that $D > 0$ for an ellipse and $D < 0$ for a hyperbola:

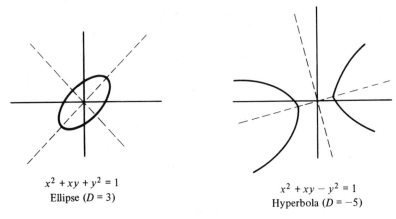

$$x^2 + xy + y^2 = 1$$
Ellipse ($D = 3$)

$$x^2 + xy - y^2 = 1$$
Hyperbola ($D = -5$)

FIGURE 7.1

In 3-space, a surface whose equation has the form $Ax^2 + By^2 + Cz^2 + Dxy + Exz + Fyz = 1$ is called a *central quadric* surface. At this point, a complete classification of such surfaces would not be particularly enlightening, but the examples of figure 7.2 should give the reader an idea of what to expect.

Note that each central conic or quadric has an equation of the form $Q(X) = 1$, where Q is a suitable quadratic form. For this reason, the study of conic sections and quadric surfaces is closely related to the theory of quadratic forms, and we shall discuss this relationship while developing the various topics of this chapter.

To facilitate our discussion of quadratic forms, we need a more simple way to represent these expressions than the double summation which appears

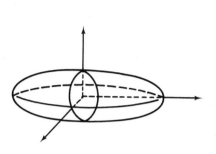

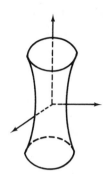

$5x^2 + 3y^2 + 2z^2 = 1$

Ellipsoid

$4x^2 + 3y^2 - z^2 = 1$

Hyperboloid of one sheet

FIGURE 7.2

in (7.1). As an illustration of the kind of representation we have in mind, note that the quadratic form $Q(X) = x_1^2 + 4x_1x_2 - 5x_2^2$ can be expressed as the matrix product

$$Q(X) = [x_1, x_2]\begin{bmatrix} 1 & 2 \\ 2 & -5 \end{bmatrix}\begin{bmatrix} x_1 \\ x_2 \end{bmatrix} = x_1^2 + 2x_1x_2 + 2x_2x_1 - 5x_2^2$$

if we agree to write the sum $2x_1x_2 + 2x_2x_1$ as $4x_1x_2$. This observation encourages us to make the following definition.

Definition 7.2. The *symmetric representative* of the real quadratic form displayed in (7.1) is the $n \times n$ real symmetric matrix $S = (s_{ij})$, where

$$s_{ii} = a_{ii}$$
$$s_{ij} = \tfrac{1}{2}a_{ij} = s_{ji} \quad \text{for } i \neq j$$

In general, if $X = (x_i)$ is a variable column n-tuple, we have

$$X^tSX = \sum_{i=1}^{n} x_i \left(\sum_{j=1}^{n} s_{ij}x_j \right) = \sum_{i=1}^{n} \sum_{j=1}^{n} s_{ij}x_ix_j$$

and if we agree to express the sum $s_{ij}x_ix_j + s_{ji}x_jx_i$ as $2 s_{ij} x_ix_j$, where $i < j$, then it follows that $Q(X) = X^tSX$. We shall use this representation throughout our work.

EXAMPLE 2.   The symmetric representative of the quadratic form

$$Q(X) = -3x_1^2 + 6x_1x_2 + 10x_1x_3 + 4x_2^2 - 4x_2x_3 - x_3^2$$

is the 3×3 matrix

$$S = \begin{bmatrix} -3 & 3 & 5 \\ 3 & 4 & -2 \\ 5 & -2 & -1 \end{bmatrix}$$

We leave it to the reader to verify that the matrix product $X^t S X$ is the same as $Q(X)$ if we express $3x_1 x_2 + 3x_2 x_1$ as $6x_1 x_2$, $5x_1 x_3 + 5x_3 x_1$ as $10x_1 x_3$, and $-2x_2 x_3 - 2x_3 x_2$ as $-4x_2 x_3$.

Next, we consider the effect of a change of variables on a given quadratic form. Let $Q(X) = X^t S X$ be the form in question, and suppose the variables $x_1, x_2, \ldots, x_n$ are related to $y_1, y_2, \ldots, y_n$ by the vector equation $X = P Y$, where P is an $n \times n$ nonsingular matrix. Then since

$$X^t S X = (P Y)^t S (P Y) = Y^t (P^t S P) Y$$

we conclude that the given quadratic form has the symmetric representative $P^t S P$ with respect to the variables $y_1, y_2, \ldots, y_n$. For this reason, it is convenient to have the terminology introduced in the following definition.

Definition 7.3. The $n \times n$ real symmetric matrix T is said to be *congruent* to the symmetric matrix S if there exists a non-singular matrix P such that $T = P^t S P$.

Note that the matrix equation $T = P^t S P$ can also be written as $S = (P^{-1})^t T (P^{-1})$, and it follows that T is congruent to S if and only if S is congruent to T. Henceforth, we shall simply say that S and T are congruent.

In the discussion preceding definition 7.3, we observed that two symmetric matrices which represent the same quadratic form with respect to different variables must be congruent. Conversely, if $S_2 = P^t S_1 P$, we find that

$$Y^t S_2 Y = Y^t (P^t S_1 P) Y = (P Y)^t S_1 (P Y) = X^t S_1 X$$

where $X = P Y$, and in this sense, congruent symmetric matrices represent the same quadratic form. Thus, we make the following definition.

Definition 7.4. The quadratic forms $X^t S_1 X$ and $Y^t S_2 Y$ are said to be *equivalent* if S_1 and S_2 are congruent and $X = P Y$, where P is a non-singular matrix which satisfies $P^t S_1 P = S_2$.

EXAMPLE 3.   The matrices

$$S_1 = \begin{bmatrix} 5 & -1 \\ -1 & 5 \end{bmatrix} \quad \text{and} \quad S_2 = \begin{bmatrix} 2 & 0 \\ 0 & 2 \end{bmatrix}$$

are congruent, for we have

$$P^t S_1 P = \begin{bmatrix} \dfrac{1}{\sqrt{6}} & -\dfrac{1}{\sqrt{6}} \\ \dfrac{1}{2} & \dfrac{1}{2} \end{bmatrix} \begin{bmatrix} 5 & -1 \\ -1 & 5 \end{bmatrix} \begin{bmatrix} \dfrac{1}{\sqrt{6}} & \dfrac{1}{2} \\ -\dfrac{1}{\sqrt{6}} & \dfrac{1}{2} \end{bmatrix} = \begin{bmatrix} 2 & 0 \\ 0 & 2 \end{bmatrix} = S_2$$

Therefore, the quadratic forms

$$X^t S_1 X = 5x_1^2 - 2x_1 x_2 + 5x_2^2$$

and

$$Y^t S_2 Y = 2y_1^2 + 2y_2^2$$

are equivalent. Note that the second form is obtained from the first by using the change of variables

$$x_1 = \frac{1}{\sqrt{6}} y_1 + \frac{1}{2} y_2$$

$$x_2 = -\frac{1}{\sqrt{6}} y_1 + \frac{1}{2} y_2$$

If the quadratic forms $X^t S_1 X$ and $Y^t S_2 Y$ are equivalent, the equations $X^t S_1 X = 1$ and $Y^t S_2 Y = 1$ represent the same surface with respect to different variables. Thus, if $X^t S_1 X$ is a complicated expression, we can hope to gain insight into the nature of the surface $X^t S_1 X = 1$ by first finding a "simple" form $Y^t S_2 Y$ which is equivalent to $X^t S_1 X$ and then analyzing the surface whose equation is $Y^t S_2 Y = 1$. However, it is important to note that even when S_1 and S_2 are congruent matrices, the corresponding surfaces, $X^t S_1 X = 1$ and $Y^t S_2 Y = 1$, are not necessarily congruent in the sense of Euclidean geometry. That is, these surfaces do not have to be identical except for their placement in space. For instance, in example 3, we found that the quadratic forms $5x_1^2 - 2x_1 x_2 + 5x_2^2$ and $2y_1^2 + 2y_2^2$ are equivalent. For this reason, we would expect the curve $5x_1^2 - 2x_1 x_2 + 5x_2^2 = 1$ to have a great deal in common with $2y_1^2 + 2y_2^2 = 1$. However, as shown in figure 7.3, the first curve is a full-fledged ellipse, while the second is a circle, and we have no way of determining such features of the ellipse as its area, the length of its axis, or the location of its foci by examining the circle.

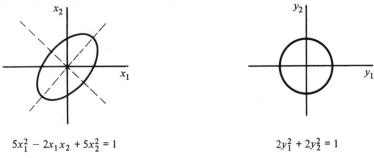

$$5x_1^2 - 2x_1 x_2 + 5x_2^2 = 1 \qquad\qquad\qquad 2y_1^2 + 2y_2^2 = 1$$

FIGURE 7.3

This example forces us to reexamine the technique of simplification we have just discussed, for this technique is valuable only if the metric features of the surface $X^t S_1 X = 1$ can be easily determined from the simplified form $Y^t S_2 Y = 1$. We can achieve this goal by insisting that the matrix P which relates the variables $x_1, \ldots, x_n$ to $y_1, \ldots, y_n$ be orthogonal, for we know that an orthogonal transformation of coordinates does preserve metric relationships (recall section 4.7). For this reason, we shall be especially interested in the concepts defined as follows.

Definition 7.5.   The $n \times n$ real symmetric matrix S_2 is said to be *orthogonally congruent* to the symmetric matrix S_1 if there exists an orthogonal matrix P such that $S_2 = P^t S_1 P$.

In this case, if $X = PY$, the quadratic forms $X^t S_1 X$ and $Y^t S_2 Y$ are said to be *orthogonally equivalent*.

If P is an orthogonal matrix, we have $P^t = P^{-1}$, and the equation $S_2 = P^t S_1 P$ can also be written as $S_2 = P^{-1} S_1 P$. Therefore, whenever S_1 and S_2 are orthogonally congruent, they are also similar and as a consequence, have the same eigenvalues, the same determinant, the same trace, and so on.

In the example which follows, we begin by observing that the quadratic form $5x_1^2 - 2x_1 x_2 + 5x_2^2$, which was analyzed in the discussion preceding definition 7.5, is orthogonally equivalent to the sum of squares form $4y_1^2 + 6y_2^2$. We then show that the curve $5x_1^2 - 2x_1 x_2 + 5x_2^2 = 1$ has the same features as the curve $4y_1^2 + 6y_2^2 = 1$, which is much easier to handle since the "cross" term, $y_1 y_2$, is absent.

EXAMPLE 4.   The symmetric matrices

$$S_1 = \begin{bmatrix} 5 & -1 \\ -1 & 5 \end{bmatrix} \quad \text{and} \quad S_2 = \begin{bmatrix} 4 & 0 \\ 0 & 6 \end{bmatrix}$$

are orthogonally congruent, for we have $P^tS_1P = S_2$ where P is the orthogonal matrix

$$P = \begin{bmatrix} \dfrac{1}{\sqrt{2}} & -\dfrac{1}{\sqrt{2}} \\[2mm] \dfrac{1}{\sqrt{2}} & \dfrac{1}{\sqrt{2}} \end{bmatrix}$$

Therefore, the quadratic forms

$$X^tS_1X = 5x_1^2 - 2x_1x_2 + 5x_2^2 \quad \text{and} \quad Y^tS_2Y = 4y_1^2 + 6y_2^2$$

are orthogonally equivalent.

Since we have $X = PY$, it follows that $Y = P^{-1}X = P^tX$, and the transformation of variables relating x_1, x_2 to y_1, y_2 can be written as

$$\begin{aligned} y_1 &= \frac{1}{\sqrt{2}}x_1 + \frac{1}{\sqrt{2}}x_2, & x_1 &= \frac{1}{\sqrt{2}}y_1 - \frac{1}{\sqrt{2}}y_2 \\[2mm] y_2 &= -\frac{1}{\sqrt{2}}x_1 + \frac{1}{\sqrt{2}}x_2, & x_2 &= \frac{1}{\sqrt{2}}y_1 + \frac{1}{\sqrt{2}}y_2 \end{aligned} \tag{7.2}$$

Note that the y_1, y_2-plane is obtained by simply rotating the x_1, x_2-plane through 45° in a counterclockwise direction about the origin (see figure 7.4). Since a rotation is a rigid motion, the rotated curve $4y_1^2 + 6y_2^2 = 1$ is congruent to $5x_1^2 - 2x_1x_2 + 5x_2^2 = 1$, which means that the metric characteristics of the two curves are identical. For example, as indicated in figure 7.4, one of the vertices of the ellipse $4y_1^2 + 6y_2^2 = 1$ is located at $(\frac{1}{2}, 0)$ in the y_1, y_2-plane,

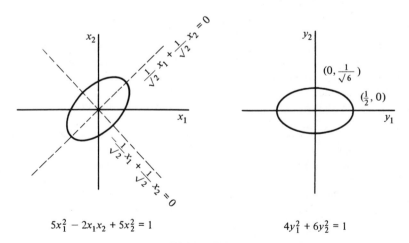

$$5x_1^2 - 2x_1x_2 + 5x_2^2 = 1 \qquad\qquad 4y_1^2 + 6y_2^2 = 1$$

FIGURE 7.4

and by substituting in equations (7.2), we find that the corresponding vertex in the x_1, x_2-plane is located at $(\sqrt{2}/4, \sqrt{2}/4)$. Similarly, the length of the major axis of $4y_1^2 + 6y_2^2 = 1$ is clearly 1, and it follows that the length of the major axis of $5x_1^2 - 2x_1x_2 + 5x_2^2 = 1$ is also 1. In exercise 10, we ask the reader to determine several more features of these curves.

To summarize, we have found that whenever the quadratic forms X^tS_1X and Y^tS_2Y are orthogonally equivalent, the surfaces $X^tS_1X = 1$ and $Y^tS_1Y = 1$ will be congruent in the usual geometric sense. Moreover, since a surface which is represented by the sum of squares equation $c_1y_1^2 + \ldots + c_ny_n^2 = 1$ is relatively easy to analyze, it would be nice to be able to show that every quadratic form is orthogonally equivalent to a sum of squares. Surprisingly, this can be done, and we shall construct a proof in the next section.

Exercise Set 7.1

1. Find the symmetric representative of each of the following quadratic forms.
   (i) $3x_1^2 - 2x_1x_2 - 5x_2^2$
   (ii) $-2x_1^2 + 6x_1x_2 - 8x_1x_3 + x_2^2 - 2x_2x_3 + 2x_3^2$
   (iii) $7x_1^2 + 4x_1x_2 + 4x_1x_3 - 4x_2x_3 - x_3^2$
2. In each of the following cases, find the quadratic form whose symmetric representative is the given matrix.

   (i) $\begin{bmatrix} -5 & 2 \\ 2 & 1 \end{bmatrix}$ (ii) $\begin{bmatrix} 0 & -3 \\ -3 & 0 \end{bmatrix}$ (iii) $\begin{bmatrix} 2 & 0 \\ 0 & -9 \end{bmatrix}$

   (iv) $\begin{bmatrix} 1 & 2 & -5 \\ 2 & 0 & 3 \\ -5 & 3 & -2 \end{bmatrix}$ (v) $\begin{bmatrix} -7 & 1 & 0 \\ 1 & 0 & 1 \\ 0 & 1 & 0 \end{bmatrix}$

3. If S is an $n \times n$ symmetric matrix, show that P^tSP is also symmetric.
4. Let S_1, S_2, and S_3 be symmetric matrices, and suppose that S_1 is congruent to S_2 and S_2 is congruent to S_3. Show that S_1 is also congruent to S_3.
5. Show that congruent matrices have the same rank.
   Hint: Use the result of exercise 8, section 4.5.
6. (i) Show that S_1 is orthogonally congruent to S_2 if and only if S_2 is orthogonally congruent to S_1.
   (ii) If S_1 is orthogonally congruent to S_2 and S_2 to S_3, show that S_1 is orthogonally congruent to S_3.
   (iii) Show that orthogonally congruent matrices are similar.
7. Let S be a symmetric matrix, and suppose that $P^tSP = D$, where P is orthogonal and $D = \text{diag}(d_1, \ldots, d_n)$.
   (i) Show that for each index j, the number d_j is an eigenvalue of S associated with the eigenvector γ_j, where γ_j is the jth column vector of P.
   (ii) If $Y = PX$, describe the quadratic form Y^tSY.
8. Let K be an $n \times n$ skew-symmetric matrix (that is, $K^t = -K$).

(i) If P is an $n \times m$ matrix, show that the $m \times m$ matrix $P^t K P$ is skew-symmetric.

(ii) If X is a column n-tuple, show that $X^t K X = 0$.

9. Note that each $n \times n$ matrix A can be written in the form $A = S + K$, where

$$S = \tfrac{1}{2}(A + A^t) \quad \text{and} \quad K = \tfrac{1}{2}(A - A^t)$$

are, respectively, symmetric and skew-symmetric matrices. Show that if X is a column n-tuple, then $X^t A X = X^t S X$.

10. This exercise refers to the situation analyzed in example 4.

(i) Verify that the ellipse $4y_1^2 + 6y_2^2 = 1$ has one focus at the point where $y_1 = \sqrt{3}/6$ and $y_2 = 0$.

(ii) The area of an ellipse is πab, where a and b are the lengths of its semi-axes. The semi-major axis of $4y_1^2 + 6y_2^2 = 1$ has length $\tfrac{1}{2}$. Find the length of its semi-minor axis and compute its area.

(iii) Find the foci of the ellipse $5x_1^2 - 2x_1 x_2 + 5x_2^2 = 1$ and compute its area.

7.2   THE PRINCIPAL AXES THEOREM

In this section, we shall prove that a real symmetric matrix S is necessarily diagonable. More specifically, we shall show that such a matrix must be orthogonally congruent to the diagonal matrix $D = \text{diag}(\lambda_1, \ldots, \lambda_n)$, where the λ's are the eigenvalues of S. Finally, as a consequence of this result, we shall then show that the quadratic form $X^t S X$ is orthogonally equivalent to the sum of squares form $\lambda_1 y_1^2 + \lambda_2 y_2^2 + \ldots + \lambda_n y_n^2$.

If S is orthogonally congruent to the diagonal matrix D, we shall say that S is *orthogonally diagonable*. If S has this property, then the diagonal entries of D must be the eigenvalues of S (see exercise 7, section 7.1). Therefore, we cannot hope to prove that each real symmetric matrix is orthogonally diagonable without first proving that such a matrix has only real eigenvalues. There are many different ways to obtain this result, all of which use the properties of complex numbers in one way or another. The particular approach we shall employ requires the notational device described in the following definition.

Definition 7.6.   The *conjugate transpose* of the $m \times n$ complex matrix $C = (c_{ij})$ is the $n \times m$ complex matrix C^* whose (i, j) entry is $\bar{c}_{ji}$, the complex conjugate of the (j, i) entry of C.

EXAMPLE 1.   If

$$C = \begin{bmatrix} -3 + i & 2i & 7 + 5i \\ 5 - 2i & -7 & 2i \end{bmatrix}$$

then

$$C^* = \begin{bmatrix} -3-i & 5+2i \\ -2i & -7 \\ 7-5i & -2i \end{bmatrix}$$

In supplementary exercises 9 and 12 of Chapter 3, we observed that the function defined by

$$(\alpha \,|\, \beta) = \sum_{i=1}^{n} a_i \bar{b}_i \quad \text{for } \alpha = \begin{bmatrix} a_1 \\ \cdot \\ \cdot \\ \cdot \\ a_n \end{bmatrix} \text{ and } \beta = \begin{bmatrix} b_1 \\ \cdot \\ \cdot \\ \cdot \\ b_n \end{bmatrix} \tag{7.3}$$

where the components a_i and b_i are complex numbers, has the following properties:

1. $(\alpha \,|\, \alpha) > 0 \quad$ if $\alpha \neq \theta$
2. $(\alpha \,|\, \beta) = \overline{(\beta \,|\, \alpha)}$
3. $(c\alpha \,|\, \beta) = c(\alpha \,|\, \beta) \quad$ and $(\alpha \,|\, c\beta) = \bar{c}(\alpha \,|\, \beta)$ (7.4)
4. $(\alpha + \beta \,|\, \gamma) = (\alpha \,|\, \gamma) + (\beta \,|\, \gamma)$
5. $(C\alpha \,|\, \beta) = (\alpha \,|\, C^*\beta) \quad$ for each $n \times n$ complex matrix C

(Note: In the list above α, β, and γ are complex n-tuples and c is a complex number.) We shall use these properties to prove our first theorem.

Theorem 7.1. Every eigenvalue of a real symmetric matrix S is real.

Proof: Let λ be an eigenvalue of S, and let X be an associated eigenvector. Then if ($\,|\,$) is the standard complex inner product [see (7.3)], we have

$$(SX \,|\, X) = (\lambda X \,|\, X) = \lambda(X \,|\, X)$$

and

$$(SX \,|\, X) = (X \,|\, S^*X) = (X \,|\, SX) = (X \,|\, \lambda X) = \bar{\lambda}(X \,|\, X)$$

where we have used the properties listed in (7.4) and the fact that $S^* = S$, since S is real and symmetric. Equating these two expressions for $(SX \,|\, X)$, we find that $\lambda(X \,|\, X) = \bar{\lambda}(X \,|\, X)$, and since $(X \,|\, X) > 0$ (why?), it follows that $\lambda = \bar{\lambda}$. Therefore, λ must be real, as claimed (why?).

The reader may be interested in working through the steps of the alternative proof for this theorem which is outlined in exercise 8. ∎

EXAMPLE 2. The characteristic polynomial of the 3×3 real symmetric matrix

$$S = \begin{bmatrix} 1 & -3 & 4 \\ -3 & 0 & 5 \\ 4 & 5 & -7 \end{bmatrix}$$

is $p_S(x) = x^3 + 6x^2 - 57x + 82$. We find that the zeroes of this polynomial are $\lambda_1 = 2$, $\lambda_2 = -4 + \sqrt{57}$, and $\lambda_3 = -4 - \sqrt{57}$. These numbers are the eigenvalues of S, and they are all real, as predicted by theorem 7.1.

We are now just about ready to show that every real symmetric matrix is orthogonally diagonable. The general proof of this result is straightforward but long, and the following lemma will help to simplify at least part of the proof.

Lemma 7.1. Let S be a real symmetric matrix with eigenvalues $\lambda_1, \ldots,$ λ_n, and let P be an orthogonal matrix whose first column vector γ_1 satisfies $S\gamma_1 = \lambda_1\gamma_1$. Then the product P^tSP has the block form

$$P^tSP = \left[\begin{array}{c|c} \lambda_1 & 0 \\ \hline 0 & S_1 \end{array}\right] \tag{7.5}$$

where S_1 is an $(n-1) \times (n-1)$ symmetric matrix with eigenvalues $\lambda_2, \ldots,$ λ_n.

Proof: If $\gamma_1, \ldots, \gamma_n$ are the column vectors of P, then the (i, j) entry of the product $T = P^tSP$ is $\gamma_i^tS\gamma_j$. In particular, for $i = 1, 2, \ldots, n$, we have

$$t_{i1} = \gamma_i^tS\gamma_1 = \gamma_i^t(\lambda_1\gamma_1) = \lambda_1(\gamma_i^t\gamma_1) = \begin{cases} \lambda_1 & \text{if } i = 1 \\ 0 & \text{if } i \neq 1 \end{cases}$$

since the γ's are mutually orthonormal. This means that the off-diagonal entries in the first column of T are all 0, and since T is symmetric (see exercise 3, section 7.1), it must have the form displayed in (7.5). Finally, since T is similar to S (why is this true?), its eigenvalues are $\lambda_1, \ldots, \lambda_n$, and since

$$p_T(x) = \det(xI_n - T) = (x - \lambda_1)\det(xI_{n-1} - S_1)$$
$$= (x - \lambda_1)p_{S_1}(x)$$

we conclude that S_1 has eigenvalues $\lambda_2, \lambda_3, \ldots, \lambda_n$. ∎

We now have all we need to prove the main result of this section.

Theorem 7.2. (The matrix form of the principal axes theorem.) If S is an $n \times n$ real symmetric matrix, there exists an orthogonal matrix P such that $P^tSP = \text{diag}(\lambda_1, \ldots, \lambda_n)$, where the λ's are the eigenvalues of S.

Proof: We shall use an inductive argument based on n, the dimension of S. This argument has a starting point since the theorem is clearly true for 1×1 matrices. For the inductive step, we assume that every $(n-1) \times (n-1)$ real symmetric matrix is orthogonally diagonable and then use this assumption to prove that the $n \times n$ real symmetric matrix S has the desired property.

Accordingly, let γ_1 be an eigenvector of S associated with λ_1, and assume that γ_1 has unit length (why does this assumption involve no loss of generality?). Next, use the Gram-Schmidt process to extend γ_1 to a basis $\{\gamma_1, \ldots, \gamma_n\}$ which is orthonormal with respect to the standard inner product. If P_1 is the $n \times n$ orthogonal matrix whose column vectors are $\gamma_1, \gamma_2, \ldots, \gamma_n$, then lemma 7.1 tells us that $P_1^t S P_1$ has the block form

$$P_1^t S P_1 = \left[\begin{array}{c|c} \lambda_1 & 0 \\ \hline 0 & S_1 \end{array}\right]$$

where S_1 is an $(n-1) \times (n-1)$ symmetric matrix with eigenvalues $\lambda_2, \ldots, \lambda_n$. According to the inductive hypothesis, there exists an orthogonal $(n-1) \times (n-1)$ matrix P_2 such that $P_2^t S_1 P_2 = \text{diag}(\lambda_2, \ldots, \lambda_n)$. Let P_3 denote the $n \times n$ matrix whose block form is

$$P_3 = \left[\begin{array}{c|c} 1 & 0 \\ \hline 0 & P_2 \end{array}\right]$$

Then P_3 is orthogonal, and we find that

$$P_3^t(P_1^t S P_1)P_3 = \text{diag}(\lambda_1, \lambda_2, \ldots, \lambda_n)$$

(verify this step in detail). Finally, since P_1 and P_3 are orthogonal, so is the product matrix $P = P_1 P_3$ (see exercise 5, section 4.7), and we have

$$P^t S P = (P_1 P_3)^t S(P_1 P_3) = P_3^t(P_1^t S P_1)P_3$$
$$= \text{diag}(\lambda_1, \lambda_2, \ldots, \lambda_n)$$

which shows that S is orthogonally diagonable. ∎

In section 6.3, we showed that if $P^{-1}AP = \text{diag}(\lambda_1, \ldots, \lambda_n)$ then the column vectors of P form a basis for R_n comprised of eigenvectors of A, and in the case where P is orthogonal, this basis is orthonormal with respect to the standard inner product (why?). Thus, we have the following result.

Corollary 7.1. If S is an $n \times n$ real symmetric matrix, there exists an orthonormal basis $\{\gamma_1, \gamma_2, \ldots, \gamma_n\}$ for R_n in which each γ_j is an eigenvector of S.

To find the orthonormal basis whose existence is guaranteed by corollary

7.1, we may proceed as follows:

1. Find a basis for the eigenspace associated with each distinct eigenvalue of S. The union of such bases is a basis for all of R_n.
2. Use the Gram-Schmidt process to orthonormalize the basis found in step 1.

The second step in this procedure promises to be tedious, but our next theorem eases the computational pain a bit by showing that eigenvectors which correspond to distinct eigenvalues of S are already mutually orthogonal.

Theorem 7.3. Eigenvectors of a real symmetric matrix which correspond to distinct eigenvalues are orthogonal.

Proof: Let X, Y be eigenvectors of the $n \times n$ symmetric matrix S which correspond to the distinct eigenvalues λ, μ, respectively. We wish to show that $(X|Y) = 0$, where $(\ |\ )$ is the standard inner product on R_n (remember, S has real eigenvalues and eigenvectors). To this end, recall that $(S\alpha | \beta) = (\alpha | S^t\beta)$ for all n-tuples α, β (see exercise 12, section 3.2). Therefore, we have

$$(SX|Y) = (\lambda X|Y) = \lambda(X|Y)$$

and

$$(SX|Y) = (X|S^tY) = (X|SY) = (X|\mu Y) = \mu(X|Y)$$

Equating these two expressions for $(SX|Y)$, we find that $\lambda(X|Y) = \mu(X|Y)$, and since $\lambda \neq \mu$, it follows that $(X|Y) = 0$, as claimed. ∎

Thanks to theorem 7.3, we know that an orthonormal basis comprised of eigenvectors of S may be constructed by simply combining orthonormal bases for the eigenspaces which correspond to the distinct eigenvalues of S. Consider the following example.

EXAMPLE 3.   The symmetric matrix

$$S = \begin{bmatrix} 2 & -1 & -1 \\ -1 & 2 & -1 \\ -1 & -1 & 2 \end{bmatrix}$$

has distinct eigenvalues $\lambda_1 = 0$ and $\lambda_2 = 3$. Following the procedure described in section 6.2, we find that $\{\eta_1\}$ and $\{\eta_2, \eta_3\}$ are bases for the eigenspaces S_{λ_1} and S_{λ_2}, respectively, where

$$\eta_1 = \begin{bmatrix} 1 \\ 1 \\ 1 \end{bmatrix} \quad \text{and} \quad \eta_2 = \begin{bmatrix} -1 \\ 0 \\ 1 \end{bmatrix}, \quad \eta_3 = \begin{bmatrix} 0 \\ -1 \\ 1 \end{bmatrix}$$

As predicted by theorem 7.3, the vector η_1 is orthogonal to both η_2 and η_3, and by applying the Gram-Schmidt process to η_2 and η_3, we find that the vector

$$\eta_3' = \eta_3 - \left[\frac{(\eta_2 \mid \eta_3)}{(\eta_2 \mid \eta_2)}\right]\eta_2 = \eta_3 - \frac{1}{2}\eta_2 = \begin{bmatrix} \frac{1}{2} \\ -1 \\ \frac{1}{2} \end{bmatrix}$$

is orthogonal to η_2. Finally, we normalize the vectors in the orthogonal basis $\{\eta_1, \eta_2, \eta_3'\}$ to obtain the orthonormal basis $\{\gamma_1, \gamma_2, \gamma_3\}$ for R_3, where

$$\gamma_1 = \begin{bmatrix} \frac{1}{\sqrt{3}} \\ \frac{1}{\sqrt{3}} \\ \frac{1}{\sqrt{3}} \end{bmatrix}, \quad \gamma_2 = \begin{bmatrix} -\frac{1}{\sqrt{2}} \\ 0 \\ \frac{1}{\sqrt{2}} \end{bmatrix}, \quad \gamma_3 = \begin{bmatrix} \frac{1}{\sqrt{6}} \\ -\frac{2}{\sqrt{6}} \\ \frac{1}{\sqrt{6}} \end{bmatrix}$$

Therefore, we have $P^tSP = \text{diag}\,(0, 3, 3)$, where

$$P = \begin{bmatrix} \frac{1}{\sqrt{3}} & -\frac{1}{\sqrt{2}} & \frac{1}{\sqrt{6}} \\ \frac{1}{\sqrt{3}} & 0 & -\frac{2}{\sqrt{6}} \\ \frac{1}{\sqrt{3}} & \frac{1}{\sqrt{2}} & \frac{1}{\sqrt{6}} \end{bmatrix} \tag{7.6}$$

is the orthogonal matrix whose column vectors are γ_1, γ_2, and γ_3.

Since our interest in orthogonal diagonability was originally motivated by the desire to obtain a simple representation for quadratic forms, it is appropriate to end the work of this section with the following result.

Theorem 7.4. (The principal axes theorem.) The quadratic form X^tSX is orthogonally equivalent to the sum of squares $\lambda_1 y_1^2 + \lambda_2 y_2^2 + \ldots + \lambda_n y_n^2$, where the λ's are the eigenvalues of S.

Proof: Let P be an orthogonal matrix such that $P^tSP = D = \mathrm{diag}\,(\lambda_1,$ $\ldots, \lambda_n)$, and let Y be chosen so that $X = PY$. Then we have

$$X^tSX = Y^t(P^tSP)Y = Y^tDY = \lambda_1 y_1^2 + \ldots + \lambda_n y_n^2$$

as desired. ∎

EXAMPLE 4.   The symmetric representative of the quadratic form

$$2x_1^2 - 2x_1x_2 - 2x_1x_3 + 2x_2^2 - 2x_2x_3 + 2x_3^2$$

is the matrix

$$S = \begin{bmatrix} 2 & -1 & -1 \\ -1 & 2 & -1 \\ -1 & -1 & 2 \end{bmatrix}$$

In example 3, we found that $P^tSP = D = \mathrm{diag}\,(0, 3, 3)$, where P is the 3×3 orthogonal matrix displayed in (7.6), and by setting $Y = P^tX$ (that is, $X = PY$), we obtain

$$X^tSX = Y^tDY = 0y_1^2 + 3y_2^2 + 3y_3^2$$

as the sum of squares expression for the given quadratic form.

Geometrically, theorem 7.4 says that it is always possible to rotate the coordinate axes of n-space to make them coincide with the principal axes of symmetry of a given surface $X^tSX = 1$. For instance, the result obtained in example 4 tells us that the quadric surface

$$2x_1^2 - 2x_1x_2 - 2x_1x_3 + 2x_2^2 - 2x_2x_3 + 2x_3^2 = 1$$

is the same as the surface

$$0y_1^2 + 3y_2^2 + 3y_3^2 = 1$$

which is a circular cylinder symmetric about the y_1-axis (see figure 7.5). This example and example 4 of the last section both illustrate the method by which the principal axes theorem may be used to analyze a given conic section or quadric surface. In the next section, we continue our study of real quadratic forms by examining a few more applications of this important theorem.

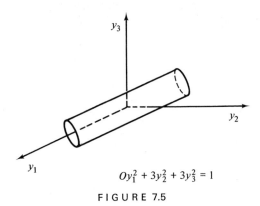

$$Oy_1^2 + 3y_2^2 + 3y_3^2 = 1$$

FIGURE 7.5

Exercise Set 7.2

1. Carry out these three steps for each of the matrices which follow:
   1. Compute the eigenvalues of the given symmetric matrix S.
   2. Construct a basis for the eigenspace associated with each distinct eigenvalue of S.
   3. Find an orthogonal matrix P such that P^tSP is diagonal.

   (i) $\begin{bmatrix} 5 & -2 \\ -2 & 2 \end{bmatrix}$
   (ii) $\begin{bmatrix} 7 & 2 \\ 2 & 4 \end{bmatrix}$
   (iii) $\begin{bmatrix} 2 & 3 \\ 3 & 2 \end{bmatrix}$

   (iv) $\begin{bmatrix} 3 & 1 & -4 \\ 1 & 3 & -4 \\ -4 & -4 & 8 \end{bmatrix}$
   (v) $\begin{bmatrix} 3 & 1 & -4 \\ 1 & 6 & -7 \\ -4 & -7 & 11 \end{bmatrix}$

   (vi) $\begin{bmatrix} -2 & 1 & 1 & 1 \\ 1 & -2 & 1 & 1 \\ 1 & 1 & -2 & 1 \\ 1 & 1 & 1 & -2 \end{bmatrix}$
   (vii) $\begin{bmatrix} 1 & 3 & -2 & 1 \\ 3 & 9 & -6 & 3 \\ -2 & -6 & 4 & -2 \\ 1 & 3 & -2 & 1 \end{bmatrix}$

   Hint: One eigenvalue is 15.

2. In each case, find a sum of squares which is orthogonally equivalent to the given quadratic form.
   (i) $2x_1^2 - 6x_1x_2 + 2x_2^2$
   (ii) $6x_1x_2 + 2x_2x_3$
   (iii) $5x_1^2 + 2x_1x_2 + 2x_1x_3 + 5x_2^2 + 2x_2x_3 + 5x_3^2$
3. Show that a non-zero symmetric matrix cannot be nilpotent.
   Hint: Can a nilpotent matrix be diagonable?
4. (i) Show that the number of non-zero eigenvalues of a real symmetric matrix equals its rank. Does the same result apply to any diagonable matrix?

(ii) Show that the result of part (i) does not necessarily apply to non-symmetric matrices. Specifically, construct a 2×2 matrix A which has two zero eigenvalues and rank 1.

5. Show that an $n \times n$ real matrix A is diagonable if and only if it is similar to a real symmetric matrix.

6. Let A be an $n \times n$ real matrix, and let $S = A^t A$.
   (i) Show that S is symmetric.
   (ii) If γ is a real n-tuple, show that $\gamma^t S \gamma$ is a non-negative real number.
   Hint: Note that $\gamma^t S \gamma = (A\gamma)^t(A\gamma)$.
   (iii) If λ is an eigenvalue of S, show that $\lambda \geq 0$.

7. Let K be an $n \times n$ skew-symmetric matrix.
   (i) Show that each non-zero eigenvalue λ of K is a pure imaginary number, that is, of the form $\lambda = ci$.
   Hint: Modify the proof of theorem 7.1 to show that $\bar{\lambda} = -\lambda$.
   (ii) Show that the eigenvectors associated with distinct eigenvalues of K must be orthogonal.

8. Let λ be an eigenvalue of the $n \times n$ real symmetric matrix S, and let $X = (x_i)$ be an associated eigenvector. Show that λ is real by completing the following steps:
   (i) Show that $\bar{X}^t S X = \lambda(\bar{X}^t X)$, where $\bar{X}$ denotes the complex column n-tuple whose ith component is $\bar{x}_i$, the complex conjugate of x_i.
   (ii) Verify that $(\overline{SX})^t = \bar{X}^t S = \bar{\lambda} \bar{X}^t$, and then show that $\bar{X}^t S X = \bar{\lambda}(\bar{X}^t X)$.
   (iii) Explain why $\bar{X}^t X$ is a positive real number, and then show that $\lambda = \bar{\lambda}$. Why does this imply that λ is real?

7.3 A CLASSIFICATION OF QUADRATIC FORMS

In this section, we shall discover that two real symmetric matrices are congruent if and only if they have the same number of positive eigenvalues and the same number that are negative. We shall then use this criterion for congruence as the basis for a general classification of all real quadratic forms. We begin by showing that each symmetric matrix is congruent to a diagonal matrix with a special arrangement of 1's, -1's, and 0's on its main diagonal.

Lemma 7.2. An $n \times n$ real symmetric matrix S with p positive eigenvalues and q that are negative is congruent to the diagonal matrix

$$D_{pq} = \text{diag}\,(\underbrace{1, \ldots, 1}_{p \text{ terms}}, \underbrace{-1, \ldots, -1}_{q \text{ terms}}, \underbrace{0, \ldots, 0}_{n\text{-}p\text{-}q \text{ terms}})$$

Proof: The matrix version of the principal axes theorem may be used to establish the existence of an orthogonal matrix P such that

$$P^t S P = D_1 = \text{diag}\,(\pi_1, \ldots, \pi_p, \nu_1, \ldots, \nu_q, 0, \ldots, 0)$$

where $\pi_1, \ldots, \pi_p$ and $\nu_1, \ldots, \nu_q$ are the positive and negative eigenvalues of S, respectively. If D_2 is the matrix

$$D_2 = \text{diag}\,(\pi_1^{-1/2}, \ldots, \pi_p^{-1/2}, (-\nu_1)^{-1/2}, \ldots, (-\nu_q)^{-1/2}, 1, \ldots, 1)$$

we find that

$$Q^t SQ = D_2^t(P^t SP)D_2 = D_2^t D_1 D_2$$
$$= \text{diag}\,(1, \ldots, 1, -1, \ldots, -1, 0, \ldots, 0) = D_{pq}$$

where $Q = PD_2$. Finally, since P and D_2 are both non-singular, so is Q, and it follows that S is congruent to D_{pq}. ∎

EXAMPLE 1.   Suppose the 5×5 symmetric matrix S has eigenvalues 1, 3, 3, -7, and 0. Then we have $p = 3, q = 1$, and lemma 7.2 tells us that S is congruent to

$$D_{31} = \text{diag}\,(1, 1, 1, -1, 0)$$

Next, let S_1 and S_2 be $n \times n$ real symmetric matrices. If they both have p positive eigenvalues and q that are negative, then according to lemma 7.2, they are both congruent to the diagonal matrix D_{pq} and hence to each other (see exercise 4, section 7.1). Conversely, if S_1 and S_2 are congruent, it can be shown that they have the same number of positive eigenvalues and the same number that are negative. Unfortunately, to prove this statement, we would have to devote a great deal of time and effort to the process of eliminating a few technical difficulties which have little to do with the main development of ideas. Instead, we shall simply state the result in question without further proof and proceed to examine its implications and applications.

Theorem 7.5.   Two $n \times n$ real symmetric matrices are congruent if and only if they have the same number of positive eigenvalues and the same number of negative eigenvalues.

In the mathematical literature, the criterion for congruence given in theorem 7.5 usually appears in a slightly different form. To state this alternative version, we need the terminology introduced in the following definition.

Definition 7.7.   If S is an $n \times n$ real symmetric matrix with p positive eigenvalues and q that are negative, then the number $s = p - q$ is called the *signature* of S.

EXAMPLE 2. Suppose S is a 5×5 symmetric matrix with eigenvalues $3, -2, -2, -1$, and 0. Then $p = 1, q = 3$, and the signature is $s = 1 - 3 = -2$.

The rank of an $n \times n$ symmetric matrix S must equal the total number of non-zero eigenvalues (see exercise 4, section 7.2). Therefore, if S has p positive eigenvalues and q that are negative, it must have rank $r = p + q$ and signature $s = p - q$. This observation may be used in conjunction with theorem 7.5 to prove the following theorem.

Theorem 7.6. Two $n \times n$ real symmetric matrices S_1 and S_2 are congruent if and only if they have the same rank and signature.

Proof: Assume that S_1 and S_2 have, respectively, p_1 and p_2 positive eigenvalues and q_1 and q_2 that are negative. Then S_1 has rank $r_1 = p_1 + q_1$ and signature $s_1 = p_1 - q_1$, while the rank and signature of S_2 are $r_2 = p_2 + q_2$ and $s_2 = p_2 - q_2$, respectively.

If S_1 and S_2 are congruent, theorem 7.5 tells us that $p_1 = p_2$ and $q_1 = q_2$, and it follows that the two matrices have the same rank and signature.

Conversely, if $r_1 = r_2$ and $s_1 = s_2$, then we have $p_1 + q_1 = p_2 + q_2$ and $p_1 - q_1 = p_2 - q_2$. Solving, we find that $p_1 = p_2$ and $q_1 = q_2$, and theorem 7.5 enables us to conclude that S_1 and S_2 are congruent. ∎

EXAMPLE 3. The matrices

$$S_1 = \begin{bmatrix} 1 & 2 \\ 2 & 1 \end{bmatrix} \quad \text{and} \quad S_2 = \begin{bmatrix} 1 & -4 \\ -4 & 1 \end{bmatrix}$$

both have rank 2 and signature 0, since S_1 has eigenvalues 3 and -1 and S_2 has eigenvalues 5 and -3. Therefore, S_1 and S_2 are congruent, but neither is congruent to the matrix

$$S_3 = \begin{bmatrix} 3 & -1 \\ -1 & 3 \end{bmatrix}$$

which has rank 2 and signature 2 (its eigenvalues are 2 and 4).

In general, it is not practical to compute the rank and signature of a given symmetric matrix S by finding all its eigenvalues and counting the number that are positive, negative, and zero. Instead, we shall find these quantities by a computational procedure based on the fact that any matrix which is congruent to S must have the same rank and signature. To be specific, observe that if E is an elementary matrix, the product ESE^t is congruent to S and may be obtained by first performing the elementary operation that corresponds to E on the rows of S and then performing the same operation on its columns. By choosing our elementary operations intelligently, we can eventually obtain a diagonal matrix D which is congruent to S, and the rank and signature of S can then be found by counting the number of positive, negative, and zero diagonal entries of D. It is important to remember that S is not necessarily similar to D, and thus, the diagonal entries of D do not have to be the eigenvalues of S. However, as long as we are only interested in finding the rank and signature of S, this is not especially important. The computational aspects of the procedure described above are illustrated in the following example.

EXAMPLE 4.   To find a diagonal matrix which is congruent to the matrix

$$S = \begin{bmatrix} -1 & 2 & -3 \\ 2 & -4 & 5 \\ -3 & 5 & 0 \end{bmatrix}$$

we begin by using elementary row operations and the corresponding elementary column operations to reduce to 0 all the off-diagonal entries in the first row and first column:

$$\begin{bmatrix} -1 & 2 & -3 \\ 2 & -4 & 5 \\ -3 & 5 & 0 \end{bmatrix} \xrightarrow[\substack{-3R_1 + R_3 \to R_3 \\ -3C_1 + C_3 \to C_3}]{\substack{2R_1 + R_2 \to R_2 \\ 2C_1 + C_2 \to C_2}} \begin{bmatrix} -1 & 0 & 0 \\ 0 & 0 & -1 \\ 0 & -1 & 9 \end{bmatrix}$$

Next, we exchange the second and third rows and the second and third columns in order to put a non-zero entry in the $(2, 2)$ position (why is this desirable?):

$$\begin{bmatrix} -1 & 0 & 0 \\ 0 & 0 & -1 \\ 0 & -1 & 9 \end{bmatrix} \xrightarrow[C_2 \longleftrightarrow C_3]{R_2 \longleftrightarrow R_3} \begin{bmatrix} -1 & 0 & 0 \\ 0 & 9 & -1 \\ 0 & -1 & 0 \end{bmatrix}$$

Finally, we use the 9 in the $(2, 2)$ position to reduce to 0 the other entries in the second row and second column:

$$\begin{bmatrix} -1 & 0 & 0 \\ 0 & 9 & -1 \\ 0 & -1 & 0 \end{bmatrix} \xrightarrow[\frac{1}{9}C_2 + C_3 \to C_3]{\frac{1}{9}R_2 + R_3 \to R_3} \begin{bmatrix} -1 & 0 & 0 \\ 0 & 9 & 0 \\ 0 & 0 & -\frac{1}{9} \end{bmatrix}$$

The diagonal matrix D produced by this reduction is congruent to S and has one positive number and two negative numbers on its main diagonal. Hence, both D and S have rank 3 and signature -1.

In general, the strategy used to carry out this method of reduction is the same as that used in the downward phase of the reduction algorithm. However, a slight modification is required in the case where S has all zero entries on its main diagonal. For instance, the procedure used in example 4 cannot be applied to the symmetric matrix

$$S = \begin{bmatrix} 0 & 1 \\ 1 & 0 \end{bmatrix}$$

since there is no way to place a non-zero entry into the $(1, 1)$ position by exchanging rows and corresponding columns (try it!). However, by adding the second row to the first and the second column to the first column, we obtain the matrix

$$\begin{bmatrix} 2 & 1 \\ 1 & 0 \end{bmatrix}$$

which can then be reduced to diagonal form as follows:

$$\begin{bmatrix} 2 & 1 \\ 1 & 0 \end{bmatrix} \xrightarrow[-\frac{1}{2}C_1 + C_2 \to C_2]{-\frac{1}{2}R_1 + R_2 \to R_2} \begin{bmatrix} 2 & 0 \\ 0 & -\frac{1}{2} \end{bmatrix}$$

Since two quadratic forms are equivalent whenever their symmetric representatives are congruent, the following result is an immediate consequence of theorem 7.7.

Corollary 7.2.   The real quadratic forms $X^t S_1 X$ and $Y^t S_2 Y$ are equivalent if and only if their symmetric representatives have the same rank and signature.

This corollary enables us to classify each real quadratic form according to the rank and signature of its symmetric representative, as described in the following definition.

Definition 7.8.   The quadratic form $X^t S X$ is said to be of type $(n; r, s)$ if its symmetric representative S is an $n \times n$ matrix with rank r and signature s.

Note that for each n, there are only finitely many different types of quadratic forms, since we must have $0 \leq r \leq n$ and $-n \leq s \leq n$. In fact, it can be shown that there are exactly $(n + 1)(n + 2)/2$ different types (see exercise 5), and in tables 7.1 and 7.2 we list the various possibilities for $n = 2$ and $n = 3$, respectively. Incidentally, as an added feature, the last column in both tables describes the kind of surface $X^t S X = 1$ which corresponds to each

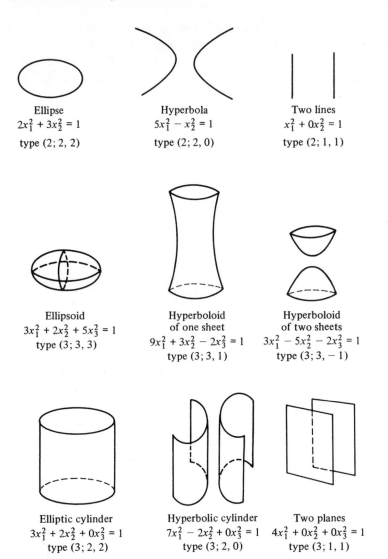

Ellipse
$2x_1^2 + 3x_2^2 = 1$
type $(2; 2, 2)$

Hyperbola
$5x_1^2 - x_2^2 = 1$
type $(2; 2, 0)$

Two lines
$x_1^2 + 0x_2^2 = 1$
type $(2; 1, 1)$

Ellipsoid
$3x_1^2 + 2x_2^2 + 5x_3^2 = 1$
type $(3; 3, 3)$

Hyperboloid
of one sheet
$9x_1^2 + 3x_2^2 - 2x_3^2 = 1$
type $(3; 3, 1)$

Hyperboloid
of two sheets
$3x_1^2 - 5x_2^2 - 2x_3^2 = 1$
type $(3; 3, -1)$

Elliptic cylinder
$3x_1^2 + 2x_2^2 + 0x_3^2 = 1$
type $(3; 2, 2)$

Hyperbolic cylinder
$7x_1^2 - 2x_2^2 + 0x_3^2 = 1$
type $(3; 2, 0)$

Two planes
$4x_1^2 + 0x_2^2 + 0x_3^2 = 1$
type $(3; 1, 1)$

type of quadratic form. For instance, in example 4, we observed that the symmetric matrix

$$S = \begin{bmatrix} -1 & 2 & -3 \\ 2 & -4 & 5 \\ -3 & 5 & 0 \end{bmatrix}$$

has rank 3 and signature -1. Therefore, the quadratic form

$$X^t S X = -x_1^2 + 4x_1x_2 - 6x_1x_3 - 4x_2^2 + 10x_2x_3$$

TABLE 7.1

Classification of Quadratic Forms in Two Variables

type of form $(2; r, s)$	eigenvalue pattern	reduced form of X^tSX	conic section associated with $X^tSX = 1$
$(2; 2, 2)$	$+ \ +$	$x_1^2 + x_2^2$	ellipse
$(2; 2, 0)$	$+ \ -$	$x_1^2 - x_2^2$	hyperbola
$(2; 2, -2)$	$- \ -$	$-x_1^2 - x_2^2$	none
$(2; 1, 1)$	$+ \ 0$	$x_1^2 + 0x_2^2$	two lines
$(2; 1, -1)$	$- \ 0$	$-x_1^2 + 0x_2^2$	none
$(2; 0, 0)$	$0 \ 0$	$0x_1^2 + 0x_2^2$	none

TABLE 7.2

Classification of Quadratic Forms in Three Variables

type of form $(3; r, s)$	eigenvalue pattern	reduced form of X^tSX	quadric surface associated with $X^tSX = 1$
$(3; 3, 3)$	$+ \ + \ +$	$x_1^2 + x_2^2 + x_3^2$	ellipsoid
$(3; 3, 1)$	$+ \ + \ -$	$x_1^2 + x_2^2 - x_3^2$	hyperboloid of one sheet
$(3; 3, -1)$	$+ \ - \ -$	$x_1^2 - x_2^2 - x_3^2$	hyperboloid of two sheets
$(3; 3, -3)$	$- \ - \ -$	$-x_1^2 - x_2^2 - x_3^2$	none
$(3; 2, 2)$	$+ \ + \ 0$	$x_1^2 + x_2^2 + 0x_3^2$	elliptic cylinder
$(3; 2, 0)$	$+ \ - \ 0$	$x_1^2 - x_2^2 + 0x_3^2$	hyperbolic cylinder
$(3; 2, -2)$	$- \ - \ 0$	$-x_1^2 - x_2^2 + 0x_3^2$	none
$(3; 1, 1)$	$+ \ 0 \ 0$	$x_1^2 + 0x_2^2 + 0x_3^2$	two planes
$(3; 1, -1)$	$- \ 0 \ 0$	$-x_1^2 + 0x_2^2 + 0x_3^2$	none
$(3; 0, 0)$	$0 \ 0 \ 0$	$0x_1^2 + 0x_2^2 + 0x_3^2$	none

is of type $(3; 3, -1)$, and according to table 7.2, the quadric surface

$$-x_1^2 + 4x_1x_2 - 6x_1x_3 - 4x_2^2 + 10x_2x_3 = 1$$

is a hyperboloid of two sheets (see page 340). On the other hand, if S happens to be a 2×2 real symmetric matrix with rank 2 and signature -2, then the equation $X^t S X = 1$ cannot be satisfied, as indicated in line 3 of table 7.1.

A quadratic form of type $(n; n, n)$ is said to be *positive definite*, and the same term is applied to the symmetric representative of such a form. In other words, a symmetric matrix is positive definite if each of its n eigenvalues is positive. According to table 7.2, if the quadratic form $X^t S X$ is positive definite, then the quadric surface $X^t S X = 1$ is an ellipsoid. In the solved problem which follows, we conclude the work of this chapter by establishing a simple relationship between the eigenvalues of S and certain features of this ellipsoid.

PROBLEM.   If $X^t S X$ is a positive definite quadratic form in three variables and $\lambda_1, \lambda_2, \lambda_3$ are the eigenvalues of S, show that the following statements can be made about the ellipsoid $X^t S X = 1$:
  (i) The semi-axes of the ellipsoid have lengths $\lambda_1^{-1/2}, \lambda_2^{-1/2}, \lambda_3^{-1/2}$
  (ii) The volume of the ellipsoid is

$$V = \frac{4}{3} \frac{\pi}{\sqrt{\det S}}$$

Solution: We shall base our solution on the fact that an ellipsoid in the standard form

$$\frac{x^2}{a^2} + \frac{y^2}{b^2} + \frac{z^2}{c^2} = 1 \quad (a, b, c \text{ all positive})$$

bounds a region with volume $\frac{4}{3}\pi abc$ and has semi-axes whose lengths are a, b, and c.

If P is an orthogonal matrix such that $P^t S P = D$, where $D = \text{diag}(\lambda_1, \lambda_2, \lambda_3)$, then the transformation of variables represented by $X = PY$ is a rigid motion. We find that

$$X^t S X = Y^t (P^t S P) Y = Y^t D Y = \lambda_1 y_1^2 + \lambda_2 y_2^2 + \lambda_3 y_3^2$$

and since a rigid motion has no effect on metric relationships, it follows that the transformed ellipsoid $\lambda_1 y_1^2 + \lambda_2 y_2^2 + \lambda_3 y_3^2 = 1$ has the same general properties as the ellipsoid $X^t S X = 1$. When the transformed ellipsoid is expressed in standard form, we have

$$\lambda_1 y_1^2 + \lambda_2 y_2^2 + \lambda_3 y_3^2 = \frac{y_1^2}{\mu_1^2} + \frac{y_2^2}{\mu_2^2} + \frac{y_3^2}{\mu_3^2} = 1$$

where $\mu_1 = \lambda_1^{-1/2}$, $\mu_2 = \lambda_2^{-1/2}$, and $\mu_3 = \lambda_3^{-1/2}$. Note that this assignment is made possible by the fact that S is positive definite (why?). From this standard form, we see that the transformed ellipsoid has semi-axes with lengths μ_1, μ_2, μ_3 and that it bounds a region whose volume is

$$V = \frac{4}{3}\pi\mu_1\mu_2\mu_3 = \frac{4}{3}\frac{\pi}{\sqrt{\lambda_1\lambda_2\lambda_3}} = \frac{4}{3}\frac{\pi}{\sqrt{\det S}}$$

Since the given ellipsoid $X^tSX = 1$ also has these characteristics, our solution is complete.

EXAMPLE 5.   Using the procedure described in example 4, it can be shown that the symmetric matrix

$$S = \begin{bmatrix} 5 & 1 & -1 \\ 1 & 2 & 2 \\ -1 & 2 & 4 \end{bmatrix}$$

is positive definite. Since $\det S = 10$, the result obtained in the solved problem above enables us to conclude that the quadric surface

$$5x_1^2 + 2x_1x_2 - 2x_1x_3 + 2x_2^2 + 4x_2x_3 + 4x_3^2 = 1$$

is an ellipsoid which bounds a region with volume

$$V = \frac{4}{3}\frac{\pi}{\sqrt{\det S}} = \frac{4}{3}\frac{\pi}{\sqrt{10}}$$

Quadratic forms and matrices which are positive definite have many interesting properties. For instance, it can be shown that a positive definite symmetric matrix S has a square root—that is, there exists a symmetric matrix S_1 such that $S_1^2 = S$. Moreover, a positive definite quadratic form X^tSX takes on only positive values; in other words, if η is a non-zero vector, then $\eta^tS\eta$ is a positive real number. These properties and others are outlined in the exercises at the end of this section and in the supplementary exercises at the end of the chapter.

This concludes our study of symmetric matrices and quadratic forms, and with the completion of this chapter, we come to the end of our survey of the fundamental concepts of linear algebra. One of the main disadvantages of this or any other survey is that it touches only lightly on many important ideas and passes over others altogether. For this reason, we strongly encourage the reader to continue his study of linear algebra and its applications by examining the material in several of the intermediate or advanced level texts listed in the bibliography.

Exercise Set 7.3

1. Find the rank and signature of each of the following matrices.

(i) $\begin{bmatrix} 1 & 2 \\ 2 & 4 \end{bmatrix}$
(ii) $\begin{bmatrix} 4 & 1 \\ 1 & 3 \end{bmatrix}$
(iii) $\begin{bmatrix} 0 & -3 \\ -3 & 0 \end{bmatrix}$

(iv) $\begin{bmatrix} 0 & 3 & 1 \\ 3 & 0 & -2 \\ 1 & -2 & 1 \end{bmatrix}$
(v) $\begin{bmatrix} 1 & -2 & 1 \\ -2 & 5 & 4 \\ 1 & 4 & 3 \end{bmatrix}$

(vi) $\begin{bmatrix} 1 & 3 & -2 & 1 \\ 3 & 9 & -6 & 3 \\ -2 & -6 & 4 & -2 \\ 1 & 3 & -2 & 1 \end{bmatrix}$
(vii) $\begin{bmatrix} 0 & -3 & 2 & 1 \\ -3 & 0 & 1 & 2 \\ 2 & 1 & 0 & -3 \\ 1 & 2 & -3 & 0 \end{bmatrix}$

2. Which of the following 3×3 matrices are congruent?

$$A_1 = \begin{bmatrix} -2 & 1 & 0 \\ 1 & 3 & 5 \\ 0 & 5 & 0 \end{bmatrix}, \quad A_2 = \begin{bmatrix} 4 & -1 & 2 \\ -1 & 2 & 1 \\ 2 & 1 & 3 \end{bmatrix}$$

$$A_3 = \begin{bmatrix} -10 & -1 & 3 \\ -1 & -1 & 6 \\ 3 & 6 & -13 \end{bmatrix}, \quad A_4 = \begin{bmatrix} 10 & 1 & 5 \\ 1 & 1 & 2 \\ 5 & 2 & -3 \end{bmatrix}$$

3. In each of the following cases, classify the given quadratic form X^tSX using definition 7.8 and then identify the corresponding surface $X^tSX = 1$.
   (i) $x_1^2 + 2x_1x_2 + 5x_2^2$
   (ii) x_1x_2
   (iii) $-x_1^2 + 2x_1x_2 - 2x_2^2$
   (iv) $2x_1x_2 - 6x_2x_3$
   (v) $7x_1^2 + 8x_1x_2 - 4x_1x_3 + 3x_2^2 + 2x_3^2$
   (vi) $-6x_1^2 + 8x_1x_2 - 24x_1x_3 - 19x_2^2 + 2x_2x_3 - 27x_3^2$

4. (i) Show that the matrix

$$S = \begin{bmatrix} 3 & -2 \\ -2 & 0 \end{bmatrix}$$

   has eigenvalues $\lambda_1 = 4$ and $\lambda_2 = -1$, and find an orthogonal matrix P such that $P^tSP = \mathrm{diag}\,(4, -1)$.
   (ii) Consider the transformation of variables $X = PY$. Explain why the conic sections $3x_1^2 - 4x_1x_2 = 1$ and $4y_1^2 - y_2^2 = 1$ have the same features.
   (iii) Find the vertices and the foci of the hyperbola $4y_1^2 - y_2^2 = 1$, and then use the formula $X = PY$ to find the vertices and foci of $3x_1^2 - 4x_1x_2 = 1$.
   (iv) The axes of symmetry of $4y_1^2 - y_2^2 = 1$ are $y_1 = 0$ and $y_2 = 0$, and the asymptotes of this hyperbola are the lines $y_2 = 2y_1$ and $y_2 = -2y_1$. Use this information to find equations for the axes of symmetry and asymptotes of $3x_1^2 - 4x_1x_2 = 1$.

5. Let $v(n)$ denote the number of different types of quadratic forms in n variables.
   (i) Show that there are $(n + 1)$ different types of quadratic forms in n variables which have rank n.
   (ii) Show that there are $v(n - 1)$ different types of quadratic forms with rank less than n.
   (iii) Combine (i) and (ii) to show that $v(n) = (n + 1) + v(n - 1)$, and then use mathematical induction to prove that

$$v(n) = \frac{(n + 1)(n + 2)}{2}$$

6. Let $Ax^2 + Bxy + Cy^2$ and $A_1x^2 + B_1xy + C_1y^2$ be quadratic forms which are orthogonally equivalent.
   (i) Show that $4AC - B^2 = 4A_1C_1 - B_1^2$.
   (ii) Show that $A + C = A_1 + C_1$.

7. Show that the quadric surface

$$19x_1^2 + 10x_1x_2 + 4x_1x_3 + 9x_2^2 + 14x_2x_3 + 6x_3^2 = 1$$

   is an ellipsoid, and then find the volume of the region which it bounds.

8. If S is a positive definite real symmetric matrix, show that $\det S > 0$.

9. The characteristic polynomial of a certain 4×4 real symmetric matrix S is

$$p_S(x) = x^4 - 3x^3 + 2x^2 - 7x - 8$$

   (i) What is the determinant of S?
   (ii) Show that S is not positive definite.

10. In answering the following questions, you may use the fact that an ellipse in standard form

$$\frac{x^2}{a^2} + \frac{y^2}{b^2} = 1$$

   bounds a region with area πab.
   (i) If λ_1, λ_2 are positive real numbers, show that the conic section $\lambda_1 y_1^2 + \lambda_2 y_2^2 = 1$ is an ellipse which bounds a region with area $A = \pi/\sqrt{\lambda_1 \lambda_2}$.
   (ii) If S is a 2×2 positive definite symmetric matrix, show that the conic section $X^t S X = 1$ is an ellipse which bounds a region with area

$$A = \frac{\pi}{\sqrt{\det S}}$$

   (iii) Find the area of the region bounded by the ellipse

$$5x_1^2 + 4x_1x_2 + x_2^2 = 1$$

11. Let S be an $n \times n$ positive definite symmetric matrix.
   (i) Show that S is congruent to I_n.
   (ii) Show that there exists a non-singular matrix P such that $S = P^t P$.

(iii) If η is an n-tuple and $\gamma = P\eta$, show that $\eta^t S \eta = \gamma^t \gamma$.

(iv) Show that $\eta^t S \eta > 0$ for each non-zero vector η.

(v) For each index j, show that $s_{jj} = \epsilon_j^t S \epsilon_j$, where ϵ_j is the jth standard basis n-tuple. Use this fact to prove that $s_{jj} > 0$, and then show that tr $(S) > 0$.

12. Let S be an $n \times n$ real symmetric matrix such that $\eta^t S \eta > 0$ for each non-zero vector η.

    (i) If λ is an eigenvalue of S associated with the eigenvector γ, show that $\gamma^t S \gamma = \lambda(\gamma^t \gamma)$

    (ii) Show that S is positive definite.

13. Let S be an $n \times n$ positive definite real symmetric matrix, and let P be an orthogonal matrix such that $P^t S P = \text{diag} (\lambda_1, \ldots, \lambda_n) = D$, where the λ's are the eigenvalues of S.

    (i) Let $D^{1/2}$ denote the matrix diag $(\sqrt{\lambda_1}, \sqrt{\lambda_2}, \ldots, \sqrt{\lambda_n})$. Show that $D = (D^{1/2})^t D^{1/2} = D^{1/2}(D^{1/2})^t$.

    (ii) Define $S^{1/2}$ to be $P D^{1/2} P^t$. Show that $(S^{1/2})^2 = S$.

    (iii) Show that $S = Q^t Q$, where Q is the non-singular matrix $(P D^{1/2})^t$.

SUPPLEMENTARY EXERCISES

1. A quadratic form $X^t S X$ is said to be *negative definite* if $\eta^t S \eta < 0$ whenever $\eta \neq \theta$. Show that $X^t S X$ is negative definite if and only if the matrix $-S$ is positive definite.

2. If $Ax^2 + Bxy + Cy^2$ is a positive definite quadratic form, show that $A^3 + C^3 > ABC$.

3. If S is a real symmetric matrix, show that there exists a real number c_0 such that the matrix $S + cI$ is positive definite whenever $c > c_0$ and is not positive definite when $c \leq c_0$.

4. Let $S = (s_{ij})$ be an $n \times n$ real symmetric matrix with the following properties:
   1. $s_{kk} > 0$ for each k

   2. $s_{kk} > \sum_{\substack{i=1 \\ i \neq k}}^{n} |s_{ki}|$ for $k = 1, 2, \ldots, n$

   Show that S must be positive definite.

   Hint: Show that the only real numbers contained in the Gerschgorin disks of S are positive (see supplementary exercises 10 and 12, chapter 6).

5. If S and T are $n \times n$ real symmetric matrices and S is positive definite, show that there exists a non-singular matrix P such that

$$P^t S P = I \quad \text{and} \quad P^t T P = D$$

   where D is a diagonal matrix.

   Hint: Let P_1 be a non-singular matrix which satisfies $S = P_1^t P_1$ and let $A = (P_1^{-1})^t T(P_1^{-1})$. Then show that $P = P_1^{-1} Q$ satisfies the desired conditions, where Q is an orthogonal matrix for which $Q^t A Q$ is diagonal.

6. Let K be an $n \times n$ skew-symmetric matrix. We have already observed that the eigenvalues of K are real (see exercise 7, section 7.2). Let $\lambda = ci$ be a non-zero eigenvalue of K, and let $\eta = X + iY$ be an associated eigenvector, where X and Y are real n-tuples.

(i) Show that $KX = -cY$ and $KY = cX$.

(ii) If (|) is the standard inner product on R_n, show that $(KX|X) = (X|K^tX) = -c(Y|X)$.

(iii) Show that X and Y are orthogonal.

(iv) Assume that $\|X\| = \|Y\| = 1$, and extend $\{X, Y\}$ to an orthonormal basis $\{X, Y, \gamma_3, \ldots, \gamma_n\}$ for R_n. Let P be the $n \times n$ matrix whose columns are $X, Y, \gamma_3, \ldots, \gamma_n$, respectively. Show that P is orthogonal and that P^tKP has the block form

$$P^tKP = \left[\begin{array}{cc|c} 0 & c & \\ -c & 0 & 0 \\ \hline & 0 & K_1 \end{array}\right]$$

where K_1 is an $(n-2) \times (n-2)$ skew-symmetric matrix.

(v) Show that the rank of K is an even number.

Hint: Use part (iv) and mathematical induction. Note that $\rho(K) = \rho(K_1) + 2$ (why?).

7. The $n \times n$ complex matrix H is said to be *hermitian* if $H = H^*$. For instance, the matrix

$$H = \begin{bmatrix} 3 & -5 + 2i & 2 - 2i \\ -5 - 2i & 1 & 4 + 2i \\ 2 + 2i & 4 - 2i & -6 \end{bmatrix}$$

has this property.

(i) Show that the eigenvalues of a hermitian matrix are all real.
Hint: Modify the proof of theorem 7.1.

(ii) Show that the eigenvectors which correspond to distinct eigenvalues of a hermitian matrix are orthogonal.

(iii) If H is hermitian, show that there exists a unitary matrix U (recall supplementary exercise 12 of chapter 3) such that U^*HU is diagonal.
Hint: Modify the proof of theorem 7.2.

*8. The following result, which is known as *Schur's theorem*, is an extremely valuable tool of matrix analysis:

Theorem.   If C is an $n \times n$ complex matrix, there exists a unitary matrix U such that U^*CU has the upper triangular form

$$U^*CU = \begin{bmatrix} \lambda_1 & & & \\ & \lambda_2 & & \\ & & \ddots & \\ & 0 & & \lambda_n \end{bmatrix}$$

where the λ_j's are the eigenvalues of C.

Prove Schur's theorem by completing the steps outlined below:

(i) Use mathematical induction. First note that the theorem is true for 1×1 matrices. For the inductive step, begin by showing that there exists a unitary matrix U_1 such that

$$U_1{}^*CU_1 = \left[\begin{array}{c|c} \lambda_1 & \alpha \\ \hline O & C_1 \end{array}\right]$$

where α is a row vector, O is a column vector of zeroes, and C_1 is an $(n-1) \times (n-1)$ matrix whose eigenvalues are $\lambda_2, \ldots, \lambda_n$. *Hint:* Modify the proof of lemma 7.1.

(ii) Using the inductive hypothesis, assert the existence of an $(n-1) \times (n-1)$ unitary matrix U_2 such that

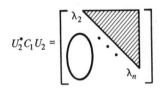

(iii) Complete the proof by combining U_1 and U_2 to form the desired unitary matrix U.
Hint: Use the proof of theorem 7.2 as a model.

*9. Let A be an $n \times n$ complex matrix. Show that there exists a unitary matrix U such that U^*AU is diagonal if and only if A satisfies $AA^* = A^*A$. (Incidentally, a matrix with this property is said to be *normal*.) *Hint:* The only real difficulty lies in showing that whenever a triangular matrix T is normal, it must actually be in diagonal form. This statement may be proved by simply comparing corresponding entries in the equation $TT^* = T^*T$.

Bibliography

ELEMENTARY AND INTERMEDIATE TEXTBOOKS

D. T. FINKBEINER, *Introduction to Matrices and Linear Transformations*, W. H. Freeman (1960)

P. R. HALMOS, *Finite-Dimensional Vector Spaces*, Van Nostrand (1958)

K. HOFFMAN AND R. KUNZE, *Linear Algebra*, Prentice-Hall (1961)

B. KOLMAN, *Elementary Linear Algebra*, Macmillan (1970)

E. D. NERING, *Linear Algebra and Matrix Theory*, Wiley (1963)

P. SHIELDS, *Elementary Linear Algebra*, Worth (1968)

R. R. STOLL AND E. T. WONG, *Linear Algebra*, Academic Press (1968)

ADVANCED TEXTBOOKS

R. BELLMAN, *Introduction to Matrix Analysis*, McGraw-Hill (1960). An excellent survey of the methods of matrix analysis. Includes a fine bibliography and is a sourcebook of solved problems on the research level.

F. R. GANTMACHER, *Theory of Matrices*, Vols. I, II, Chelsea (1959). Probably the best single source of information regarding matrices.

349

I. M. GEL'FAND, *Lectures on Linear Algebra*, Interscience (1961)

G. HADLEY, *Linear Algebra*, Addison-Wesley (1961)

N. JACOBSON, *Lectures in Abstract Algebra*, Vol. II, Van Nostrand (1953). A thorough treatment of the algebraic aspects of linear algebra.

TEXTS CONTAINING APPLICATIONS OF LINEAR ALGEBRA
AND MATRICES

R. G. D. ALLEN, *Mathematical Economics*, St. Martins (1951)

N. R. AMUNDSON, *Mathematical Methods in Chemical Engineering*, Prentice-Hall (1966)

D. K. FADDEEV AND V. N. FADDEEVA, *Computational Methods of Linear Algebra*, W. H. Freeman (1963)

D. GALE, *The Theory of Linear Economic Models*, McGraw-Hill (1960)

S. I. GASS, *Linear Programming*, McGraw-Hill (1964)

E. A. GUILLEMIN, *The Mathematics of Circuit Analysis*, Wiley (1949)

A. S. HOUSEHOLDER, *The Theory of Matrices in Numerical Analysis*, Random (1964)

J. G. KEMENY AND J. L. SNELL, *Finite Markov Chains*, Van Nostrand (1960)

D. L. KREIDER, R. G. KULLER, D. R. OSTBERG, AND F. W. PERKINS, *Introduction to Linear Analysis*, Addison-Wesley (1966)

B. NOBLE, *Applications of Undergraduate Mathematics in Engineering*, Macmillan (1967)

B. NOBLE, *Applied Linear Algebra*, Prentice-Hall (1969)

S. R. SEARLE, *Matrix Algebra for the Biological Sciences*, Wiley (1966)

D. VANDERMEULEN, *Linear Economic Theory*, Prentice-Hall (1971)

R. VARGA, *Matrix Iterative Analysis*, Prentice-Hall (1962)

Answers and Suggestions
for Selected Exercises

CHAPTER 1

Section 1.1 (*pp. 8–10*)

1. (i) unique solution, $(0, 0)$ (ii) inconsistent   (iii) unique solution $(3, 2, 4)$
(iv) $x_1 = 3 - 5t, x_2 = 7t, x_3 = t, x_4 = 8$

2. (i) $a_{24} = -6$ and $a_{32} = 8$ (ii) $\begin{bmatrix} -3 & 3 & -6 \\ 7 & 5 & 10 \end{bmatrix}$ (iii) B_3 is not.

3. (i) $\begin{bmatrix} 1 & -1 & 0 & 1 \\ 0 & 6 & 7 & -3 \\ 0 & 4 & 2 & 11 \end{bmatrix}$ (ii) $\begin{bmatrix} 1 & -3 & 4 & 2 \\ 0 & 1 & -13 & -11 \\ 0 & 0 & 0 & 0 \end{bmatrix}$

4. (iii) $x_1 = 4 + 2t_1 - t_2, x_2 = -3t_1 - t_2, x_3 = t_1, x_4 = t_2$

5. inconsistent

6. The solution is $(-2, 5, 7)$.

7. 4 pennies, 3 nickels, and 7 dimes

8. $a = -\frac{1}{2}, b = 1, c = \frac{13}{2}$

Section 1.2 (*pp. 18–20*)

1. (i) $x_1 = 1 + 2t_1 + 3t_2, x_2 = t_1, x_3 = 8 - 4t_2, x_4 = t_2, x_5 = t_2$
(ii) $x_1 = 9 - 3t_1 + 5t_2, x_2 = t_1, x_3 = t_2$ (iii) inconsistent

(iv) $x_1 = -1$, $x_2 = 2 - 2t$, $x_3 = t$

(v) $x_1 = t_1$, $x_2 = -2 + 3t_2$, $x_3 = t_2$, $x_4 = 7$

(vi) $x_1 = 5$, $x_2 = 0$, $x_3 = 2$

2. (1) is equivalent to (c).

(2) and (3) are equivalent to (d).

(4) is equivalent to (a).

3. (i) $x_1 = 10 - 2t$, $x_2 = -9 + t$, $x_3 = -17 + 4t$, $x_4 = t$

(ii) inconsistent

(iii) $x_1 = 4 + t_1 - 3t_2$, $x_2 = t_1$, $x_3 = 1 + 2t_2$, $x_4 = t_2$

(iv) unique solution $(-3, 2, 6, 7)$

5. unique solution if t is not -12.

6. (i) true   (ii) false   (iii) true   (iv) false   (v) false

7. (i) $x_1 = 2 + 5t_1 - 3t_2$, $x_2 = t_1$, $x_3 = -1 + 7t_2$, $x_4 = t_2$, $x_5 = 4$

(ii) η_1 and η_3 are solutions.

Section 1.3 (pp. 28–30)

1. (i) $r = 2$; basic columns are 1 and 3.

(iii) $r = 3$; basic columns are 2, 4, and 6.

(ii) and (iv) are not in reduced row echelon form.

2. (i) $\begin{bmatrix} 1 & 0 & 10.5 & 0 \\ 0 & 1 & .5 & 0 \\ 0 & 0 & 0 & 1 \\ 0 & 0 & 0 & 0 \end{bmatrix}$ (ii) $\begin{bmatrix} 1 & 0 & 0 & 0 \\ 0 & 1 & 0 & 0 \\ 0 & 0 & 1 & 0 \\ 0 & 0 & 0 & 1 \end{bmatrix}$ (iii) $\begin{bmatrix} 1 & 0 & -3 & 0 \\ 0 & 1 & 5 & 0 \\ 0 & 0 & 0 & 1 \end{bmatrix}$

(iv) $\begin{bmatrix} 1 & 0 & 5 & 2 & 0 & 0 \\ 0 & 1 & -3 & 1 & 0 & 0 \\ 0 & 0 & 0 & 0 & 1 & 1 \\ 0 & 0 & 0 & 0 & 0 & 0 \end{bmatrix}$

3. (i) inconsistent

(ii) $x_1 = -7s + 5t$, $x_2 = 3s - 2t$, $x_3 = s$, $x_4 = t$

(iii) $x_1 = 5 - 2s - 3t$, $x_2 = 1 + s - 7t$, $x_3 = s$, $x_4 = t$, $x_5 = 2$

(iv) $x_1 = 4 - 2s$, $x_2 = s$, $x_3 = 3$

4. (i) Unique solution $(7, -3, 1)$

(ii) Solution is not unique.

6. A_1, A_3, and A_4 are all row equivalent to $\begin{bmatrix} 1 & 0 & -3 & 2 \\ 0 & 1 & 5 & -1 \\ 0 & 0 & 0 & 0 \end{bmatrix}$.

7. $x_1 = -2 - 7s - 3t$, $x_2 = 5 + 2s - 9t$, $x_3 = s$, $x_4 = t$

8. (iii) $\begin{bmatrix} 1 & 0 & 0 \\ 0 & 1 & 0 \\ 0 & 0 & 1 \\ 0 & 0 & 0 \end{bmatrix}$

10. $\begin{bmatrix} 1 & 0 & -2 & 1 & 0 \\ 0 & 1 & 5 & 7 & 0 \\ 0 & 0 & 0 & 0 & 1 \\ 0 & 0 & 0 & 0 & 0 \end{bmatrix}$

12. Solution is unique whenever λ is not 1, 8, or 9.

Section 1.4 (pp. 36–39)

1. $A + B$ and BC are meaningless.

(v) $\quad AB = \begin{bmatrix} -5 & 2 \\ -4 & 16 \\ 33 & 2 \end{bmatrix}$ (vii) $\quad CA = \begin{bmatrix} -49 & 11 & 51 \\ -12 & 0 & 36 \\ 49 & -7 & -35 \end{bmatrix}$

(ix) $\quad C(AB) = \begin{bmatrix} 247 & 66 \\ 84 & 84 \\ -231 & -14 \end{bmatrix}$

2. (i) true for all real numbers

(ii) $y = 0$, $x = 0$; $y = 0$, $x = \frac{1}{2}$; $y = -2$, $x = -1$

(iii) $x = y = 0$

3. The solution is $(-26, 46, 15)$.

7. (i) $\quad T = \begin{bmatrix} 6 & 19 & 18 & 12 \\ 27 & 13 & 20 & 19 \\ 23 & 18 & 16 & 24 \end{bmatrix}$

(ii) Francine and Ben are made for each other.

(iii) Ben and Francine; Arthur with Gertrude; Charles and Inger; Hattie loses.

Section 1.5 (pp. 45–47)

2. (i) $2AB - BC$ (ii) $-4A^2 - 10AB + 3B^2$

(iii) $2(A^2B + ABA + BA^2 + B^3)$ (iv) $-A^2 - BA + 2AB - B^2 - 3B - A^2B$

(v) $AB^2 + ACB - ABC - AC^2$

3. (i) $X = \frac{2}{3}A - \frac{7}{3}B$ (ii) no solution

(iii) $X = \begin{bmatrix} 5 & 0 \\ 2 & 0 \end{bmatrix}$ is one solution.

4. (i), (iii), (iv), and (vi) are true.

(ii) is false; for example, if $X = \begin{bmatrix} 0 & 1 \\ 1 & 0 \end{bmatrix}$, then $X^2 = I$.

(v) is false; for example, consider $A = \begin{bmatrix} 1 & 0 \\ 0 & 0 \end{bmatrix}$, $B = \begin{bmatrix} 0 & 1 \\ 0 & 0 \end{bmatrix}$.

Section 1.6 (pp. 53–56)

1. (iii) and (v) are singular. The inverses of the others are as follows:

(i) $\begin{bmatrix} \frac{1}{19} & 0 & 0 \\ 0 & -\frac{1}{5} & 0 \\ 0 & 0 & \frac{1}{3} \end{bmatrix}$ (ii) $\begin{bmatrix} \frac{1}{5} & \frac{3}{5} & \frac{13}{30} \\ 0 & 1 & \frac{2}{3} \\ 0 & 0 & -\frac{1}{6} \end{bmatrix}$

(iv) $\begin{bmatrix} 7 & -3 & -17 \\ 14 & -6 & -35 \\ -2 & 1 & 5 \end{bmatrix}$ (vi) $\begin{bmatrix} -48 & 18 & 9 & -7 \\ 82 & -31 & -15 & 12 \\ 49 & -18 & -9 & 7 \\ 27 & -10 & -5 & 4 \end{bmatrix}$

2. The matrices in question are non-singular except when ...
   (i) λ is -31 (ii) λ is -2 or 1 (iii) λ is -2 (iv) λ is 5 or -9
3. Non-singular, since B is non-singular
4. $a = 5; b = 7; c = -1; d = -1; e = -63; f = 17$
5. (i) $A^{-1} = \begin{bmatrix} 1 & 3 & -1 \\ -11 & -34 & 13 \\ -5 & -16 & 6 \end{bmatrix}$ (iii)

a	b	c	solution
0	0	0	$(0, 0, 0)$
1	0	0	$(1, -11, -5)$
2	-1	0	$(-20, 229, 108)$

8. (i) false   (ii) true   (iii) true   (iv) true   (v) true   (vi) true
   (vii) true
10. $A^{-1} = \begin{bmatrix} -7 & -8 & -3 \\ 12 & 14 & 5 \\ 41 & 47 & 17 \end{bmatrix}$
   (i) $X = \frac{1}{2}\begin{bmatrix} -13 & -15 \\ 22 & 26 \\ 75 & 88 \end{bmatrix}$ (ii) $X = (A^{-1})^t A = \begin{bmatrix} -419 & 124 & -108 \\ -480 & 143 & -124 \\ -174 & 52 & -45 \end{bmatrix}$

Section 1.7 (pp. 61–62)

3. $A^{-1} = \begin{bmatrix} -\frac{1}{3} & 4 & 0 \\ 0 & 0 & \frac{1}{3} \\ 0 & 1 & 0 \end{bmatrix}$

4. $B = E_1^{-1}E_2^{-1}E_3^{-1}A = \begin{bmatrix} -7 & 2 & -5 & 8 \\ 6 & 2 & 4 & -2 \\ -26 & 11 & -25 & 41 \end{bmatrix}$

5. (ii) $P_1 = \frac{1}{5}\begin{bmatrix} 2 & 1 & 0 \\ 1 & -2 & 0 \\ 5 & 5 & 5 \end{bmatrix}$, $P_2 = \frac{1}{14}\begin{bmatrix} 4 & 2 & 0 \\ -1 & 3 & 0 \\ 37 & -13 & 14 \end{bmatrix}$
   (iii) $P = P_1^{-1}P_2 = \frac{1}{14}\begin{bmatrix} 7 & 7 & 0 \\ 6 & -4 & 0 \\ 24 & -16 & 14 \end{bmatrix}$

Section 1.8 (pp. 70–71)

1. (i) $I_1 = -1; I_2 = 2; I_3 = 3$
   (ii) yes; $E_1 = 8$ and $E_2 = 6.5$
2. (i) Annabelle on Monday, $X_1 = (.37, .3, .33)$
      Christine on Tuesday, $X_2 = (.37, .23, .40)$
   (ii) Annabelle, $Y = (.43, .19, .38)$

3. (i)
$$(I_3 - P)^{-1} = \tfrac{10}{49}\begin{bmatrix} 24 & 27 & 29 \\ 37 & 60 & 59 \\ 23 & 32 & 38 \end{bmatrix}$$

(ii)
$$\begin{bmatrix} 24.2 \\ 48.9 \\ 28.5 \end{bmatrix}$$

Supplementary exercises (pp. 71–73)

1. (ii) *Hint:* Let $C = AB$. Then $\sum_{j=1}^{n} c_{ij} = \sum_{j=1}^{n}\left(\sum_{k=1}^{n} a_{ik}b_{kj}\right) = \sum_{k=1}^{n} a_{ik}\left(\sum_{j=1}^{n} b_{kj}\right).$

2. (i) $a = -\tfrac{4}{5}$, $b = \tfrac{12}{5}$, $c = 0$

3. $d = 10$ ft/sec^2; $v_0 = 30$ ft/sec; $s_1 = 45$ ft; $s_2 = 125$ ft

6. $A^{-1} = -\tfrac{1}{2}A^2 + \tfrac{7}{2}A - \tfrac{1}{2}I$

8. (i) $X^2 = \begin{bmatrix} t^2 + e^t & 2te^t \\ 2t & t^2 + e^t \end{bmatrix}$ $X^{-1} = \dfrac{1}{t^2 - e^t}\begin{bmatrix} t & -e^t \\ -1 & t \end{bmatrix}$

   (iii) no

9. (i) $A^{-1} = \begin{bmatrix} 3 & -5 \\ -1 & 2 \end{bmatrix}$ **(ii)** $x_1 = \tfrac{3}{2}t^2 - 35 \sin t$
$x_2 = -\tfrac{1}{2}t^2 + 14 \sin t$

CHAPTER 2

Section 2.1 (pp. 80–81)

7. α is longer since $\alpha = 3\beta$.

8. (i) $\alpha_3 = -25\beta_2 - 24\beta_3$
   (ii) If $\alpha_3 = \theta$, then $\beta_2 = -\tfrac{24}{25}\beta_3$ and $\beta_1 = -\tfrac{1}{25}\beta_3$.

Section 2.2 (pp. 87–88)

1. (i) $(\tfrac{22}{5}, \tfrac{2}{5})$ **(iv)** yes, on the segment between 0 and P_2
   (v) P_5 is inside; P_6 is outside.

4. $(34, -19, -20)$

5. The line contains the points $P_2(1, -1, 0)$ and $P_3(3, 1, 3)$. One parameterization of the plane is
$$x = -1 + 2s + 4t, \qquad y = 3 - 4s - 2t, \qquad z = 7 - 7s - 4t$$
The equation of the plane is $x - 10y + 6z = 11$.

6. $\dfrac{x}{2} = \dfrac{y}{-1} = \dfrac{z}{-7}$

7. $(-3, 13)$ and $(6, -10)$

8. (i) $\alpha_1 = 2\beta_1 + 3\beta_2$ and $\alpha_2 = 3\beta_1 + 5\beta_2$ **(ii)** $(-9, -12)$

Section 2.3 (pp. 94–96)

4.

					Property					
System	1	2	3	4	5	6	7	8	9	10
S_1	N	Y	Y	N	N	N	Y	Y	Y	Y
S_2	N	Y	Y	Y	N	N	Y	Y	Y	Y
S_3	N	Y	Y	N	Y	N	Y	Y	Y	Y
S_4	Y	Y	Y	Y	N	N	Y	Y	Y	Y
S_5	N	Y	Y	Y	Y	Y	Y	Y	Y	Y

9. $\alpha_1 = -2\beta_1 - \beta_2$ and $\alpha_2 = -5\beta_1 - 2\beta_2$

10. The system satisfies axioms 1, 2, 3, 6, 7, 9, 10. Note that $(s + t) \odot \alpha = \alpha$, but $(s \odot \alpha) \oplus (t \odot \alpha)$ is either 0 or 1.

Section 2.4 *(pp. 104–106)*

1. (i) yes   (ii) no; not closed under addition   (iii) yes   (iv) no

2. S is a subspace of P_2.

4. (ii) the line $\dfrac{5x_1 - 6}{14} = \dfrac{5x_2 + 16}{11} = x_3$

5. (ii) the plane $7x + 10y - z = 0$
(iii) all of 3-space

6. β is not contained in Sp $\{\alpha_1, \alpha_2, \alpha_3\}$ since the system

$$
\begin{aligned}
s_1 - 3s_2 - 7s_3 &= 13 \\
7s_1 + s_2 + 17s_3 &= -7 \\
-2s_1 + 4s_2 + 8s_3 &= 2
\end{aligned}
$$

is inconsistent. However, $\gamma = \alpha_1 - 4\alpha_2$.

7. (i) $\gamma = -8\alpha_1 + 4\alpha_2 - 29\alpha_3$
(ii) $s_1\beta_1 + s_2\beta_2 = (2s_1 - 2s_2)\alpha_1 + (-s_1 + s_2)\alpha_2 + (2s_1 - 5s_2)\alpha_3$
(iii) $x\alpha_1 + y\alpha_2 + z\alpha_3$ lies in Sp $\{\beta_1, \beta_2\}$ if and only if $x + 2y = 0$. An example of a linear combination not in Sp $\{\beta_1, \beta_2\}$ is $\alpha_1 + \alpha_2 + \alpha_3$.

11. (i) yes   (ii) yes   (iii) yes   (iv) no

14. (i) S and T are both planes
(ii) $S \cap T$ is the line $-\frac{7}{3}x = 7y = z$.
$S \cup T$ is the collection of all points in both planes.
$S + T$ is all of 3-space.

Section 2.5 *(pp. 115–117)*

1. (i) dependent; $\alpha_3 = \alpha_1 + 2\alpha_2$ (ii) dependent; $p_3 = p_1 + 2p_2$
(iii) dependent: $\alpha_4 = 2\alpha_1 + \alpha_2 - \alpha_3$.

2. (i) basis   (ii) not a basis; the vectors do not span R_3
(iii) not a basis; the γ's are dependent since $\gamma_3 = 4\gamma_1 + 3\gamma_2$

3. (i) $\{\alpha_1, \alpha_2, \alpha_4\}$ (ii) $\beta = -3\alpha_1 - 3\alpha_2 + \alpha_4$

4. $\{\alpha_1, \alpha_2, \epsilon_1, \epsilon_3\}$ is one such basis. Note that $\epsilon_2 = 3\alpha_1 + \alpha_2 + 4\epsilon_1$.

5. $s_1 = 2$ and $s_2 = 1$

10. (i) $\gamma = -6\alpha_1 - 25\alpha_2 + 33\alpha_4$

13.
$$
E = \begin{bmatrix}
1 & -4 & 0 & -4 & 0 & -14 \\
0 & 0 & 1 & 1 & 0 & 0 \\
0 & 0 & 0 & 0 & 1 & 1 \\
0 & 0 & 0 & 0 & 0 & 0 \\
0 & 0 & 0 & 0 & 0 & 0
\end{bmatrix}
$$

14. *Suggestions:* (i) If $s_1\alpha_1 + \ldots + s_k\alpha_k = s_{k+1}\alpha_{k+1} + \ldots + s_n\alpha_n$, then each coefficient s_j must be 0 since the α's are linearly independent.
(iii) If $\sigma_1 + \tau_1 = \sigma_2 + \tau_2$, then $\sigma_1 - \sigma_2$ and $\tau_2 - \tau_1$ are in $S \cap T$.

Section 2.6 (pp. 124–126)

1. (i) 2   (ii) 3   (iii) 2

2. (i)
$$\eta = \tfrac{1}{7}\begin{bmatrix} -10 \\ -1 \\ 7 \end{bmatrix}$$
(ii)
$$\eta_1 = \begin{bmatrix} -14 \\ -5 \\ 1 \\ 0 \end{bmatrix}, \qquad \eta_2 = \begin{bmatrix} -31 \\ -11 \\ 0 \\ 1 \end{bmatrix}$$

(iii)
$$\eta_1 = \tfrac{1}{7}\begin{bmatrix} 1 \\ -2 \\ 7 \\ 0 \\ 0 \end{bmatrix}, \qquad \eta_2 = \begin{bmatrix} 2 \\ 0 \\ 0 \\ 1 \\ 0 \end{bmatrix}, \qquad \eta_3 = \begin{bmatrix} -2 \\ 1 \\ 0 \\ 0 \\ 1 \end{bmatrix}$$

3. (i) Plane; a basis is $\{\eta_1, \eta_2\}$, where $\eta_1 = (5, 3, 0)$, $\eta_2 = (-1, 0, 3)$.
(ii) Line; a basis is $\{\eta\}$, where $\eta = (7, -3, 4)$.
(iii) Line; a basis is $\{\eta\}$, where $\eta = (-2, 5, 1)$.
(iv) Line; basis is $\{\eta\}$, where $\eta = (7, 9, 5)$.

4. (i)
$$\eta_1 = \begin{bmatrix} 3 \\ 1 \\ 0 \\ 0 \\ 0 \end{bmatrix}, \qquad \eta_2 = \begin{bmatrix} 5 \\ 0 \\ -7 \\ 1 \\ 0 \end{bmatrix}, \qquad \eta_3 = \begin{bmatrix} -2 \\ 0 \\ 1 \\ 0 \\ 1 \end{bmatrix}$$

(ii) yes; $\eta = 2\eta_1 - \eta_2 - \eta_3$
(iii) no, since $p - p_0 = (-7, 4, 1, -8, 6)$ does not satisfy the homogeneous system.

7. (ii) p_1 and p_2 are independent; $p_3 = \tfrac{1}{2}(3p_1 + p_2)$ and $p_4 = \tfrac{1}{2}(5p_1 + p_2)$.
(iii) dim $V = 2$

8. The dimension is 2; the subspace is spanned by A_1, A_2, and we have $A_3 = 3A_1 + 12A_2$ and $A_4 = 7A_1 + 33A_2$.

9.
$$\begin{bmatrix} 1 & 0 & 7 & -5 & 0 & -1 \\ 0 & 1 & -3 & 1 & 0 & 3 \\ 0 & 0 & 0 & 0 & 1 & 2 \\ 0 & 0 & 0 & 0 & 0 & 0 \\ 0 & 0 & 0 & 0 & 0 & 0 \end{bmatrix}$$

Section 2.7 (pp. 133–136)

1. (i) $\gamma = (-3, -11, 28)$ (ii) $[p]_T = (-186, 109, -17)$
(iii)
$$\begin{bmatrix} 76 & -9 & -5 \\ -44 & 6 & 3 \\ 9 & 1 & 0 \end{bmatrix}$$
(iv) $[p]_S = (-2, 1, 5)$

2. (i) $[\gamma_1]_S = (2, 0, 1)$, $[\gamma_2]_S = (18, -44, 37)$, $[\gamma_3]_S = (20, -15, 8)$
(ii) $2\beta_1 - 3\beta_2 - \beta_3 = \theta$ (iii) Sp $\{\beta_1, \beta_2, \beta_3\}$ has dimension 2.

3. (i) $S = \{p_1, p_2\}$ is a basis for V.
(ii) $[q_1]_S = (-2, 3)$ and $[q_4]_S = (2, -1)$; q_2 and q_3 are not in V.

4. $\{A_1, A_2\}$ is a basis for $\text{Sp}\{A_1, A_2, A_3, A_4\}$.

5. The fundamental solutions are $\eta_1 = \frac{1}{2}(1, 1, 2, 0)$ and $\eta_2 = \frac{1}{2}(-3, -1, 0, 2)$.

   (i) $[\beta_1]_S = (1, 1)$ and $[\beta_2]_S = (9, -5)$.

   (ii) T is a basis since the solution space has dimension 2 and β_1, β_2 are linearly independent. The T to S and S to T change of basis matrices are, respectively,

   $$\begin{bmatrix} 1 & 9 \\ 1 & -5 \end{bmatrix} \quad \text{and} \quad \frac{1}{14}\begin{bmatrix} 5 & 9 \\ 1 & -1 \end{bmatrix}$$

6. (i) $[f_1]_S = (2, 0, 0)$, $[f_2]_S = (0, 1, 1)$, $[f_3]_S = (1, 0, -1)$

   (iii) $$\frac{1}{2}\begin{bmatrix} 1 & -1 & 1 \\ 0 & 2 & 0 \\ 0 & 2 & -2 \end{bmatrix}, \quad [g]_T = (-\tfrac{1}{2}, 4, 6)$$

7. Since $\beta_2 = \alpha_1 - \alpha_2 - \alpha_3$, we have $\alpha_3 = \alpha_1 - \alpha_2 - \beta_2 = (-11, 4, 8)$. Similarly, $\beta_1 = (-9, -6, 11)$ and $\beta_3 = (4, 23, 4)$.

Section 2.8 (pp. 142–144)

1. A_2 and A_3 have rank 2, and A_1 has rank 3.

2. A suitable basis for the row space of A_1 is either the first two row vectors or $\{(1, 0, -28, -31), (0, 1, -11, -11)\}$. For the row space of A_2, a basis is formed by either the first two row vectors or $\{(1, 0, -5, 3), (0, 1, 4, -1)\}$.

3. (iii) $m - r$

6. (iii) Columns 1, 2, and 4 form a suitable basis.

12. (i) $p_1 = (4, -20, -1, -13)$, $p_2 = (2, -10, 7, 1)$

   (ii) $\alpha_1 = 1 \cdot p_1 + 0 \cdot p_2$

   $\alpha_2 = 0 \cdot p_1 + 1 \cdot p_2 \qquad \gamma_1 = \begin{bmatrix} 1 \\ 0 \\ \frac{5}{6} \end{bmatrix}, \qquad \gamma_2 = \begin{bmatrix} 0 \\ 1 \\ -\frac{1}{6} \end{bmatrix}$

   $\alpha_3 = \frac{5}{6}p_1 - \frac{1}{6}p_2$

   (iii) $\beta_1 = 4\gamma_1 + 2\gamma_2$, $\quad \beta_2 = -20\gamma_1 - 10_2$

   $\beta_3 = -\gamma_1 + 7\gamma_2$, $\quad \beta_4 = -13\gamma_1 + \gamma_2$

   (iv) $\beta = 2\gamma_1 + 4\gamma_2$. The system $AX = \beta$ is consistent since β is contained in the column space of A.

13. (i) If the p's were linearly dependent, the rank of A would be less than 3.

   (ii) $\gamma_1 = (-7, 4, 9, -2)$, $\gamma_2 = (2, 0, 1, 1)$, $\gamma_3 = (-5, 1, -2, 3)$

   (iii) The system $AX = \beta$ is inconsistent since β cannot be expressed as a linear combination of the γ's.

Supplementary exercises (pp. 144–148)

1. (i) Since $\overrightarrow{P_1 C_1} = \frac{1}{3}(\alpha_1 + \alpha_2 + \alpha_3)$, it follows that C_1 has S-coordinates $(\frac{1}{3}, \frac{1}{3}, \frac{1}{3})$.

   (ii) P has S-coordinates $(\frac{1}{4}, \frac{1}{4}, \frac{1}{4})$.

3. (i) yes (ii) yes (iii) no (iv) yes (v) yes (vi) yes

   (vii) yes (viii) no

4. (i) $p_1 = (1, 3, 0)$, $p_2 = (4, 0, 3)$

   (ii) $p_3 = (1, -2, 3)$

(iv) $P = \begin{bmatrix} 1 & 4 & 1 \\ 3 & 0 & -2 \\ 0 & 3 & 3 \end{bmatrix}$ and $P^{-1} = -\frac{1}{21}\begin{bmatrix} 6 & -9 & -8 \\ -9 & 3 & 5 \\ 9 & -3 & -12 \end{bmatrix}$

9. (iv) $\gamma = \begin{bmatrix} 1 \\ 1 \\ 1 \end{bmatrix}$, $N\gamma = \begin{bmatrix} -12 \\ -4 \\ 4 \end{bmatrix}$, $N^2\gamma = \begin{bmatrix} 8 \\ -24 \\ 8 \end{bmatrix}$

$[\epsilon_1]_S = \frac{1}{64}(8, -4, 1)$ $[\epsilon_2]_S = \frac{1}{32}(8, 0, -1)$ $[\epsilon_3]_S = \frac{1}{64}(40, 4, 1)$

12. (i) $A(v) = k_v \begin{bmatrix} 1 & \dfrac{-v}{c^2} \\ -v & 1 \end{bmatrix}$ (ii) $A^{-1}(v) = A(-v) = k_v \begin{bmatrix} 1 & \dfrac{v}{c^2} \\ v & 1 \end{bmatrix}$

(iii) $v_3 = \dfrac{(v_1 + v_2)c^2}{v_1 v_2 + c^2}$

CHAPTER 3

Section 3.1 (pp. 157–159)

1. (i) $||\alpha|| = 13$, $\alpha \cdot \beta = -15$, $\cos \varphi = -\frac{3}{13}$, $||\beta|| = 5$, $d(\alpha, \beta) = 4\sqrt{14}$

(ii) $||\alpha|| = \sqrt{11}$, $\alpha \cdot \beta = 0$, $\cos \varphi = 0$, $||\beta|| = \sqrt{30}$, $d(\alpha, \beta) = \sqrt{41}$

(iii) $||\alpha|| = 37$, $\alpha \cdot \beta = 96$, $\cos \varphi = \frac{96}{629}$, $||\beta|| = 17$, $d(\alpha, \beta) = \sqrt{1466}$

2. $42 + \sqrt{346}$

3. The right angle is at P_2; the area is 25.

4. $\beta = (2\sqrt{47}, -2, 2)$ is one solution.

6. (ii) The sum of the squares of the sides of a parallelogram is equal to the sum of the squares of its diagonals.

9. (i) $t = \frac{2}{7}$ (ii) $\gamma = \alpha - \frac{2}{7}\beta$; $s_1 = 1$, $s_2 = -\frac{2}{7}$

10. The distance is $\left| \dfrac{-57}{\sqrt{169}} \right| = \dfrac{57}{13}$.

11. $-2x + y + 5z = 33$

14. *Suggestion:* Observe that the area is $\frac{1}{2}||\alpha|| \, ||\beta|| |\sin \varphi|$

15. The area is $\frac{5}{2}\sqrt{5}$.

Section 3.2 (pp. 165–167)

1. $(\alpha | \beta) = -27$, $||\alpha|| = 6$, $||\beta|| = 9$, $d(\alpha, \beta) = \sqrt{171}$, $\cos \varphi = -\frac{1}{2}$; $\varphi = 120°$ or $2\pi/3$ radians

2. $||p|| = \sqrt{\dfrac{13}{3}}$, $d(p, q) = \sqrt{\dfrac{161}{30}}$, $(p|q) = \dfrac{10}{3}$

5. Only (|)$_1$ is an inner product. Note that if $\alpha = (1, 1)$ and $\beta = (1, 0)$, then $(\alpha|\alpha)_2 = (\beta|\beta)_4 = 0$ and $(\alpha|\beta)_3 \neq (\beta|\alpha)_3$.

12. (iii) If $A = -A^t$, then $(AX|Y) = -(X|AY)$

Section 3.3 (pp. 173–174)

1. There are many possibilities, among which are the following:

(i) $\{(5, 2, -1), (-5, 8, -9), (1, -5, -5)\}$

(ii) $\{(1, 1, -1), (2, -1, 1), (0, 1, 1)\}$ (iii) $\left\{\dfrac{1}{\sqrt{5}}(1, 2, 0), \dfrac{1}{\sqrt{105}}(-8, 4, 5)\right\}$

2. $a = 5, b = 2, c = -13$

3. (i) $\{3t - 1, 1 - t\}$ is one possibility.   (ii) $\{1, t\}$

4. $p(t) = \sqrt{\tfrac{3}{43}}(7 - 13t)$; that is, $a = -13\sqrt{\tfrac{3}{43}}, b = 7\sqrt{\tfrac{3}{43}}$

5. (i) $\left\{(1, 0, 0), \dfrac{1}{\sqrt{2}}(1, 1, 0), (-3, -1, 1)\right\}$

(ii) $(\sigma \mid \tau)_* = 61, \|\sigma\|_* = 13, d(\sigma, \tau)_* = 13$

8. $\gamma_1 = (1, 1, 0), \gamma_2 = (-3, 3, 2), \gamma_3 = (0, 0, 0)$

9. (iii) An orthogonal basis is
$$\{(-3, 1, 1, 2), (1, 1, 0, 1), (1, -2, 3, 1)\}$$

Section 3.4 (pp. 181–182)

1. (i) $\alpha_{S_1} = -\tfrac{2}{7}(3, -2, 1)$ (ii) $\alpha_{S_2} = \tfrac{1}{7}(-17, 26, 20)$

(iii) $\alpha_{S_3} = \alpha$, since α lies in S_3.

2. $\alpha_S = \tfrac{1}{3}(3, -1, 2, 4)$ and the distance from α to S is $\tfrac{1}{3}\sqrt{6}$.

3. $(x - 1)^2 + (y + 3)^2 + (z - 2)^2 = 9$

4. $q_S(t) = 18t - 6$; the distance is $\tfrac{11}{3}\sqrt{3}$.

5. (i) $\alpha_S = 3\gamma_1 - 5\gamma_2$ (ii) The distance is 3.

6. $\tfrac{3}{2}t^2 - \tfrac{3}{5}t + \tfrac{1}{20}$

7. (i) $\alpha_S = (2, 1, 2)$ (ii) $\alpha_S = (2, 1, 2)$

10. (i) $\{\gamma_1, \gamma_2\}$ is an orthonormal basis for the plane in question, where $\gamma_1 = \dfrac{1}{\sqrt{2}}(1, 1, 0)$ and $\gamma_2 = \dfrac{1}{\sqrt{6}}(1, -1, -2)$. A suitable orthonormal basis for all of R_3 is $\{\gamma_1, \gamma_2, \gamma_3\}$, where $\gamma_3 = \dfrac{1}{\sqrt{3}}(1, -1, 1)$.

(ii) An orthonormal basis for $S^\perp$ is $\{\gamma_3\}$.

Section 3.5 (pp. 190–191)

2. The normal equations are

$$4\bar{k}_1 - \bar{k}_2 = 6$$
$$-\bar{k}_1 + 4\bar{k}_2 = 6$$

The best approximations are $\bar{k}_1 = 2, \bar{k}_2 = 2$.

3. The normal equations are

$$731\bar{a} + 83\bar{b} - 27\bar{c} = 139$$
$$83\bar{a} + 11\bar{b} - 3\bar{c} = 19$$
$$-27\bar{a} - 3\bar{b} + 4\bar{c} = -2$$

Solving, we find that $a = 0, b = 2, c = 1$, and it follows that the best-fitting curve of the desired type is $y = 2x + 1$.

Supplementary exercises (pp. 191–194)

1. *Hint:* Apply the Cauchy-Schwarz inequality with $\alpha = (a_1, \ldots, a_n, b_1, \ldots, b_n)$ and $\beta = (\cos x, \ldots, \cos nx, \sin x, \ldots, \sin nx)$.

4. $2 \sin x$

5. (i) $||\beta_n||^2 = ||\alpha||^2 - \sum_{j=1}^{n} (\alpha | \gamma_j)^2$

(ii) Observe that $||\beta_n||^2 \geq 0$ for all n.

(iii) The series has non-negative terms, and its sequence of partial sums is bounded by $||\alpha||^2$.

7. (ii) Note that $(\alpha_i | \beta) = \sum_{j=1}^{n} c_j (\alpha_i | \alpha_j)$.

CHAPTER 4

Section 4.1 (pp. 201–203)

3.
$$A = \begin{bmatrix} -2 & 1 \\ 3 & -7 \\ 5 & -1 \end{bmatrix}$$

4. (i) yes   (ii) no   (iii) no   (iv) yes   (v) yes

6. (ii) $\gamma = 5\alpha_1 + 3\alpha_2 - 7\alpha_3$ (iii) $L(\gamma) = (1, -4, 2, -4)$

7. $(7, -2) = -59(-1, 4) + 26(-2, 9)$

(i) $L(7, -2) = -59t^2 + 26t + 203$ (ii) $\alpha = (1, -6)$

(iii) $p(t) = at^2 + bt + c$ has no preimage if $3a - b + c \neq 0$.

10. (iii) the point $(1, 1, 1)$

Section 4.2 (pp. 209–210)

1. (i) The null space contains only $(0, 0, 0)$.   (ii) $\{(2, 1, 0), (-6, 0, 1)\}$

(iii) $\{(1, -2, 1)\}$

2. (i) A basis for $\mathfrak{N}_L$ is $\{(-2, 0, 1, 0,), (1, \frac{2}{3}, 0, 1)\}$; A basis for R_4 is $\{(-2, 0, 1, 0), (1, \frac{2}{3}, 0, 1), (1, 0, 0, 0), (0, 1, 0, 0)\}$.

(ii) The range of L is spanned by $L(\epsilon_1) = (2, 1, 0)$ and $L(\epsilon_2) = (-3, 0, -3)$.

(iii) $\alpha = (-2, -\frac{19}{3}, 0, 0)$

3. (i) $\mathfrak{N}_L = \text{Sp}\{2t^2 - \frac{3}{2}t + 1\}$ and $\mathfrak{R}_L = \text{Sp}\{-t^2 + 3t + 1, 4t + 2\}$

(ii) $\mathfrak{N}_L = \text{Sp}\{\frac{1}{3}t^2 + \frac{8}{3}t + 1\}$ and $\mathfrak{R}_L = \text{Sp}\{(1, -3, 0), (-2, 0, -3)\}$

4. (ii) $x + y - z = 0$

7. (i) non-singular   (ii) singular   (iii) non-singular

Section 4.3 (pp. 217–219)

1. (i) $\begin{bmatrix} -6 & -95 & -42 \\ 13 & 214 & 91 \end{bmatrix}$ (ii) $\begin{bmatrix} -1 & 21 & 2 \\ 3 & -44 & -12 \end{bmatrix}$

(iii) $\begin{bmatrix} -15 & -122 & 119 \\ 35 & 270 & -261 \end{bmatrix}$

2. (i) $\begin{bmatrix} 1 & -2 \\ 3 & 1 \\ -7 & 1 \end{bmatrix}$ (ii) $\begin{bmatrix} -11 & 11 \\ 2 & 5 \\ 12 & -25 \end{bmatrix}$ (iii) $\begin{bmatrix} 23 & 96 \\ 109 & 490 \\ -8 & -43 \end{bmatrix}$

3. $L[(a_1, a_2, a_3)] = (-a_1 + 6a_2 + 8a_3, -a_1 + 3a_2 + 3a_3)$

4. (i) $L(5, 3) = (-2, 1, -3)$

(ii) $(1, 0) = 3(2, 1) - (5, 3)$ and $(0, 1) = -5(2, 1) + 2(5, 3)$

(iii) $\begin{bmatrix} 53 & -89 \\ 29 & -48 \\ 48 & -81 \end{bmatrix}$

5. $\begin{bmatrix} -12 & -19 & 55 \\ 20 & 37 & -109 \end{bmatrix}$

6. (i) dim $U = 5$ and dim $V = 4$ (ii) $\rho(L) = 3$, $\nu(L) = 2$

(iii) $[L(\alpha)]_T = (-22, 0, -22, -44)$

7. (i)
$$A = \begin{bmatrix} 1 & -1 & 3 \\ 2 & 0 & 1 \\ 2 & 3 & -6 \end{bmatrix}$$
(ii)
$$A^{-1} = \begin{bmatrix} -3 & 3 & -1 \\ 14 & -12 & 5 \\ 6 & -5 & 2 \end{bmatrix}$$

8. (ii) $\{(4, 1, 0), (-1, 0, 1)\}$ (iv) diag $(2, 2, -1)$

10. (i)
$$A = \begin{bmatrix} 7 & 1 & 0 & 0 & 0 \\ -1 & 7 & 0 & 0 & 0 \\ 0 & 0 & 0 & 0 & 0 \\ 0 & 0 & 0 & 0 & 0 \\ 0 & 0 & 0 & 0 & 0 \end{bmatrix}$$

(ii) $\mathfrak{N}_L = $ Sp $\{e^{-x}, e^{2x}, xe^{2x}\}$ and $\mathfrak{R}_L = $ Sp $\{\sin x, \cos x\}$

(iii) The solutions are functions of the form
$$f(x) = 3 \sin x - 2 \cos x + C_1 e^{-x} + C_2 e^{2x} + C_3 x e^{2x}$$
where C_1, C_2, and C_3 are constants.

Section 4.4 (pp. 226–227)

1. (i) $(L_1 + L_2)(a_1, a_2) = (-3a_1 + 6a_2, 2a_1 + 4a_2, -2a_1 - 2a_2)$

(ii) $(3L_1)(a_1, a_2) = (-9a_1 + 15a_2, 3a_2, 3a_1 - 21a_2)$

(iii) $(3L_1 - 2L_2)(a_1, a_2) = (-9a_1 + 13a_2, -4a_1 - 3a_2, 9a_1 - 31a_2)$

matrix representatives:

(i) $\begin{bmatrix} -3 & 6 \\ 2 & 4 \\ -2 & -2 \end{bmatrix}$ (ii) $\begin{bmatrix} -9 & 15 \\ 0 & 3 \\ 3 & -21 \end{bmatrix}$ (iii) $\begin{bmatrix} -9 & 13 \\ -4 & -3 \\ 9 & -31 \end{bmatrix}$

2. (i) $(L_1 \circ L_1)(c_1, c_2) = (c_1 - 21c_2, 7c_1 + 22c_2)$

$(L_2 \circ L_1)(c_1, c_2) = (-13c_2, 5c_1 + 12c_2, 9c_1 - 46c_2)$

(ii) $\mu(L_1) = \begin{bmatrix} 2 & -3 \\ 1 & 5 \end{bmatrix}$ $\mu(L_2) = \begin{bmatrix} 1 & -2 \\ 1 & 3 \\ 7 & -5 \end{bmatrix}$

$\mu(L_1 \circ L_1) = \begin{bmatrix} 1 & -21 \\ 7 & 22 \end{bmatrix}$ $\mu(L_2 \circ L_1) = \begin{bmatrix} 0 & -13 \\ 5 & 12 \\ 9 & -46 \end{bmatrix}$

3. (i)
$$\mu(L_1) = \begin{bmatrix} 3 & -5 & 3 \\ -1 & -2 & 3 \\ 0 & 1 & -1 \end{bmatrix} \quad \mu(L_2) = \begin{bmatrix} 1 & 2 & 9 \\ 1 & 3 & 12 \\ 1 & 3 & 11 \end{bmatrix}$$

(iii) $L_1 \circ L_2 = I = L_2 \circ L_1$

4. (ii) $\mu(L_1) = \begin{bmatrix} -2 & 8 & 13 \\ 0 & 13 & 7 \end{bmatrix}$ and $\mu(L_2) = \begin{bmatrix} 3 & -13 & -19 \\ 2 & -19 & -9 \end{bmatrix}$

(iii) $\mu^*(L_1) = \begin{bmatrix} -36 & 61 & -63 \\ -23 & 41 & -43 \end{bmatrix}$ and $\mu^*(L_2) = \begin{bmatrix} 58 & -88 & 95 \\ 37 & -60 & 65 \end{bmatrix}$

7. $L_1(a_1, a_2) = (a_1 - a_2, 2a_1 + a_2)$
$L_2(a_1, a_2) = (3a_1 + a_2, a_1 + 5a_2)$

8. (i) *Suggestion:* Show that if $L_1(\alpha) = \theta_V$, then $(L_2 \circ L_1)(\alpha) = \theta_W$

(ii) Recall corollary 4.1.

(iii) *Suggestion:* If $(L_2 \circ L_1)(\alpha) = \beta$, show that $L_2(\gamma) = \beta$, where $\gamma = L_1(\alpha)$.

(iv) Not generally true; for example, if $L_1(a_1, a_2) = (0, 0, a_1 - a_2)$ and $L_2(b_1, b_2, b_3) = (0, 0, b_3)$, then $v(L_2) = 2$ and $v(L_2 \circ L_1) = 1$.

9. *Suggestion:* Let β be any non-zero vector in $\mathcal{R}_L$. Then $\{\beta\}$ is a basis for $\mathcal{R}_L$ (why?), and $L(\beta) = s\beta$ for some scalar s. Next, let α be a typical vector in V, and show that $(L \circ L)(\alpha) = sL(\alpha)$.

Section 4.5 (pp. 234–236)

1. (i)
$$A = \begin{bmatrix} -1 & 1 & 1 \\ 1 & 0 & -2 \\ -2 & -1 & 0 \end{bmatrix}, \quad B = \begin{bmatrix} 0 & -2 & -3 \\ 3 & 3 & 1 \\ 1 & -2 & 0 \end{bmatrix} \quad \text{(iii) yes}$$

2. (i) $L^{-1}(a_1, a_2) = (-a_1 + 4a_2, 2a_1 - 7a_2)$ (ii) singular

(iii) $L^{-1}(a_1, a_2, a_3)$
$$= (-3a_1 + 2a_2 - 8a_3, -5a_1 + 3a_2 - 12a_3, 2a_1 - a_2 + 5a_3)$$

3. (i)
$$A = \begin{bmatrix} 1 & -1 & 0 \\ 0 & 2 & 1 \\ 2 & -1 & 1 \end{bmatrix} \quad \text{(ii)} \quad A^{-1} = \begin{bmatrix} 3 & 1 & -1 \\ 2 & 1 & -1 \\ -4 & -1 & 2 \end{bmatrix}$$

(iii) $L^{-1}(t^2) = 3t^2 + 2t - 4$
$L^{-1}(t) = t^2 + t - 1$
$L^{-1}(1) = -t^2 - t + 2$

(iv) $L^{-1}(p) = (3a + b - c)t^2 + (2a + b - c)t + (-4a - b + 2c)$, where $p(t) = at^2 + bt + c$

6. (i) $(2, 0, 1)$ (ii)
$$A = \begin{bmatrix} -3 & 2 & 0 \\ 1 & 0 & 1 \\ 7 & 1 & 8 \end{bmatrix}, \quad A^{-1} = \begin{bmatrix} -1 & -16 & 2 \\ -1 & -24 & 3 \\ 1 & 17 & -2 \end{bmatrix}$$

(iii) $\alpha = (-3, 0, 4)$

13. (ii) No. For example, consider
$$A = \begin{bmatrix} 1 & 3 \\ 2 & 5 \end{bmatrix} \text{ and } B = \begin{bmatrix} 1 & 1 \\ 0 & 0 \end{bmatrix}$$

Section 4.6 (pp. 241–242)

1. (i)

$$A = \begin{bmatrix} 2 & -1 & 1 \\ 1 & 0 & 5 \\ 1 & 3 & -2 \end{bmatrix} \qquad B = \begin{bmatrix} -11 & 24 & 14 \\ -16 & 21 & 8 \\ 20 & -26 & -10 \end{bmatrix}$$

(ii)

$$P = \begin{bmatrix} 1 & 3 & 3 \\ 1 & -2 & -1 \\ 0 & 1 & 1 \end{bmatrix} \qquad P^{-1} = \begin{bmatrix} 1 & 0 & -3 \\ 1 & -1 & -4 \\ -1 & 1 & 5 \end{bmatrix}.$$

2. (i)

$$\begin{bmatrix} -16 & -5 & 17 \\ 52 & 5 & -34 \\ -12 & -6 & 17 \end{bmatrix} \qquad \text{(ii)} \quad \begin{bmatrix} 7 & 0 & 0 \\ 0 & 0 & 0 \\ 0 & 0 & -1 \end{bmatrix}$$

3.

$$\begin{bmatrix} 15 & 2 & -8 \\ -2 & 2 & 0 \\ 23 & 3 & -12 \end{bmatrix}$$

Section 4.7 (pp. 248–250)

1. A_1, A_2, A_3, A_5 are orthogonal.

2. (i) and (iii) are orthogonal.

3. (i) $\frac{1}{2}\begin{bmatrix} \sqrt{3} & 1 \\ 1 & -\sqrt{3} \end{bmatrix}$ (ii) L is orthogonal.

4. $a = 0$, $b = 1/\sqrt{7}$, $c = -1/\sqrt{42}$, $d = 0$

8. (i) rotation through $60°$ (ii) reflection in the line $y = \frac{1}{2}x$
   (iii) reflection in the line $y = -\frac{1}{5}x$ (iv) rotation through $225°$

9. $x_1 = 5 + \frac{5}{2}\sqrt{2}$, $x_2 = 5 + 4\sqrt{2}$, $x_3 = -2\sqrt{2}$, $x_4 = 5 - \frac{13}{2}\sqrt{2}$

Supplementary exercises (pp. 250–254)

1. $\overrightarrow{OQ} = \frac{1}{3}\alpha + \frac{1}{3}\beta$, where $\alpha = \overrightarrow{P_1P_2} = (5, -1, 0)$ and $\beta = \overrightarrow{P_1P_3} = (2, 0, 0)$. Thus, Q is $\frac{1}{3}(7, -1, 0)$, and $L(Q) = \frac{1}{3}(5, 22, 4)$.

2. (i) $\alpha = (-9, 7, 0, 0)$ (ii) $(9, 5, 1)$ and $(-5, 3, -1)$
   (iii) $\begin{bmatrix} 5 & 7 & 9 & -5 \\ -3 & 1 & 5 & 3 \\ 1 & 1 & 1 & -1 \end{bmatrix}$

3. (i) A and C are similar. Thus, they both have rank n.   (ii) $A^{-1}CA = B$

4. (i) It suffices to show that the β's are linearly independent.
   (ii) $B = P^{-1}AP$, where

$$P = \begin{bmatrix} 2 & 1 & -1 \\ 1 & 3 & 0 \\ -1 & 1 & 1 \end{bmatrix}$$

Therefore,

$$B = \begin{bmatrix} -15 & -3 & 9 \\ 5 & 1 & -3 \\ -21 & -3 & 13 \end{bmatrix}$$

5. *Suggestion for (ii):* Let $\{\alpha_1, \ldots, \alpha_r\}$ be a basis for $\mathfrak{R}_L$ and $\{\alpha_{r+1}, \ldots, \alpha_n\}$ be a basis for $\mathfrak{N}_L$. Show that $\alpha_1, \ldots, \alpha_n$ are linearly independent.

(iii) $\begin{bmatrix} I_r & 0 \\ 0 & 0 \end{bmatrix}$

6. *Suggestion for (i):* Let $\beta = s_1\gamma + s_2N(\gamma) + \ldots + s_kN^{(k-1)}(\gamma)$. If $\beta = \theta$, note that $N(\beta) = N^2(\beta) = \ldots = N^{(k-1)}(\beta) = \theta$, and use this fact to show that each s_j is 0.

(ii) For example, in the case where $n = 4$, N is represented by

$$\begin{bmatrix} 0 & 1 & 0 & 0 \\ 0 & 0 & 1 & 0 \\ 0 & 0 & 0 & 1 \\ 0 & 0 & 0 & 0 \end{bmatrix}$$

7. $F(x, y, z) = \sqrt[3]{xyz}$ is such a function.

9. (i) $\mathfrak{N}_{L_A}$ contains more than O; for instance, $L_A(A) = O$.

(ii) $\nu(L) = 2; \rho(L) = 2$

10. *Suggestions:*

(i) Show that each vector in $\mathfrak{R}_{L_1+L_2}$ is also contained in $\mathfrak{R}_{L_1} + \mathfrak{R}_{L_2}$. Then use the formula established in supplementary exercise 8 of chapter 2.

(ii) Let $\{\alpha_1, \ldots, \alpha_k\}$ be a basis for $\mathfrak{N}_{L_1}$ and extend it to a basis $\{\alpha_1, \ldots, \alpha_p\}$ for $\mathfrak{N}_{L_2 \cdot L_1}$. Show that the vectors $L_1(\alpha_{k+1}), \ldots, L_1(\alpha_p)$ are linearly independent members of $\mathfrak{N}_{L_2}$ (this is the hard part). Conclude that $\nu(L_2) \geq p - k$

12. (iv) $f_1(x, y, z) = -x + y - z$

$f_2(x, y, z) = 3x - y + 2z$

$f_3(x, y, z) = 2x - y + z$

13. (i) *Suggestion:* If β is contained in $\mathfrak{N}_{L^*}$, then $(L(\alpha) \mid \beta) = (\alpha \mid L^*(\beta)) = 0$ for each α in V, and β must be orthogonal to each vector in $\mathfrak{R}_L$. On the other hand, if γ is in $\mathfrak{R}_L^\perp$, then $(L^*(\gamma) \mid L^*(\gamma)) = (L \circ L^*(\gamma) \mid \gamma) = 0$ since $(L \circ L^*)(\gamma)$ is a vector in $\mathfrak{R}_L$.

CHAPTER 5

Section 5.1 (pp. 262–265)

1. (i) 2   (ii) -1 (iii) 0   (iv) -12

2. (i) 0   (ii) 1   (iii) 0   (iv) -1

3. (i) -3 or 2   (ii) 5 or -4 (iii) 0, 5, or -3 (iv) -1 or 3

13. -20

15. (i) 25   (ii) -32

Section 5.2 (pp. 270–271)

2. (i) 1   (ii) $-5, 3, 1$ (iii) $0, -3$ (iv) $1, -7$
3. (i) non-singular   (ii) singular   (iii) non-singular
9. $\det Q = 17$, $\det P = -5$ and $\det B = (17)(5)(-5) = -425$
10. 3

Section 5.3 (pp. 279–281)

1. (i) -56 (ii) $(-1)^{2+3}(117) = -117$ (iii) -175
2. (i) -127 (ii) -142 (iii) 12
3. (i) $\begin{bmatrix} 3 & -4 \\ 2 & 7 \end{bmatrix}$ (ii) $\begin{bmatrix} -3 & -14 & 15 \\ -3 & -10 & 11 \\ 2 & 8 & -10 \end{bmatrix}$ (iii) $\begin{bmatrix} 2 & -1 & -7 \\ -7 & 3 & 25 \\ -9 & 4 & 31 \end{bmatrix}$

(iv) $\begin{bmatrix} -17 & 34 & 17 \\ 9 & -18 & -9 \\ -10 & 20 & 10 \end{bmatrix}$ (v) the zero matrix

4. For (i), (ii), and (iii) see exercise 3, above.

(iv) $\dfrac{1}{105}\begin{bmatrix} 0 & 0 & 21 \\ 35 & 0 & 0 \\ 0 & -15 & 0 \end{bmatrix}$ (v) $\begin{bmatrix} 20 & 4 & -52 & 15 \\ 35 & 7 & -91 & 26 \\ 104 & 21 & -273 & 78 \\ -40 & -8 & 105 & -30 \end{bmatrix}$

5. (i) $x_1 = -2$ (ii) $x_1 = -3$ (iii) $x_1 = 1$ (iv) $x_1 = 2$
$x_2 = 1$ $x_2 = -2$ $x_2 = 5$ $x_2 = 1$
$x_3 = 4$ $x_3 = 2$ $x_3 = 3$
$x_4 = -1$

9. Non-singular. Note that $\det (\operatorname{adj} A) \neq 0$.
11. $\det B < 0$ but if $B = \operatorname{adj} A$, we would have $\det B = (\det A)^2$.
12. $\det A = 1$ and $\operatorname{adj} A = A^{-1}$
13. *Suggestion for (ii):* For the "only if" part, use the fact that whenever A^{-1} is a matrix of integers, $\det A^{-1}$ is an integer.
(iii) $x = -3$, $y = 2$, $z = 1$ is a suitable choice.

Supplementary exercises (pp. 281–283)

8. yes
9. $\det H_6 = a^6 - 5(bc)a^4 + 6(bc)^2 a^2 - (bc)^3$
10. (i) $Q_j(c_{11}a_j + c_{12}b_j, c_{21}a_j + c_{22}b_j)$ for $j = 1, 2, 3$. They cannot be collinear since L is non-singular.

(ii) *Suggestion:* Multiply $\begin{bmatrix} a_1 & b_1 & 1 \\ a_2 & b_2 & 1 \\ a_3 & b_3 & 1 \end{bmatrix}$ by $\begin{bmatrix} c_{11} & c_{21} & 0 \\ c_{12} & c_{22} & 0 \\ 0 & 0 & 1 \end{bmatrix}$

11. (i) $L(x, y) = \left(\dfrac{1}{7}x + \dfrac{1}{70}y, \dfrac{\sqrt{3}}{10}y \right)$

CHAPTER 6

Section 6.1 (*pp. 290–292*)

1. (i) $t^2 - 6t + 8$; 4 and 2   (ii) $t^2 - 12t + 36$; 6 and 6
   (iii) $t^2 - 2t - 15$; 5 and -3 (iv) $t^3 - 14t^2 + 64t - 96$; 4, 4, and 6
   (v) $t^3 + 2t^2 - 8t$; 0, -4, and 2
2. (i) $\begin{bmatrix} 2 & 1 & 0 \\ 0 & 2 & 1 \\ 6 & -1 & -2 \end{bmatrix}$ (ii) 0, -1, 3
3. 0, -1, -1
4. (i) $\begin{bmatrix} 2 & -2 & 0 & 0 \\ 0 & 4 & 0 & 0 \\ 0 & 0 & 0 & 0 \\ 0 & 0 & 0 & 0 \end{bmatrix}$ (ii) 2, 2, 0, 0
6. (iii) no
12. $-3, 7, 5, 0$
15. *Suggestion:* Form the product $A\gamma$, where $\gamma = (c_j)$ is the n-tuple with $c_j = 1$ for all j.
16. (i) $p_A(t) = t^3 + t^2 - 2$ (ii) $\lambda_3 = -1 + i$ and $\lambda_2 = -1 - i$
    (iii) no   (iv) yes

Section 6.2 (*pp. 298–300*)

1. (i) $\lambda_1 = 0, \lambda_2 = 2, \lambda_3 = 3$
   bases: $\{(1, 0, 1)\}$ for S_{λ_1}; $\{(3, 2, 1)\}$ for S_{λ_2}; $\{(1, -1, 1)\}$ for S_λ
   (ii) $\lambda_1 = 1, \lambda_2 = -5$
   bases: $\{(1, 0, 1)\}$ for S_{λ_1} and $\{(1, -1, 1)\}$ for S_{λ_2}
   (iii) $\lambda_1 = 1, \lambda_2 = -2$
   bases: $\{(2, -1, 1)\}$ for S_{λ_1} and $\{(0, 1, 0), (1, 0, 1)\}$ for S_{λ_2}
   (iv) $\lambda_1 = 1, \lambda_2 = -2, \lambda_3 = 3$
   bases: $\{(1, -1, 1)\}$ for S_{λ_1}; $\{(1, 0. 1)\}$ for S_{λ_2}; $\{(2, -1, 1)\}$ for S_{λ_3}
2. (i) $\begin{bmatrix} 7 & 0 & -10 \\ -5 & -3 & 5 \\ 5 & 0 & -8 \end{bmatrix}$ (ii) $\lambda_1 = 2, \lambda_2 = -3 = \lambda_3$
   (iii) bases: $\{(2, -1, 1)\}$ for S_{λ_1} and $\{(1, 0, 1), (0, 1, 0)\}$ for S_λ
   (iv) $\{2t^2 - t + 1, t^2 + 1, t\}$; $D = \text{diag}(2, -3, -3)$
3. (i) $\lambda_1 = 0$; $\{(t^2 - t + 1)\}$ is a basis for S_{λ_1}
   (ii) $\lambda_1 = 1, \lambda_2 = 4$
   bases: $\{(1, -1, 1)\}$ for S_{λ_1} and $\{(1, 0, 1)\}$ for S_{λ_2}
   (iii) $\lambda_1 = 1, \lambda_2 = -2$
   bases: $\{(1, 1, 1, 1)\}$ for S_{λ_1} and $\{(-1, 2, -4, 8)\}$ for S_{λ_2}
4. (i) $\lambda_2 = -3$ bases: $\{(-1, 0, 0, 2), (0, 1, 1, 0)\}$ for S_{λ_1} and $\{(1, 1, 0, 1)\}$ for S_{λ_2}
   (ii) no
7. *Hint:* Use the fact that similar matrices have the same trace.
8. (ii) $[L(\gamma)]_S = (d_1 c_1, d_2 c_2, \ldots, d_n c_n)$

9. (i)

$$\begin{bmatrix} 1 & -2 & 2 & 0 & 0 & 0 \\ 0 & 1 & -4 & 0 & 0 & 0 \\ 0 & 0 & 1 & 0 & 0 & 0 \\ 0 & 0 & 0 & 4 & 0 & 0 \\ 0 & 0 & 0 & 0 & 0 & 2 \\ 0 & 0 & 0 & 0 & -2 & 0 \end{bmatrix}$$

(ii) The differential equation is equivalent to the functional relationship $L(f) = \lambda f$. The solutions are as follows:

for $\lambda_1 = 1$, the constant functions $f(x) = C$
for $\lambda_2 = 4$. functions of the form $f(x) = Ce^{-x}$.

Section 6.3 (*pp. 306–308*)

1. (i) yes (ii) yes (iii) no (iv) yes
2. (i) yes (ii) no (iii) yes
3. (i) $p_A(x) = x^3 + 169x$ (ii) no
4. (i) $\lambda_1 = -4, \lambda_2 = 1, \lambda_3 = 3$
(ii) bases: $\{(36, 31, 1)\}$ for S_{λ_1}, $\{(1, 1, 1)\}$ for S_{λ_2}, $\{(1, 3, 1)\}$ for S_{λ_3}
(iii)

$$P = \begin{bmatrix} 36 & 1 & 1 \\ 31 & 1 & 3 \\ 1 & 1 & 1 \end{bmatrix} \qquad P^{-1} = -\frac{1}{70} \begin{bmatrix} -2 & 0 & 2 \\ -28 & 35 & -77 \\ 30 & -35 & 5 \end{bmatrix}$$

5. $\{t^2 - t + 1, -t^2 + t, -t^2 + 1\}$
7. (ii) $m(x) = x^2 - x - 2$
(iii) $A^2 = A + 2I, \qquad A^3 = 3A + 2I, \qquad A^4 = 5A + 6I,$
$A^5 = 11A + 10I, \qquad A^{-1} = \frac{1}{2}A - \frac{1}{2}I$

Section 6.4 (*pp. 313–314*)

1. (i) $x_1 = C_1e^{3t} + C_2e^t$ (ii) $x_1 = -2C_1e^{3t} + C_2e^{2t}$
  $x_2 = C_1e^{3t} - C_2e^t$ $x_2 = 3C_1e^{3t} - C_2e^{2t}$
(iii) $x_1 = \qquad C_2e^{2t} + 3C_3e^{4t}$
  $x_2 = -C_1e^t \qquad + 4C_3e^{4t}$
  $x_3 = \quad C_1e^t \qquad + 2C_3e^{4t}$
(iv) $x_1 = \quad C_1e^{2t} + 3C_2e^{2t} - \quad C_3e^{-7t}$
  $x_2 = 2C_1e^{2t} \qquad\qquad + C_3e^{-7t}$
  $x_3 = \qquad\qquad 2C_2e^{2t} + 2C_3e^{-7t}$
2. (i), (ii), and (iv) are in Jordan form.
3. $x_1 = C_3e^{2t} + C_2te^{2t} + \frac{1}{2}C_1t^2e^{2t}$
  $x_2 = C_2e^{2t} + C_1te^{2t}$
  $x_3 = C_1e^{2t}$
4. (i)

$$P^{-1} = \begin{bmatrix} -1 & 0 & 1 \\ 4 & -1 & -3 \\ -2 & 1 & 2 \end{bmatrix} \quad \text{and} \quad J = \begin{bmatrix} 2 & 1 & 0 \\ 0 & 2 & 1 \\ 0 & 0 & 2 \end{bmatrix}$$

(iii) $x_1 = \qquad -te^{2t} + t^2e^{2t}$
  $x_2 = \qquad 6te^{2t} - 2t^2e^{2t}$
  $x_3 = e^{2t} - 4te^{2t} + 2t^2e^{2t}$

5. (i) $\begin{bmatrix} 0 & 1 \\ -2 & 3 \end{bmatrix}$ (ii) $\begin{bmatrix} 0 & 1 \\ 9 & 0 \end{bmatrix}$ (iii) $\begin{bmatrix} 0 & 1 & 0 \\ 0 & 0 & 1 \\ -6 & 7 & 0 \end{bmatrix}$

$x^2 - 3x + 2$ $x^2 - 9$ $x^3 - 7x + 6$

$\{e^{2t}, e^t\}$ $\{e^{3t}, e^{-3t}\}$ $\{e^t, e^{2t}, e^{-3t}\}$

Supplementary exercises (*pp. 315–318*)

1. (i) $p_{E_n}(x) = x^{n-1}(x - n)$

2. (i) $\lambda_3 = 1$ (use property 2) and $\lambda_4 = 2$ (use property 3)

  (ii) $p_A(x) = p_{A^t}(x) = (x - 1)(x - 2)^2(x - 3)$

  (iii) Eigenvalues of A^{-1} are $1, \frac{1}{2}, \frac{1}{2}, \frac{1}{3}$. Therefore,

$$p_{A^{-1}}(x) = (x - 1)(x - \tfrac{1}{2})^2(x - \tfrac{1}{3})$$

3. Since $A^3 - 7A^2 + 5A - 9I = 0$, we have $A^{-1} = \frac{1}{9}[A^2 - 7A + 5I]$ and $\det A = 9$. Thus, adj $A = (\det A)A^{-1} = A^2 - 7A + 5I$.

4 (i) $\begin{bmatrix} 0 & 2 & -2 & 0 \\ 2 & 0 & 0 & -2 \\ -2 & 0 & 0 & 2 \\ 0 & -2 & 2 & 0 \end{bmatrix}$

  (ii) Distinct eigenvalues are $\lambda_1 = 0, \lambda_2 = 4, \lambda_3 = -4$.

  bases: $\left\{ \begin{bmatrix} 0 & 1 \\ 1 & 0 \end{bmatrix}, \begin{bmatrix} 1 & 0 \\ 0 & 1 \end{bmatrix} \right\}$ for C_{λ_1}; $\left\{ \begin{bmatrix} -1 & -1 \\ 1 & 1 \end{bmatrix} \right\}$ for C_{λ_2} and $\left\{ \begin{bmatrix} -1 & 1 \\ -1 & 1 \end{bmatrix} \right\}$ for C_{λ_3}

  (iii) yes

5. (i) $L(\alpha_j) = \alpha_j$ for $j = 1, 2, \ldots, k$; $L(\alpha_j) = -\alpha_j$ for $j = k + 1, \ldots, n$. The matrix representative is diag $(1, \ldots 1, -1, \ldots, -1)$.   (ii) 1 and -1

7. (i) $p_{A_1}(x) = x^3 + 8x^2 + 8$; $p_{A_2}(x) = x^3 - 16x + 40$

8. (ii) Since dim $S_\lambda = 1$, each eigenvector associated with λ must be a scalar multiple of X.

9. (i) If $C = \begin{bmatrix} C_1 & C_2 \\ C_3 & C_4 \end{bmatrix}$ show that CJ_m and J_mC both have the characteristic polynomial $p(x) = x^{n-m} \det (xI_m - C_1)$.

  (ii) Let $C = P^{-1}BQ^{-1}$ and use part (ii).

11. 7 lies outside the union of the four disks.

12. Since 0 lies outside the union of the disks, it cannot be an eigenvalue. Hence, H is non-singular.

CHAPTER 7

Section 7.1 (*pp. 326–327*)

1. (i) $\begin{bmatrix} 3 & -1 \\ -1 & -5 \end{bmatrix}$ (ii) $\begin{bmatrix} -2 & 3 & -4 \\ 3 & 1 & -1 \\ -4 & -1 & 2 \end{bmatrix}$ (iii) $\begin{bmatrix} 7 & 2 & 2 \\ 2 & 0 & -2 \\ 2 & -2 & -1 \end{bmatrix}$

2. (i) $-5x_1^2 + 4x_1x_2 + x_2^2$ (ii) $-6x_1x_2$ (iii) $2x_1^2 - 9x_2^2$

  (iv) $x_1^2 + 4x_1x_2 - 10x_1x_3 + 6x_2x_3 - 2x_3^2$ (v) $-7x_1^2 + 2x_1x_2 + 2x_2x_3$

8. (ii) *Hint:* Show that $s = (X^t K X)$ is a scalar which satisfies $s = -s$.

10. (ii) The semi-minor axis has length $1/\sqrt{6}$. The area of the ellipse is $\pi(\tfrac{1}{2})(1/\sqrt{6})$
$= \dfrac{\pi\sqrt{6}}{12}$.

(iii) The foci of $4y_1^2 + 6y_2^2 = 1$ are located at $(\sqrt{3}/6, 0)$ and $(-\sqrt{3}/6, 0)$ in the y_1, y_2-plane. Using (7.2), we find that the foci of the unrotated ellipse are at $(\sqrt{6}/12, \sqrt{6}/12)$ and $(-\sqrt{6}/12, -\sqrt{6}/12)$. The area is $\pi\sqrt{6}/12$.

Section 7.2 (*pp. 334–335*)

1. (i) $\lambda_1 = 6, \lambda_2 = 1; P = \dfrac{1}{\sqrt{5}}\begin{bmatrix} -2 & 1 \\ 1 & 2 \end{bmatrix}$

(ii) $\lambda_1 = 8, \lambda_2 = 3; P = \dfrac{1}{\sqrt{5}}\begin{bmatrix} 2 & 1 \\ 1 & -2 \end{bmatrix}$

(iii) $\lambda_1 = 5, \lambda_2 = -1; P = \dfrac{1}{\sqrt{2}}\begin{bmatrix} 1 & 1 \\ 1 & -1 \end{bmatrix}$

(iv) $\lambda_1 = 2, \lambda_2 = 12, \lambda_3 = 0; P = \begin{bmatrix} -\dfrac{1}{\sqrt{2}} & -\dfrac{1}{\sqrt{6}} & \dfrac{1}{\sqrt{3}} \\[2mm] \dfrac{1}{\sqrt{2}} & -\dfrac{1}{\sqrt{6}} & \dfrac{1}{\sqrt{3}} \\[2mm] 0 & \dfrac{2}{\sqrt{6}} & \dfrac{1}{\sqrt{3}} \end{bmatrix}$

(v) $\lambda_1 = 3, \lambda_2 = 17, \lambda_3 = 0; P = \begin{bmatrix} -\dfrac{5}{\sqrt{42}} & -\dfrac{1}{\sqrt{14}} & \dfrac{1}{\sqrt{3}} \\[2mm] \dfrac{4}{\sqrt{42}} & -\dfrac{2}{\sqrt{14}} & \dfrac{1}{\sqrt{3}} \\[2mm] \dfrac{1}{\sqrt{42}} & \dfrac{3}{\sqrt{14}} & \dfrac{1}{\sqrt{3}} \end{bmatrix}$

(vi) $\lambda_1 = -3, \lambda_2 = 1; P = \begin{bmatrix} -\dfrac{1}{\sqrt{2}} & -\dfrac{1}{\sqrt{6}} & -\dfrac{1}{2\sqrt{3}} & \dfrac{1}{2} \\[2mm] \dfrac{1}{\sqrt{2}} & -\dfrac{1}{\sqrt{6}} & -\dfrac{1}{2\sqrt{3}} & \dfrac{1}{2} \\[2mm] 0 & \dfrac{2}{\sqrt{6}} & -\dfrac{1}{2\sqrt{3}} & \dfrac{1}{2} \\[2mm] 0 & 0 & \dfrac{3}{2\sqrt{3}} & \dfrac{1}{2} \end{bmatrix}$

(vii) $\lambda_1 = 0, \lambda_2 = 15; P = \begin{bmatrix} -\dfrac{1}{\sqrt{2}} & \dfrac{1}{\sqrt{3}} & \dfrac{9}{\sqrt{551}} & \dfrac{1}{\sqrt{15}} \\[2mm] 0 & \dfrac{1}{\sqrt{3}} & \dfrac{10}{\sqrt{551}} & \dfrac{3}{\sqrt{15}} \\[2mm] 0 & 0 & \dfrac{12}{\sqrt{551}} & -\dfrac{2}{\sqrt{15}} \\[2mm] \dfrac{1}{\sqrt{2}} & \dfrac{1}{\sqrt{3}} & -\dfrac{15}{\sqrt{551}} & \dfrac{1}{\sqrt{15}} \end{bmatrix}$

2. (i) $5y_1^2 - y_2^2$ (ii) $0y_1^2 + \sqrt{10}\,y_2^2 - \sqrt{10}\,y_3^2$ (iii) $7y_1^2 + 4y_2^2 + 4y_3^2$

Section 7.3 (pp. 344–346)

1. (i) $r = 1$ (ii) $r = 2$ (iii) $r = 2$ (iv) $r = 3$ (v) $r = 3$
   $s = 1$ $s = 2$ $s = 0$ $s = 1$ $s = 1$
   (vi) $r = 1$ (vii) $r = 3$
   $s = 1$ $s = 1$

2. Only A_1 and A_3 have the same rank and signature.

$$\begin{array}{cccc} A_1 & A_2 & A_3 & A_4 \\ r = 3 & r = 3 & r = 3 & r = 3 \\ s = -1 & s = 3 & s = -1 & s = 1 \end{array}$$

3. (i) $(2; 2, 2)$ ellipse   (ii) $(2; 2, 0)$ hyperbola   (iii) $(2; 2, -2)$ no locus
   (iv) $(3; 2, 0)$ hyperbolic cylinder   (v) $(3; 3, 1)$ hyperboloid of one sheet
   (vi) $(3; 2, -2)$ no locus

4. (i) $P = \dfrac{1}{\sqrt{5}} \begin{bmatrix} -2 & 1 \\ 1 & 2 \end{bmatrix}$

   (iii) and (iv)

	in the y_1, y_2-plane	in the x_1, x_2-plane
vertices	$(\tfrac{1}{2}, 0), (-\tfrac{1}{2}, 0)$	$(-\sqrt{5}/5, \sqrt{5}/10), (\sqrt{5}/5, -\sqrt{5}/10)$
foci	$(\sqrt{5/4}, 0), (-\sqrt{5/4}, 0)$	$(-1, \tfrac{1}{2}), (1, -\tfrac{1}{2})$
axes	$y_1 = 0, y_2 = 0$	$x_1 = -2x_2, x_2 = 2x_1$
asymptotes	$2y_1 = y_2, 2y_1 = -y_2$	$x_1 = 0, x_2 = \dfrac{3}{4}x_1$

6. *Suggestion:* Note that $AC - \tfrac{1}{4}B^2$ and $A + C$ are the determinant and trace, respectively, of the symmetric representative of $AX^2 + Bxy + Cy^2$. Then use the fact that similar matrices have the same trace and determinant.

7. The symmetric representative of the related quadratic form is

$$S = \begin{bmatrix} 19 & 5 & 2 \\ 5 & 9 & 7 \\ 2 & 7 & 6 \end{bmatrix}$$

   Since S is positive definite, $X^t S X = 1$ is an ellipsoid. Its volume is

$$\frac{4}{3} \frac{\pi}{\sqrt{\det S}} = \frac{4\pi}{21}$$

8. The determinant of S is the product of its eigenvalues, all of which are positive.

9. (i) $\det S = -8$ (See exercise 20, section 6.1)
   (ii) *Hint:* Use part (i) and the result of exercise 8.

10. (iii) The ellipse has area π.

Supplementary exercises (pp. 346–348)

2. *Hint:* Let $x = A$ and $y = -C$.
3. *Hint:* If λ is an eigenvalue of S, then $\lambda + c$ is an eigenvalue of $S + cI$.

Location of
Key Definitions

Index